Oscillation and Dynamics
in Delay Equations

Recent Titles in This Series

(Continued in the back of this publication)

CONTEMPORARY MATHEMATICS

129

Oscillation and Dynamics in Delay Equations

Proceedings of an AMS Special Session
held January 16–19, 1991

John R. Graef
Jack K. Hale
Editors

American Mathematical Society
Providence, Rhode Island

The AMS Special Session on Oscillation Dynamics in Delay Equations was held at the 863rd Meeting of the American Mathematical Society in San Francisco, California, on January 16–19, 1991.

1991 *Mathematics Subject Classification*. Primary 34-06, 34K05, 34K10, 34K15, 34K25, 34K30, 34K40, 34A05, 34C10, 34C11, 34C15, 34E05, 34E10, 11N25, 35B05, 35B25, 35B40, 35L10, 39A12, 45E10, 58F09.

Library of Congress Cataloging-in-Publication Data

Oscillation and dynamics in delay equations: proceedings of an AMS special session held January 16–19, 1991/John R. Graef, Jack K. Hale, editors.
    p. cm.—(Contemporary mathematics, ISSN 0271-4132; v. 129)
    ISBN 0-8218-5140-3 (alk. paper)
    1. Delay differential equations—Congresses. 2. Oscillations—Congresses.
3. Differentiable dynamical systems—Congresses. I. Graef, John R., 1942- . II. Hale, Jack K. III. American Mathematical Society, Meeting, (863rd: 1991: San Francisco, Calif.)
IV. Series: Contemporary mathematics (American Mathematical Society); v. 129
QA372.O82 1992   92-12229
515′.352—dc20   CIP

Contents

Preface

Oscillation theory and dynamical systems have long been rich and active areas of mathematical research. The papers in this volume are based on presentations at the Special Session of the same title held at the 97th Annual Meeting of the American Mathematical Society, in San Francisco, CA, in January 1991. With a special emphasis on delay equations, the papers cover a broad range of topics in ordinary, partial, and difference equations, and include applications to problems in commodity prices, biological modeling, and number theory. All the papers in this volume are in final form, and no versions of them will be submitted for publication elsewhere. The editors wish to express their appreciation to the authors of each paper for their assistance with the preparation of these proceedings.

John R. Graef

Jack K. Hale

Contemporary Mathematics
Volume **129**, 1992

Oscillations of Second-Order Neutral Differential Equations with Deviating Arguments

S. J. BILCHEV, M. K. GRAMMATIKOPOULOS,
AND I. P. STAVROULAKIS

ABSTRACT. Consider the second-order neutral differential equation

(E) $$\frac{d^2}{dt^2}\left[x(t) + \sum_{\mathcal{I}} p_i x\left(t - \tau_i\right)\right] + \sum_{\mathcal{K}} q_k x\left(t - \sigma_k\right) = 0,$$

where $\mathcal{I}$ and $\mathcal{K}$ are initial segments of natural numbers and p_i, τ_i, q_k, $\sigma_k \in \mathbb{R}$ for $i \in \mathcal{I}$ and $k \in \mathcal{K}$. Then a necessary and sufficient condition for the oscillation of all solutions of (E) is that its characteristic equation

$$\lambda^2 + \lambda^2 \sum_{\mathcal{I}} p_i e^{-\lambda \tau_i} + \sum_{\mathcal{K}} q_k e^{-\lambda \sigma_k} = 0$$

has no real roots. The method of proof has the advantage that it results in easily verifiable sufficient conditions (in terms of the coefficients and the arguments only) for the oscillation of all solutions of Eq. (E). Such equations were encountered in the study of vibrating masses attached to an elastic bar and also as the Euler equations in some variational problems.

1. Introduction

Consider the second-order neutral differential equation

(E) $$\frac{d^2}{dt^2}\left[x(t) + \sum_{\mathcal{I}} p_i x\left(t - \tau_i\right)\right] + \sum_{\mathcal{K}} q_k x\left(t - \sigma_k\right) = 0$$

where $\mathcal{I}$, $\mathcal{K}$, are initial segments of natural numbers and p_i, τ_i, q_k, $\sigma_k \in \mathbb{R}$ for $i \in \mathcal{I}$ and $k \in \mathcal{K}$. In the case where $\mathcal{I} = \emptyset$ Eq. (E) reduces to the (non-neutral)

1991 *Mathematics Subject Classification*. Primary 34K40; Secondary 34K15, 34C10.

equation

$$(E_1) \qquad \ddot{x}(t) + \sum_{\mathcal{K}} q_k x\left(t - \sigma_k\right) = 0,$$

while when $\mathcal{K} = \emptyset$ Eq. (E) yields

$$(E_2) \qquad \frac{d^2}{dt^2}\left[x(t) + \sum_{\mathcal{I}} p_i x\left(t - \tau_i\right)\right] = 0,$$

which admits non-oscillatory solutions. Thus we assume that $\mathcal{K} \neq \emptyset$. When $p_i > 0$ or $p_i < 0$ for $i \in \mathcal{I}$ Eq. (E) leads, respectively, to

$$\frac{d^2}{dt^2}\left[x(t) + \sum_{\mathcal{I}} p_i x\left(t - \tau_i\right)\right] + \sum_{\mathcal{K}} q_k x\left(t - \sigma_k\right) = 0,$$

or

$$\frac{d^2}{dt^2}\left[x(t) - \sum_{\mathcal{I}} |p_i| x\left(t - \tau_i\right)\right] + \sum_{\mathcal{K}} q_k x\left(t - \sigma_k\right) = 0,$$

while in all other cases Eq. (E) can be written in the form

$$(E_3) \quad \frac{d^2}{dt^2}\left[x(t) + \sum_{\mathcal{I}} p_i x\left(t - \tau_i\right) - \sum_{\mathcal{J}} r_j x\left(t - \rho_j\right)\right] + \sum_{\mathcal{K}} q_k x\left(t - \sigma_k\right) = 0,$$

where $p_i > 0$ and $r_j > 0$ for $i \in I$, $j \in J$. Observe that the former two equations are special cases of the latter one and therefore it suffices to study Eq. (E_3).

It is easy to see (cf. [9, 11]) that in the case where

$$I_1 = \{i \in I : \tau_i < 0\} \subseteq I, \quad J_1 = \{j \in J : \rho_j < 0\} \subseteq J$$

are nonempty, by taking

$$\tau = \max_{i \in I_1} |\tau_i| \text{ and } \rho = \max_{j \in J_1} |\rho_j|$$

Eq. (E_3) leads to an equation of the same form with $\tau_i > 0$ and $\rho_j > 0$ for $i \in I$ and $j \in J$. So in the sequel we will assume $\tau_i > 0$ and $\rho_j > 0$ for $i \in I$ and $j \in J$. Also, because q_k, $\sigma_k \in \mathbb{R}$, (E_3) can be written in the following form

$$(1) \quad \frac{d^2}{dt^2}\left[x(t) + \sum_{I} p_i x\left(t - \tau_i\right) - \sum_{J} r_j x\left(t - \rho_j\right)\right]$$
$$+ \sum_{K_1} q_k x\left(t - \sigma_k\right) + \sum_{K_2} \hat{q}_k x\left(t + \hat{\sigma}_k\right) - \sum_{S_1} g_s x\left(t - \mu_s\right) - \sum_{S_2} \hat{g}_s x\left(t + \hat{\mu}_s\right) = 0,$$

where I, J, K_1, K_2, S_1, S_2 are initial segments of natural numbers, p_i, τ_i, r_j, ρ_j, q_k, $\hat{q}_k$, g_s, $\hat{g}_s \in (0, \infty)$ and σ_k, $\hat{\sigma}_k$, μ_s, $\hat{\mu}_s \in [0, \infty)$ for $i \in I$, $j \in J$, $k \in K_1 \cup K_2$, $s \in S_1 \cup S_2$.

Our aim in this paper is to obtain a necessary and sufficient condition under which all solutions of Eq. (1) oscillate. Indeed, we prove that every solution of Eq. (1) oscillates if and only if its characteristic equation

$$(2) \quad F(\lambda) \equiv \lambda^2 + \lambda^2 \sum_I p_i e^{-\lambda \tau_i} - \lambda^2 \sum_J r_j e^{-\lambda \rho_j} + \sum_{K_1} q_k e^{-\lambda \sigma_k}$$
$$+ \sum_{K_2} \hat{q}_k e^{\lambda \hat{\sigma}_k} - \sum_{S_1} g_s e^{-\lambda \mu_s} - \sum_{S_2} \hat{g}_s e^{\lambda \hat{\mu}_s} = 0$$

has no real roots. That is, the oscillatory character of the solutions is determined by the roots of the characteristic equation. This is in contrast with the fact that the stability character is not determined by the characteristic roots. Some of these differences as well as some applications of neutral differential equations are discussed in [2, 4, 5, 6, 7, 14, 15, 26, 27]. Especially, second-order neutral differential equations were encountered in the study of vibrating masses attached to an elastic bar and also as the Euler equations in some variational problems (see Hale [15, p. 7]].

Necessary and sufficient conditions (in terms of the characteristic equation) for the oscillation of all solutions of first order neutral differential equations have been obtained by Sficas and Stavroulakis [25], Grove, Ladas and Meimaridou [13], Kulenoviĉ, Ladas and Meimaridou [17], Schultz [24], Grammatikopoulos, Sficas and Stavroulakis [10], Farrell [8], and Grammatikopoulos and Stavroulakis [11, 12]. Necessary and sufficient conditions for the oscillation of second-order neutral equations were recently obtained by Ladas, Partheniadis and Sficas [19, 20], Partheniadis [23], and Kulenoviĉ, Ladas, and Sficas [18]. For higher-order equations see [21, 16, 28, 3]. See also Arino and Győri [1].

Let $T = \max_{i,j,k,s} \{r_i, p_j, \sigma_k, \hat{\sigma}_k, \mu_s, \hat{\mu}_s\}$. We say that $x(t)$ is a solution of Eq. (1) provided there exists $t_0 \in \mathbb{R}$ such that $x \in C([t_0 - T, \infty), \mathbb{R})$,

$$x(t) + \sum_I p_i x(t - \tau_i) - \sum_J r_j x(t - \rho_j) \in C^2[t_0, \infty)$$

and Eq. (1) holds for $t \geq t_0$.

As is customary, a solution is called *oscillatory* if it has arbitrarily large zeros. Otherwise it is called *nonoscillatory*; that is, if it is eventually positive or eventually negative.

In the sequel all functional inequalities that we write are assumed to hold eventually; that is, for sufficiently large t.

In the case where I, J, K_1, K_2, S_1, S_2 are nonempty we can assume, without loss of generality, that

$$I = \{1, 2, \ldots, l\}, J = \{1, 2, \ldots, m\}, K_1 = \{1, 2, \ldots, n_1\}, K_2 = \{1, 2, \ldots, n_2\},$$
$$S_1 = \{1, 2, \ldots, \nu_1\}, S_2 = \{1, 2, \ldots, \nu_2\},$$
$$0 < \tau_1 < \cdots < \tau_l, 0 < \rho_1 < \cdots < \rho_m, 0 \leq \sigma_1 < \cdots < \sigma_{n_1}, 0 \leq \hat{\sigma}_1 < \cdots < \hat{\sigma}_{n_2}$$
$$0 \leq \mu_1 < \cdots < \mu_{\nu_1}, 0 \leq \hat{\mu}_1 < \cdots < \hat{\mu}_{\nu_2},$$

and

$$\tau_i \neq \rho_j,\ i \in I,\ j \in J; \sigma_k \neq \mu_s,\ k \in K_1,\ s \in S_1; \hat{\sigma}_k \neq \hat{\mu}_s,\ k \in K_2,\ s \in S_2,$$

since otherwise the terms in Eq. (1) can be abbreviated. Also for convenience we use the following notations

$$P = \sum_I p_i,\ R = \sum_J r_j,\ Q_1 = \sum_{K_1} q_k,$$

$$Q_2 = \sum_{K_2} \hat{q}_k,\ G_1 = \sum_{S_1} g_s,\ G_2 = \sum_{S_2} \hat{g}_s$$

and

$$Q = Q_1 + Q_2,\ \ G = G_1 + G_2.$$

Note that $F(0) = Q - G$ and when $Q - G = 0$ the characteristic equation (2) has a real root. Thus, we will assume that $Q - G \neq 0$. In the sequel we will consider Eq. (1) in the case where $Q - G > 0$ since the case $Q - G < 0$ can be treated similarly (cf. [11, 12, 3]).

2. Preliminaries

In this section we establish some useful lemmas which will be used in the proof of our main theorem (cf. [10, 11, 12, 3]).

LEMMA 1. *Consider Eq. (1). Then the following inequalities*

(3) $$\hat{\sigma}_{n_2} > \hat{\mu}_{\nu_2},$$

(4) $\max\{\tau_l, \sigma_{n_1}\} > \max\{\rho_m, \mu_{\nu_1}\}$ *or* $\tau_l \geqq \mu_{\nu_1}$ *when*

$$\tau_l = \max\{\tau_l, \sigma_{n_1}\}\ \ and\ \ \mu_{\nu_1} = \max\{p_m, \mu_{\nu_1}\},$$

and

(5) $$m = \min_{\lambda \in \mathbb{R}} F(\lambda) > 0$$

are necessary conditions for the characteristic equation (2) to have no real roots.

PROOF. As $F(0) = Q - G > 0$ and Eq. (2) has no real roots it follows that $F(\lambda) > 0$ for $\lambda \in \mathbb{R}$. Thus $F(+\infty)$ must be positive or $+\infty$. But when $\hat{\sigma}_{n_2} < \hat{\mu}_{\nu_2}$ it follows that $F(+\infty) = -\infty$; that is, Eq. (2) has a real root. This is impossible and thus condition (3) must hold. Also $F(-\infty)$ must be positive or $+\infty$. But when $\max\{\tau_l, \sigma_{n_1}\} < \max\{\rho_m, \mu_{\nu_1}\}$ or $\sigma_{n_1} = \max\{\tau_l, \sigma_{n_1}\} \leqq \max\{\rho_m, \mu_{\nu_1}\} = \rho_m$ or $\tau_l = \max\{\tau_l, \sigma_{n_1}\} < \max\{\rho_m, \mu_{\nu_1}\} = \mu_{\nu_1}$ it follows that $F(-\infty) = -\infty$. This is impossible and thus condition (4) must hold. Finally, since Eq. (2) has no real roots and $F(+\infty) = F(-\infty) = +\infty$ it follows that (5) holds. The proof of the lemma is complete.

From (5) it follows that for all $\lambda \in \mathbb{R}$

(6) $$F(\lambda) \geqq m\ (\text{equivalently } F(-\lambda) \geqq m).$$

LEMMA 2. *Let $x(t)$ be a solution of Eq.(1) and let a, b and c be real numbers. Then each one of the functions*

$$x(t-a), \ \int_{t-b}^{t-c} x(\xi)d\xi, \ and \ \int_{t}^{\infty} x(\xi)d\xi \ for \ x(t) \in L^{1}[t_{0}, \infty) \ and \ \lim_{t\to\infty} x(t) = 0$$

is a solution of (1).

PROOF. The conclusion follows easily from the linearity of Eq. (1) and its autonomous nature.

DEFINITION. Consider the equation (1). Then the set of all solutions $w(t)$ of Eq. (1) that are at least μ-times ($\mu \geq 2$) continuously differentiable and such that

$$(-1)^{\nu} w^{(\nu)}(t) > 0, \ \nu = 0, 1, \ldots, \mu, \ and \ \lim_{t\to\infty} w^{(\nu)}(t) = 0, \ \nu = 0, 1, \ldots, \mu - 1$$

is called Class I_{μ}, while the set of all solutions $w(t)$ of Eq. (1) that are at least μ-times continuously differentiable and such that

$$w^{(\nu)}(t) > 0, \ \nu = 0, 1, \ldots, \mu \ and \ \lim_{t\to\infty} w^{(\nu)}(t) = \infty, \ \nu = 0, 1, \ldots, \mu - 1$$

is called Class II_{μ}.

LEMMA 3. *Let $x(t)$ be a nonoscillatory solution of Eq. (1). Then Eq. (1) also has a nonoscillatory solution $w(t)$ such that either $w(t) \in$ Class I_4 or $w(t) \in$ Class II_4.*

PROOF. Without loss of generality $x(t)$ can be considered eventually positive. Set

$$z(t) = (\Phi_1 x)(t) = \frac{d}{dt}[\zeta_x(t)]$$

$$+ \sum_{K_1} q_k \int_{t-T}^{t-\sigma_k} x(\xi)d\xi + \sum_{K_2} \hat{q}_k \int_{t-T}^{t+\hat{\sigma}_k} x(\xi)d\xi$$

$$- \sum_{S_1} g_s \int_{t-T}^{t-\mu_s} x(\xi)d\xi - \sum_{S_2} \hat{g}_s \int_{t-T}^{t+\hat{\mu}_s} x(\xi)d\xi,$$

$$u(t) = (\Phi_1 z)(t), \ v(t) = (\Phi_1 u)(t) \ and \ w(t) - (\Phi_1 v)(t),$$

where

(7) $$\zeta_x(t) \equiv x(t) + \sum_I p_i x(t - \tau_i) - \sum_J r_j x(t - \rho_j)$$

and

(8) $$T = \max\{\tau_l, \rho_m, \sigma_{n_1}, \hat{\sigma}_{n_2}, \mu_{\nu_1}, \hat{\mu}_{\nu_2}\}.$$

Then it is easy to see that each one of the functions $z(t)$, $u(t)$, $v(t)$, $w(t)$ is a solution of Eq. (1). Furthermore, $z \in C^1[t_0, \infty)$, $u \in C^2[t_0, \infty)$, $v \in$

$C^3[t_0, \infty)$, $w \in C^4[t_0, \infty)$ and they are eventually strictly monotone functions. We have

$$\dot{z}(t) = -(Q - G)x(t - T) < 0, \tag{9}$$

and

$$\begin{cases} u^{(\nu)}(t) = -(Q - G)z^{(\nu-1)}(t - T), & \nu = 1, 2 \\ v^{(\nu)}(t) = -(Q - G)u^{(\nu-1)}(t - T), & \nu = 1, 2, 3 \\ w^{(\nu)}(t) = -(Q - G)v^{(\nu-1)}(t - T), & \nu = 1, 2, 3, 4. \end{cases} \tag{10}$$

Since $\dot{z}(t) < 0$, $z(t)$ is strictly decreasing. So either $\lim_{t\to\infty} z(t) = -\infty$ or $\lim_{t\to\infty} z(t) = l \in \mathbb{R}$. In the former case it follows that $\lim_{t\to\infty} \dot{u}(t) = \infty$, which implies that $u(t) > 0$ and $\lim_{t\to\infty} u(t) = \infty$. Consequently, we find

$$\lim_{t\to\infty} v(t) = \lim_{t\to\infty} \dot{v}(t) = \lim_{t\to\infty} \ddot{v}(t) = -\infty$$

and finally

$$\lim_{t\to\infty} w^{(\nu)}(t) = \infty, \ \nu = 0, 1, 2, 3,$$

that is, $w(t) \in$ Class II_4. In the latter case, we claim that $l = 0$. Suppose that $l \neq 0$. Then, integrating (9) over the interval $[t_0, \infty)$, we see that

$$z(t_0) - l = (Q - G) \int_{t_0}^{\infty} x(\xi - T)\, d\xi,$$

which shows that $x \in L^1[t_0, \infty)$ and therefore $z \in L^1[t_0, \infty)$. Since $z(t)$ is strictly decreasing it follows that $\lim_{t\to\infty} z(t) = 0$, which is a contradiction. Thus we have established that $\dot{z}(t) < 0$ and $\lim_{t\to\infty} z(t) = 0$. This implies that $z(t) > 0$. Now, in view of (10), we see that

$$\lim_{t\to\infty} w^{(\nu)}(t) = 0 \text{ and } (-1)^\nu w^{(\nu)}(t) > 0, \ \nu = 0, 1, 2, 3.$$

Moreover, $w^{(4)}(t) > 0$. Otherwise, $w^{(4)}(t) < 0$ which, together with $w^{(3)}(t) < 0$, contradicts the positivity of $w(t)$. Therefore in this case $w(t)$ belongs to Class I_4. The proof is complete.

LEMMA 4.

(a) *Let $x(t) \in$ Class I_4. Then for any $\omega > 0$*

$$(-1)^\nu x^{(\nu)}(t) < \frac{1}{w^\nu} x(t - \nu\omega), \ \nu = 1, 2, 3.$$

(b) *Let $x(t) \in$ Class II_4. Then for any $\omega > 0$*

$$x^{(\nu)}(t) < \frac{1}{w^\nu} x(t + \nu\omega), \ \nu = 1, 2, 3.$$

PROOF. It follows easily by applying the mean-value theorem to the functions $x(t)$, $\dot{x}(t)$, $\ddot{x}(t)$.

LEMMA 5.

(a) *Let $x(t) \in$ Class I_4. Then there exists a solution $w(t)$ of Eq. (1) which belongs to Class I_4, such that*

$$\Lambda(w) = \{\lambda > 0 : -\ddot{w}(t) + \lambda^2 w(t) \leqq 0\} \neq \emptyset.$$

(b) *Let $x(t) \in$ Class II_4. Then there exists a solution $w(t)$ of Eq. (1) which belongs to Class II_4, such that $\Lambda(w) \neq \emptyset$.*

PROOF. (a)   Let $x(t) \in$ Class I_4. Set

$$(11) \quad z(t) = (\Phi_2 x)(t) = \frac{d}{dt}[\zeta_x(t)]$$

$$+ \sum_{K_1} q_k \int_{t-2T}^{t-\sigma_k} x(\xi)d\xi + \sum_{K_2} \hat{q}_k \int_{t-2T}^{t+\hat{\sigma}_k} x(\xi)d\xi$$

$$- \sum_{S_1} g_s \int_{t-2T}^{t-\mu_s} x(\xi)d\xi - \sum_{S_2} \hat{g}_s \int_{t-2T}^{t+\hat{\sigma}_k} x(\xi)d\xi,$$

and

$$(12) \qquad\qquad w(t) = (\Phi_2 z)(t),$$

where $\zeta_x(t)$ and T are as in (7) and (8). It is easy to see that $z(t)$ and $w(t)$ both belong to Class I_4 and that

$$-\ddot{w}(t) + (Q - G)^2 x(t - 4T) = 0.$$

Since $x(t)$ and $z(t)$ belong to Class I_4, in view of Lemma 4(a), we have

$$-\dot{x}(t) < \frac{1}{T}x(t - T)$$

and

$$-\dot{z}(t) < \frac{1}{T}z(t - T).$$

Thus, (11) and (12) imply that

$$z(t) < \left(\frac{R}{T} + 3QT\right)x(t - 2T)$$

and

$$w(t) < \left(\frac{R}{T} + 3QT\right)z(t - 2T) < \left(\frac{R}{T} + 3QT\right)^2 x(t - 4T) \equiv A^2 x(t - 4T).$$

Therefore

$$0 = -\ddot{w}(t) + (Q - G)^2 x(t - 4T) > -\ddot{w}(t) + \left(\frac{Q - G}{A}\right)^2 w(t),$$

which says that

$$0 < \frac{Q - G}{A} \in \Lambda(w),$$

that is, $\Lambda(w) \neq \emptyset$.

(b)   Let $x(t) \in$ Class II_4. Set

$$(11') \quad z(t) = (\Phi_3 x)(t) = -\frac{d}{dt}[\zeta_x(t)]$$

$$+ \sum_{K_1} q_k \int_{t-\sigma_k}^{t+2T} x(\xi)d\xi + \sum_{K_2} \hat{q}_k \int_{t+\hat{\sigma}_k}^{t+2T} x(\xi)d\xi$$

$$- \sum_{S_1} g_s \int_{t-\mu_s}^{t+2T} x(\xi)d\xi - \sum_{S_2} \hat{g}_s \int_{t+\hat{\mu}_s}^{t+2T} x(\xi)d\xi,$$

and

$$(12') \qquad\qquad w(t) = (\Phi_3 z)(t),$$

where $\zeta_x(t)$ and T are as in (7) and (8). It is easy to see that $z(t)$ and $w(t)$ both belong to Class II_4 and that

$$-\ddot{w}(t) + (Q-G)^2 x(t+4T) = 0.$$

Since $x(t)$ and $z(t)$ belong to Class II_4, in view of Lemma 4(b), we have

$$\dot{x}(t) < \frac{1}{T}x(t+T)$$

and

$$\dot{z}(t) < \frac{1}{T}z(t+T).$$

Thus, $(11')$ and $(12')$ imply that

$$z(t) < \left(\frac{R}{T} + 3QT\right)x(t+2T)$$

and

$$w(t) < \left(\frac{R}{T} + 3QT\right)z(t+2T) < \left(\frac{R}{T} + 3QT\right)^2 x(t+4T) \equiv B^2 x(t+4T).$$

Therefore

$$0 = -\ddot{w}(t) + (Q-G)^2 x(t+4T) > -\ddot{w}(t) + \left(\frac{Q-G}{B}\right)^2 w(t),$$

which says that

$$0 < \frac{Q-G}{B} \in \Lambda(w),$$

that is, $\Lambda(w) \neq \emptyset$. The proof of the lemma is complete.

LEMMA 6. *Assume that (3) and (4) hold. Then we have the following.*

(a) *Let $x(t) \in$ Class I_4 for which $\Lambda(x) \neq \emptyset$. Then the set $\Lambda(x)$ has an upper bound which is independent of x.*

(b) *Let $x(t) \in$ Class II_4 for which $\Lambda(x) \neq \emptyset$. Then the set $\Lambda(x)$ has an upper bound which is independent of x.*

PROOF. (a) Let $x(t) \in$ Class I_4. Set

$$(13) \quad w(t) = (\Phi_4 x)(t) = \frac{d}{dt} [\zeta_x(t)]$$

$$- \sum_{K_1} q_k \int_{t-\sigma_k}^{t} x(\xi)d\xi + \sum_{K_2} \hat{q}_k \int_{t}^{t+\hat{\sigma}_k} x(\xi)d\xi$$

$$+ \sum_{S_1} g_s \int_{t-\mu_s}^{t} x(\xi)d\xi - \sum_{S_2} \hat{g}_s \int_{t}^{t+\hat{\mu}_s} x(\xi)d\xi,$$

where $\zeta_x(t)$ is given by (7). It is easy to see that $w(t)$ belongs to Class I_4. Since (4) holds, if we set $\alpha = \max\{\rho_m, \mu_{\nu_1}\}$ the following cases are possible.

$$\sigma_{n_1} > \alpha \text{ or } \tau_l > \alpha \text{ or } \tau_l \geq \alpha = \mu_{\nu_1} \text{ when } \tau_l = \max\{\tau_l, \sigma_{n_1}\}.$$

1. $\sigma_{n_1} > \alpha$. As $w(t) > 0$, from (13), we have

$$- \sum_{J} r_j \dot{x}(t - \rho_j) + \sum_{K_2} \hat{q}_k \int_{t}^{t+\hat{\sigma}_k} x(\xi)d\xi + \sum_{S_1} g_s \int_{t-\mu_s}^{t} x(\xi)d\xi$$

$$> -\dot{x}(t) - \sum_{I} p_i \dot{x}(t - \tau_i) + \sum_{K_1} q_k \int_{t-\sigma_k}^{t} x(\xi)d\xi + \sum_{S_2} \hat{g}_s \int_{t}^{t+\hat{\mu}_s} x(\xi)d\xi$$

from which it follows that

$$- R\dot{x}(t - \alpha) + (Q_2 \hat{\sigma}_{n_2} + G_1 \mu_{\nu_1}) x(t - \alpha) > q_{n_1} \int_{t-\sigma_{n_1}}^{t} x(\xi)d\xi.$$

Set $\omega_1 = \sigma_{n_1} - \alpha > 0$. Then we see that

$$- R\dot{x}(t) + (Q_2 \hat{\sigma}_{n_2} + G_1 \mu_{\nu_1}) x(t)$$

$$> q_{n_1} \int_{t-\sigma_{n_1}+\alpha}^{t+\alpha} x(\xi)d\xi > q_{n_1} \int_{t-\omega_1}^{t} x(\xi)d\xi$$

$$> q_{n_1} \int_{t-\omega_1}^{t-\frac{1}{2}\omega_1} x(\xi)d\xi > \frac{1}{2}\omega_1 q_{n_1} x\left(t - \frac{1}{2}\omega_1\right).$$

Integrating the inequality

$$- R\dot{x}(t) + (Q_2 \hat{\sigma}_{n_2} + G_1 \mu_{\nu_1}) x(t) > \frac{1}{2}\omega_1 q_{n_1} x\left(t - \frac{1}{2}\omega_1\right)$$

over the interval $[t - \frac{1}{4}\omega_1, t]$, we obtain

$$\left[R + \frac{1}{4}\omega_1 (Q_2 \hat{\sigma}_{n_2} + G_1 \mu_{\nu_1})\right] x\left(t - \frac{1}{4}\omega_1\right) > \frac{\omega_1^2}{8} q_{n_1} x\left(t - \frac{1}{2}\omega_1\right)$$

or

$$\frac{2}{\omega_1^2 q_{n_1}} [4R + \omega_1 (Q_2 \hat{\sigma}_{n_2} + G_1 \mu_{\nu_1})] x(t) > x\left(t - \frac{1}{4}\omega_1\right)$$

Applying Lemma 4(a) for $\omega = \frac{1}{8}\omega_1$, the last inequality yields

$$- \ddot{x}(t) + A_1 x(t) > 0,$$

where

$$A_1 = \frac{128}{\omega_1^4 q_{n_1}} \left(4R + \omega_1 \left(Q_2 \hat{\sigma}_{n_2} + G_1 \mu_{\nu_1} \right) \right).$$

Thus, the positive number $\lambda_1 = A_1^{\frac{1}{2}}$ is an upper bound of the set $\Lambda(x)$.

2. $\tau_l > \alpha$. As previously, from (13) it follows that

$$-R\dot{x}(t) + \left(Q_2 \hat{\sigma}_{n_2} + G_1 \mu_{\nu_1} \right) x(t) > -p_l \dot{x} \left(t - \omega_2 \right),$$

where $\omega_2 = \tau_l - \alpha > 0$. Integrating over the interval $[t - \omega_2, t]$, we obtain

$$\left[R + p_l + \omega_2 \left(Q_2 \hat{\sigma}_{n_2} + G_1 \mu_{\nu_1} \right) \right] x(t) > p_l x \left(t - \omega_2 \right)$$

and applying Lemma 4(a) for $\omega = \frac{1}{2} \omega_2$, we obtain

$$-\ddot{x}(t) + A_2 x(t) > 0,$$

where

$$A_2 = \frac{4}{\omega_2^2 p_l} \left[R + p_l + \omega_2 \left(Q_2 \hat{\sigma}_{n_2} + G_1 \mu_{\nu_1} \right) \right].$$

Thus, the positive number $\lambda_2 = A_2^{\frac{1}{2}}$ is an upper bound of the set $\Lambda(x)$.

3. $\tau_l = \alpha = \mu_{v_1}$. In this case we want to show that the positive number

$$\lambda_3 \equiv \max \left\{ \left(\frac{Q}{p_l} \right)^{\frac{1}{2}}, \left(\frac{2^9 R}{(Q - G)\omega_3^4} \right)^{\frac{1}{2}} \right\} \notin \Lambda(x), \text{ where } \omega_3 = (\tau_l - \rho_m) > 0.$$

Otherwise $\lambda_3 \in \Lambda(x)$. Since $\Lambda(x)$ is an interval this implies that

$$\left(\frac{Q}{P_l} \right)^{\frac{1}{2}} \in \Lambda(x) \text{ and } \left(\frac{2^9 R}{(Q - G)\omega_3^4} \right)^{\frac{1}{2}} \in \Lambda(x).$$

Since

$$\left(\frac{Q}{p_l} \right)^{\frac{1}{2}} \in \Lambda(x),$$

it follows that

$$-\ddot{x} \left(t - \tau_l \right) + \frac{Q}{p_l} x \left(t - \tau_l \right) \leqq 0.$$

This implies that

$$(14) \qquad X \left(t - \tau_l \right) \equiv -\dot{x} \left(t - \tau_l \right) - \frac{Q}{p_l} \int_{t - \tau_l}^{\infty} x(\xi) d\xi \geqq 0$$

since

$$\dot{X} \left(t - \tau_l \right) = -\ddot{x} \left(t - \tau_l \right) + \frac{Q}{p_l} x \left(t - \tau_l \right) \leqq 0$$

and $\lim_{t \to \infty} X \left(t - \tau_l \right) = 0$. Set

$$w(t) = \frac{d}{dt} \left[\zeta_x(t) \right] + \sum_{K_1} q_k \int_{t - \tau_l}^{t - \sigma_k} x(\xi) d\xi + \sum_{K_2} \hat{q}_k \int_{t - \tau_l}^{t + \hat{\sigma}_k} x(\xi) d\xi,$$

where $\zeta_x(t)$ is given by (7). Using (14) we see that

$$w(t) = \dot{x}(t) + \sum_I p_i \dot{x}(t - \tau_i) - \sum_J r_j \dot{x}(t - \rho_j)$$

$$+ \left(Q_1 \int_{t-\tau_l}^{\infty} x(\xi)d\xi - \sum_{K_1} q_k \int_{t-\sigma_k}^{\infty} x(\xi)d\xi \right)$$

$$+ \left(Q_2 \int_{t-\tau_l}^{\infty} x(\xi)d\xi - \sum_{K_2} \hat{q}_k \int_{t+\hat{\sigma}_k}^{\infty} x(\xi)d\xi \right)$$

$$< \left[p_l \dot{x}(t - \tau_l) + Q \int_{t-\tau_l}^{\infty} x(\xi)d\xi \right] - \sum_J r_j \dot{x}(t - \rho_j) < -R\dot{x}(t - \rho_m).$$

Taking into account that $w(t) \in$ Class I_4, if we integrate the inequality

$$w(t) < -R\dot{x}(t - \rho_m)$$

over the interval $\left[t - \frac{1}{2}\omega_3, t\right]$, we obtain

$$(15) \qquad w(t) < \frac{2R}{\omega_3} x\left(t - \rho_m - \frac{1}{2}\omega_3\right).$$

Furthermore

$$\dot{w}(t) = -Qx(t - \tau_l) + \sum_{S_1} g_s x(t - \mu_s) + \sum_{S_2} \hat{g}_s x(t + \hat{\mu}_s)$$

$$< -Qx(t - \tau_l) + Gx(t - \tau_l) = -(Q - G)x(t - \tau_l),$$

and integrating over the interval $\left[t - \frac{1}{4}\omega_3, t\right]$, we obtain

$$(16) \qquad w(t) > \frac{1}{4}(Q - G)\omega_3 x\left(t - \tau_l + \frac{1}{4}\omega_3\right).$$

Combining the inequalities (15) and (16) and applying Lemma 4(a) for $\omega = \frac{1}{8}\omega_3$, we see that

$$-\ddot{x}(t) + \frac{2^9 R}{(Q - G)\omega_3^4} x(t) > 0.$$

This implies that

$$\left(\frac{2^9 R}{(Q - G)\omega_3^4} \right)^{\frac{1}{2}} \notin \Lambda(x)$$

which is a contradiction.

(b) Let $x(t) \in$ Class II_4. Set

$$(13') \qquad w(t) = -(\Phi_4 x)(t).$$

It is easy to see that $w(t) \in$ Class II_4. As $w(t) > 0$, from (13'), we have

$$\sum_J r_j \dot{x}(t - \rho_j) + \sum_{K_1} q_k \int_{t-\sigma_k}^t x(\xi)d\xi + \sum_{S_2} \hat{g}_s \int_t^{t+\hat{\mu}_s} x(\xi)d\xi$$

$$> \dot{x}(t) + \sum_I p_i \dot{x}(t - \tau_i) + \sum_{K_2} \hat{q}_k \int_t^{t+\hat{\sigma}_k} x(\xi)d\xi + \sum_{S_1} g_s \int_{t-\mu_s}^t x(\xi)d\xi$$

from which it follows that

$$R\dot{x}(t) + (Q_1\sigma_{n_1} + G_2\hat{\mu}_{\nu_2})\, x(t + \hat{\mu}_{\nu_2}) > \hat{q}_{n_2} \int_t^{t+\hat{\sigma}_{n_2}} x(\xi)d\xi$$

Set $\omega_4 = \hat{\sigma}_{n_2} - \hat{\mu}_{\nu_2} > 0$. Then we see that

$$R\dot{x}(t) + (Q_1\sigma_{n_1} + G_2\hat{\mu}_{\nu_2})\, x(t) > R\dot{x}(t - \hat{\mu}_{\nu_2}) + (Q_1\sigma_{n_1} + G_2\hat{\mu}_{\nu_2})\, x(t)$$

$$> \hat{q}_{n_2} \int_{t-\hat{\mu}_{\nu_2}}^{t+\hat{\sigma}_{n_2}-\hat{\mu}_{\nu_2}} x(\xi)d\xi$$

$$> \hat{q}_{n_2} \int_t^{t+\omega_4} x(\xi)d\xi > \hat{q}_{n_2} \int_{t+\frac{1}{2}\omega_4}^{t+\omega_4} x(\xi)d\xi$$

$$> \frac{1}{2}\omega_4 \hat{q}_{n_2} x\left(t + \frac{1}{2}\omega_4\right).$$

Integrating the inequality

$$R\dot{x}(t) + (Q_1\sigma_{n_1} + G_2\hat{\mu}_{\nu_2})\, x(t) > \frac{1}{2}\omega_4 \hat{q}_{n_2} x\left(t + \frac{1}{2}\omega_4\right)$$

over the interval $\left[t,\, t + \frac{1}{4}\omega_4\right]$ and applying Lemma 4(b) for $\omega = \frac{1}{8}\omega_4$, we obtain

$$-\ddot{x}(t) + A_4 x(t) > 0,$$

where

$$A_4 = \frac{128}{\omega_4^4 \hat{q}_{n_2}} \left[4R + \omega_4 \left(Q_1\sigma_{n_1} + G_2\hat{\mu}_{\nu_2}\right)\right].$$

Thus, the positive $\lambda_4 = A_4^{\frac{1}{2}}$ is an upper bound of the set $\Lambda(x)$.

The proof of the lemma is complete.

3. Main Result

Our main result is the following.

THEOREM. *Consider the second-order neutral differential equation*

$$(1) \quad \frac{d^2}{dt^2} \left[x(t) + \sum_I p_i x(t - \tau_i) - \sum_J r_j x(t - \rho_j)\right]$$

$$+ \sum_{K_1} q_k x(t - \sigma_k) + \sum_{K_2} \hat{q}_k x(t + \hat{\sigma}_k)$$

$$- \sum_{S_1} g_s x(t - \mu_s) - \sum_{S_2} \hat{g}_s x(t + \hat{\mu}_s) = 0,$$

where I, J, K_1, K_2, S_1, S_2 are initial segments of natural numbers, p_i, τ_i, r_j, ρ_j, q_k, $\hat{q}_k$, g_s, $\hat{g}_s \in (0, \infty)$ and σ_k, $\hat{\sigma}_k$, μ_s, $\hat{\mu}_s \in [0, \infty)$ for $i \in I$, $j \in J$, $k \in K_1 \cup K_2$, $s \in S_1 \cup S_2$. Then a necessary and sufficient condition for the oscillation of all solutions of Eq. (1) is that its characteristic equation

$$(2) \quad \lambda^2 + \lambda^2 \sum_I p_i e^{-\lambda \tau_i} - \lambda^2 \sum_J r_j e^{-\lambda \rho_j}$$
$$+ \sum_{K_1} q_k e^{-\lambda \sigma_k} + \sum_{K_2} \hat{q}_k e^{\lambda \hat{\sigma}_k}$$
$$- \sum_{S_1} g_s e^{-\lambda \mu_s} - \sum_{S_2} \hat{g}_s e^{\lambda \hat{\mu}_s} = 0$$

has no real roots.

PROOF. The theorem will be proved in the contrapositive form: There is a nonoscillatory solution of (1) if and only if the characteristic equation (2) has a real root. Assume first that (2) has a real root λ. Then (1) has the nonoscillatory solution $x(t) = e^{\lambda t}$.

Assume, conversely, that there is a nonoscillatory solution of (1) and, for the sake of contradiction, that Eq. (2) has no real roots. Then by Lemma 3, Eq. (1) also has a nonoscillatory solution $x(t)$ which belongs either to Class I_4 or to Class II_4. Consider the following cases:

(a) $x(t) \in$ Class I_4. For this solution $x(t)$, by Lemma 5(a), we can assume, without loss of generality, that $\Lambda(x) \neq \emptyset$. Let $\lambda_5 \in \Lambda(x)$. Also, by Lemma 6(a), there exists a positive number, say λ_0, such taht $\Lambda(x)$ is bounded above by λ_0.

For $\lambda \in \Lambda(x)$ consider the functions

$$y(t) = (\Psi_1 x)(t) \equiv (\Phi_1 x)(t),$$
$$z(t) = (\Psi_1 y)(t),$$
$$w(t) = (\Psi_2 z)(t) = \ddot{z}(t) - \lambda \dot{z}(t)$$

and

$$(17) \quad u(t) = (\Psi_3 w)(t) \frac{d}{dt}[\zeta_w(t)] + \sum_{K_1} q_k \int_{t-2T}^{t-\sigma_k} w(\xi)d\xi$$
$$+ \sum_{K_2} \hat{q}_k \int_{t-2T}^{t+\hat{\sigma}_k} w(\xi)d\xi + \sum_{S_1} g_s \int_{t-\mu_s}^{\infty} w(\xi)d\xi$$
$$+ \sum_{S_2} \hat{g}_s \int_{t+\hat{\mu}_s}^{\infty} w(\xi)d\xi + \lambda^2(1+P) \int_{t-2T}^{\infty} w(\xi)d\xi,$$

where $(\Phi_1 x)(t)$, $\zeta_x(t)$, T are as in Lemma 3. It is easy to see that $y(t)$, $z(t)$, $w(t)$, and $u(t)$ belong to Class I_4.

Since Eq. (2) has no real roots, then, by Lemma 1, the inequalities (5) and (6) hold. We shall show that $\left(\lambda^2 + m_0\right)^{\frac{1}{2}} \in \Lambda(u)$, where $m_0 = m/N_1 > 0$ with

$$N_1 = \left(1 + P + R + \frac{Q+G}{\lambda_5^2}\right) e^{2\lambda_0 T}.$$

To this end it suffices to show that

$$-\ddot{u}(t) + \left(\lambda^2 + m_0\right) u(t) \leqq 0.$$

Define $\phi(t) = e^{\lambda t} \left(-\dot{w}(t)\right)$. Then $\phi(t) > 0$ and for $\lambda \in \Lambda(x)$ with $\lambda > \lambda_5$, we have

$$\dot{\phi} = e^{\lambda t} \left[-\ddot{w}(t) - \lambda \dot{w}(t)\right]$$
$$= e^{\lambda t} \left[-z^{(4)}(t) + \lambda^2 \ddot{z}(t)\right]$$
$$= (Q-G)^2 e^{\lambda t} \left[-\ddot{x}(t-2T) + \lambda^2 x(t-2T)\right] \leqq 0,$$

that is, $\phi(t)$ is decreasing. Since $w(t) \in$ Class I_4 and

$$(18) \qquad\qquad -\dot{w}(t) = e^{-\lambda t}\phi(t),$$

we see that

$$(19) \qquad\qquad w(t) = \int_t^\infty -\dot{w}(\xi)d\xi \leqq \frac{1}{\lambda}e^{-\lambda t}\phi(t)$$

and therefore for any $\omega \leqq 2T$

$$(20) \qquad\qquad \int_{t-2T}^{t-\omega} w(\xi)d\xi \leqq \frac{1}{\lambda^2}e^{-\lambda t}\phi(t-2T)\left[e^{2\lambda T} - e^{\lambda\omega}\right].$$

In view of (18), (19), it is easy to see that for any $\omega \in \mathbb{R}$

$$(21) \qquad\qquad \dot{w}(t-\omega) + \lambda^2 \int_{t-\omega}^\infty w(\xi)d\xi \leqq 0$$

and, in view of (20), for any $\omega \leqq 2T$

$$(22) \quad \dot{w}(t-\omega) + \lambda^2 \int_{t-2T}^\infty w(\xi)d\xi$$
$$\leqq \lambda^2 \int_{t-2T}^{t-\omega} w(\xi)d\xi \leqq e^{-\lambda t}\phi(t-2T)\left[e^{2\lambda T} - e^{\lambda\omega}\right].$$

Now, from (17), in view of (18), we obtain

$$(23) \quad -\ddot{u}(t) = \left(\lambda^2 + \lambda^2 P + Q\right) \dot{w}(t-2T)$$
$$= -\left(\lambda^2 + \lambda^2 P + Q\right) e^{2\lambda T} e^{-\lambda t}\phi(t-2T).$$

Also, from (17), in view of (18), (19), (20), (21), and (22), we obtain

$$
\begin{aligned}
u(t) &\leqq e^{-\lambda t}\phi(t-2T)\left(e^{2\lambda T}-1\right)\\
&\quad + e^{-\lambda t}\phi(t-2T)\sum_I p_i\left[e^{2\lambda T}-e^{\lambda\tau_i}\right] + \sum_J r_j e^{-\lambda(t-\rho_j)}\phi\left(t-\rho_j\right)\\
&\quad + \frac{1}{\lambda^2}e^{-\lambda t}\phi(t-2T)\left(\sum_{K_1} q_k\left[e^{2\lambda T}-e^{\lambda\sigma_k}\right] + \sum_{K_2}\hat{q}_k\left[e^{2\lambda T}-e^{-\lambda\hat{\sigma}_k}\right]\right)\\
&\quad + \frac{1}{\lambda^2}e^{-\lambda t}\phi(t-2T)\left(\sum_{S_1} g_s e^{\lambda\mu_s} + \sum_{S_2}\hat{g}_s e^{-\lambda\hat{\mu}_s}\right),
\end{aligned}
$$

which, in view of (2), yields

$$
(24) \qquad u(t) \leqq e^{-\lambda t}\phi(t-2T)\left[-\frac{1}{\lambda^2}F(-\lambda) + \left(1 + P + \frac{Q}{\lambda^2}\right)e^{2\lambda T}\right].
$$

This inequality implies

$$
(25) \quad u(t) \leqq e^{-\lambda t}\phi(t-2T)\left(1 + P + R + \frac{Q+G}{\lambda^2}\right)e^{2\lambda T} \leqq e^{-\lambda t}\phi(t-2T)N_1.
$$

Finally, in view of (23), (24), (25) and (6), we obtain

$$
\begin{aligned}
-\ddot{u}(t) + \left(\lambda^2 + m_0\right)u(t) &= \left[-\ddot{u}(t) + \lambda^2 u(t)\right] + m_0 u(t)\\
&\leqq e^{-\lambda t}\phi(t-2T)\left(-m + m_0 N_1\right) = 0,
\end{aligned}
$$

as was to be shown. Now set

$$
x_0 \equiv x,\ x_1 = \Psi x_0 = \Psi_3\left(\Psi_2\left(\Psi_1\left(\Psi_1 x\right)\right)\right) \equiv u,\ x_2 = \Psi x_1,
$$

and in general

$$
x_\nu = \Psi x_{\nu-1},\ \nu = 1, 2, \ldots
$$

and observe that $x_\nu \in$ Class I_4 with $\Lambda\left(x_\nu\right) \neq \emptyset$, and for

$$
\lambda \in \Lambda(x) \equiv \Lambda\left(x_0\right) \Rightarrow \left(\lambda^2 + m_0\right)^{\frac{1}{2}} \in \Lambda(u) \equiv \Lambda\left(x_1\right)
$$

and after ν steps

$$
\left(\lambda^2 + \nu m_0\right)^{\frac{1}{2}} \in \Lambda\left(x_\nu\right),\ \nu = 1, 2, \ldots,
$$

which is a contradiction since λ_0 is a common upper bound for all $\Lambda\left(x_\nu\right)$. This completes the proof when $x(t) \in$ Class I_4.

(b) $x(t) \in$ Class II_4. By Lemma 5(b) we can assume that $\Lambda(x) \neq \emptyset$. Let $\lambda_6 \in \Lambda(x)$. Also, by Lemma 6(b), there exists a positive number, say λ_0, such that $\Lambda(x)$ is bounded above by λ_0. For $\lambda \in \Lambda(x)$ consider the functions

$$y(t) = (\Phi_5 x)(t) = -\frac{d}{dt}[\zeta_x(t)]$$
$$+ \sum_{K_1} q_k \int_{t-\sigma_k}^{t+T} x(\xi)d\xi + \sum_{K_2} \hat{q}_k \int_{t+\hat{\sigma}_k}^{t+T} x(\xi)d\xi$$
$$- \sum_{S_1} g_s \int_{t-\mu_s}^{t+T} x(\xi)d\xi - \sum_{S_2} \hat{g}_s \int_{t+\hat{\mu}_s}^{t+T} x(\xi)d\xi,$$

$z(t) = (\Phi_5 y)(t)$, $w(t) = \ddot{z}(t) + \lambda \dot{z}(t)$ and

$$u(t) = -\frac{d}{dt}[\zeta_w(t)] + \sum_{K_1} q_k \int_{t-\sigma_k}^{t+2T} w(\xi)d\xi$$
$$+ \sum_{S_2} \hat{q}_k \int_{t+\hat{\sigma}_k}^{t+2T} w(\xi)d\xi + \sum_{S_1} g_s \int_{-\infty}^{t-\mu_s} w(\xi)d\xi$$
$$+ \sum_{S_2} \hat{g}_s \int_{-\infty}^{t+\hat{\mu}_s} w(\xi)d\xi + \lambda^2(1+P) \int_{-\infty}^{t+2T} w(\xi)d\xi,$$

where $\zeta_x(t)$ and T are given by (7) and (8) respectively. It is easy to see that $y(t)$, $z(t)$, $w(t)$, and $u(t)$ belong to Class II_4.

Since Eq. (2) has no real roots inequality (6) holds. We shall show that $\left(\lambda^2 + m_0\right)^{\frac{1}{2}} \in \Lambda(u)$, where $m_0 = m/N_2 > 0$ with

$$N_2 = \left(1 + P + R + \frac{Q+G}{\lambda_6^2}\right)e^{2\lambda_0 T}.$$

To this end it suffices to show that

$$-\ddot{u}(t) + \left(\lambda^2 + m_0\right)u(t) \leqq 0.$$

Define $\phi(t) = e^{-\lambda t}\dot{w}(t)$. Then $\phi(t) > 0$ and for $\lambda \in \Lambda(x)$ with $\lambda > \lambda_6$, we have

$$\dot{\phi}(t) = e^{-\lambda t}[\ddot{w}(t) - \lambda\dot{w}(t)] = e^{-\lambda t}\left[z^{(4)}(t) - \lambda^2 \ddot{z}(t)\right] =$$
$$- (Q-G)^2 e^{-\lambda t}\left[-\ddot{x}(t+2T) + \lambda^2 x(t+2T)\right] \geqq 0,$$

that is, $\phi(t)$ is increasing. Now, as in [28], we extend the definition of $w^{(\nu)}(t)$, $\nu = 0, 1, 2$ such that $w^{(\nu)}(t)$ is continuous, positive and increasing on $(-\infty, \infty)$ and $\lim_{t \to -\infty} w^{(\nu)}(t) = 0$, $\nu = 0, 1$. Then, in view of the fact that

(18')
$$\dot{w}(t) = e^{\lambda t}\phi(t),$$

we see that

(19')
$$w(t) = \int_{-\infty}^{t} \dot{w}(\xi)d\xi \leqq \frac{1}{\lambda}e^{\lambda t}\phi(t)$$

and therefore for any $\omega \leqq 2T$

$$(20') \qquad \int_{t-\omega}^{t+2T} w(\xi)\xi \leqq \frac{1}{\lambda^2}e^{\lambda t}\phi(t+2T)\left[e^{2\lambda T} - e^{-\lambda\omega}\right].$$

Also, in view of $(18')$, $(19')$, it is easily seen that for any $\omega \in \mathbb{R}$

$$(21') \qquad -\dot{w}(t-\omega) + \lambda^2\int_{-\infty}^{t-\omega} w((\xi)d\xi \leqq 0$$

and, in view of $(20')$, for any $\omega \leqq 2T$

$$(22') \quad -\dot{w}(t-\omega) + \lambda^2\int_{-\infty}^{t+2T} w(\xi)d\xi \leqq$$

$$\lambda^2\int_{t-\omega}^{t+2T} w(\xi)d\xi \leqq e^{\lambda t}\phi(t+2T)\left[e^{2\lambda T} - e^{-\lambda\omega}\right].$$

Next, as in case (a), we obtain the following relations analogous to (23), (24), and (25),

$$(23') \qquad -\ddot{u}(t) = -\left(\lambda^2 + \lambda^2 P + Q\right)e^{2\lambda T}e^{\lambda t}\phi(t+2T),$$

$$(24') \qquad u(t) \leqq e^{\lambda t}\phi(t+2T)\left[-\frac{1}{\lambda^2}F(\lambda) + \left(1 + P + \frac{Q}{\lambda^2}\right)e^{2\lambda T}\right],$$

and

$$(25') \qquad u(t) \leqq e^{\lambda t}\phi(t+2T)\left(1 + P + R + \frac{Q+G}{\lambda^2}\right)e^{2\lambda t} \leqq e^{\lambda t}\phi(t+2T)N_2.$$

Finally

$$-\ddot{u}(t) + \left(\lambda^2 + m_0\right)u(t) = \left[-\ddot{u}(t) + \lambda^2 u(t)\right] + m_0 u(t)$$
$$\leqq e^{\lambda t}\phi(t+2T)\left(-m + m_0 N_2\right) = 0,$$

as was to be shown. Now, as in case (a), we are led to a contradiction.

The proof of the theorem is complete.

4. Applications and Examples

In this section we apply our main theorem and obtain some useful corollaries. It is to be emphasized that Corollaries 2 and 3 below, concerning non-neutral differential equations, are new results.

COROLLARY 1.. *Consider the second-order neutral differential equation with positive and negative coefficients*

$$\frac{d^2}{dt^2}\left[x(t) + px(t-\tau) - rx(t-\rho)\right] + qx(t-\sigma) - gx(t-\mu) = 0,$$

where p, r, q, g, τ, ρ, σ, μ are positive constants. Then all solutions of this equation oscillate if and only if its characteristic equation

$$\lambda^2 + \lambda^2 pe^{-\lambda\tau} - \lambda^2 re^{-\lambda\rho} + qe^{-\lambda\sigma} - ge^{-\lambda\mu} = 0$$

has no real roots.

COROLLARY 2.. *Consider the second-order delay differential equation with positive and negative coefficients*

$$\ddot{x}(t) + qx(t - \sigma) - gx(t - \mu) = 0,$$

where q, g, σ, μ are positive constants. Then all solutions of this equation oscillate if and only if its characteristic equation

$$\lambda^2 + qe^{-\lambda\sigma} - ge^{-\lambda\mu} = 0$$

has no real roots.

COROLLARY 3.. *Consider the mixed type differential equation*

$$\ddot{x}(t) + \delta \left(\sum_{K_1} q_k\, x\, (t - \sigma_k) + \sum_{K_2} \hat{q}_k\, (t + \hat{\sigma}_k) \right) = 0,$$

where $\delta = \pm 1$, q_k, $\hat{q}_k$, σ_k, $\hat{\sigma}_k$ are positive constants for $k \in K_1 \cup K_2$. Then all solutions of this equation oscillate if and only if its characteristic equation

$$\lambda^2 + \delta \left(\sum_{K_1} q_k e^{-\lambda\sigma\cdot} + \sum_{K_2} \hat{q}_k e^{\lambda\hat{\sigma}_k} \right) = 0$$

has no real roots.

Observe that in the case of the mixed type equations (cf. [22])

$$\ddot{x}(t) + \delta \left(\sum_{K_1} q_k x\, (t - \sigma_k) - \sum_{K_2} \hat{q}_k x\, (t + \hat{\sigma}_k) \right) = 0, \quad \delta = \pm 1$$

the characteristic equation is

$$f(\lambda) \equiv \lambda^2 + \delta \left(\sum_{K_1} q_k e^{-\lambda\sigma_k} - \sum_{K_2} \hat{q}_k e^{\lambda\hat{\sigma}_k} \right)$$

and

$$f(+\infty)f(-\infty) < 0.$$

Therefore equations of the above form admit non-oscillatory solutions.

The method of proof which we used to establish our main result is short (cf. [23]) and also has the advantage that it results in easily verifiable sufficient conditions for the oscillation of solutions to Eq. (1). Indeed, this is derived by comparing elements of the set $\Lambda(x)$ in each case. Observe that we found points λ_a and λ_b such that $\lambda_a \in \Lambda(x)$, while λ_b is an upper bound of $\Lambda(x)$. Thus, if we assume

$$\lambda_a \gneq \lambda_b,$$

we are led to a contradiction. Utilizing this idea we can obtain several sufficient conditions (in terms of the coefficients and the arguments only) for the oscillation of solutions of Eq. (1). The advantage of working with these sufficient conditions rather than the characteristic equation (2) directly is that the said conditions are explicit, while determining whether or not a real root to Eq. (2) exists may be quite a problem in itself. Thus, using Lemmas 5 and 6, one can draw a number of corollaries. We confine ourselves to the following:

COROLLARY 4.. *Consider Eq (1) and assume that*

$$(26) \quad \frac{T(Q-G)}{R+3QT^2} \geqq$$

$$\left(\frac{128}{q_{n_1}\left(\sigma_{n_1} - \alpha\right)^4}\left[4R + \left(\sigma_{n_1} - \alpha\right)\left(Q_2\hat{\sigma}_{n_2} + G_1\mu_{\nu_1}\right)\right]\right)^{\frac{1}{2}}$$

$$when \ \sigma_{n_1} > \alpha = \max\left\{\rho_m, \mu_{\nu_1}\right\}.$$

Then Eq. (1) has no (non-oscillatory) solutions of Class I_4.

COROLLARY 5.. *Consider Eq. (1) and assume that*

$$(27) \quad \frac{T(Q-G)}{R+3QT^2} \geqq$$

$$\left(\frac{128}{\hat{q}_{n_2}\left(\hat{\sigma}_{n_2} - \hat{\mu}_{\nu_2}\right)^4}\left[4R + \left(\hat{\sigma}_{n_2} - \hat{\mu}_{\nu_2}\right)\left(Q_1\sigma_{n_1} + G_2\hat{\mu}_{\nu_2}\right)\right]\right)^{\frac{1}{2}}, \ \ \hat{\sigma}_{n_2} > \hat{\mu}_{\nu_2}$$

Then Eq. (1) has no (non-oscillatory) solutions of Class II_4.

REMARK 1.. Observe that if there exists a bounded non-oscillatory solution of Eq. (1) then Class I_4 is not empty.

REMARK 2.. It is clear that when Eq. (1) has no (non-oscillatory) solutions of Class I_4 and Class II_4, then all solutions of (1) oscillate.

EXAMPLE 1.. Consider the neutral differential equation

$$(28) \quad \frac{d^2}{dt^2}\left[x(t) - x(t - \pi)\right] + 3x\left(t - 10^4\pi\right) - x\left(t - 2\pi\right) = 0$$

and observe that $\sigma > \alpha = \mu > \rho$. Also condition (26) is satisfied. Therefore, by Corollary 4, Eq. (28) has no (non-oscillatory) solutions of Class I_4 and, in view of Remark 1, *all bounded solutions* of (28) oscillate. For example, $\sin t$ and $\cos t$ are bounded oscillatory solutions of (28).

EXAMPLE 2.. Consider the mixed type differential equation

$$(29) \quad \ddot{x}(t) + 3e^{\frac{\pi}{2}}x\left(t - \frac{\pi}{2}\right) + e^{-\frac{\pi}{2}}x\left(t + \frac{\pi}{2}\right) = 0.$$

Observe that this equation has no (non-oscillatory) solutions of Class I_4 and Class II_4 and, in view of Remark 2, *all solutions* oscillate. For example $e^t\sin t$ and $e^t\cos t$ are (non-bounded) oscillatory solutions of Eq. (29).

REFERENCES

1. O. Arino and I. Győri, *Necessary and sufficient condition for oscillation of neutral differential system with several delays*, J. Differential Equations **81** (1989), 98–105.

2. R. Bellman and K. L. Cooke, *Differential - Difference Equations*, Academic Press (1963), New York.

3. S. J. Bilchev, M. K. Grammatikopoulos, and I. P. Stavroulakis, *Oscillations of higher - order neutral differential equations*, J. Austral. Math. Soc. Ser. A. (to appear).

4. R. K. Brayton and R. A. Willoughby, *On the numerical integration of a symmetric system of difference - differential equations of neutral type*, J. Math. Anal. Appl. **18** (1967), 182–189.

5. W. E. Brumley, *On the asymptotic behavior of solutions of differential - difference equations of neutral type*, J. Differential Equatitons **7** (1970), 175–188.

6. R. D. Driver, *Existence and continuous dependence of solutions of a neutral functional - differential equation*, Arch. Rational Mech. Anal. **19** (1965), 149–166.

7. R. D. Driver, *A mixed neutral system*, Nonlinear Anal. **8** (1984), 155–158.

8. K. Farrell, *Necessary and sufficient conditions for oscillation of neutral equations with real coefficients*, J. Math. Anal. Appl. **140** (1989), 251–261.

9. M. K. Grammatikopoulos, E. A. Grove, and G. Ladas, *Oscillation and asymptotic behavior of neutral differential equations with deviating arguments*, Applicable Anal. **22** (1986), 1–19.

10. M. K. Grammatikopoulos, Y. G. Sficas, and I. P. Stavroulakis, *Necessary and sufficient conditions for oscillations of neutral equations with several coefficients*, J. Differential Equations **76** (1988), 294–311.

11. M. K. Grammatikopoulos and I. P. Stavroulakis, *Necessary and sufficient conditions for oscillation of neutral equations with deviating arguments*, J. London Math. Soc. (2) **41** (1990), 244–260.

12. M. K. Grammatikopoulos and I. P. Stavroulakis, *Oscillations of neutral differential equations*, Radovi Matematički **7** (1991), 47–71.

13. E. A. Grove, G. Ladas, and A. Meimaridou, *A necessary and sufficient condition for the oscillation of neutral equations*, J. Math. Anal. Appl. **126** (1987), 341–354.

14. J. K. Hale, *Stability of linear systems with delays*, Stability Problems C.I.M.E. no. June 1974 (1974), 19–35, Ed. Cremoneze, Roma.

15. J. Hale, *Theory of functional differential equations*, Springer-Verlag (1977), New York.

16. J. Jaroš and T. Kusano, *Sufficient conditions for oscillations in higher-order linear functional differential equations of neutral type*, Japan. J. Math **15** (1989), 415–432.

17. M. R. S. Kulenouič, G. Ladas, and A. Meimaridou, *Necessary and sufficient condition for oscillations of neutral differential equations*, J. Austral. Math. Soc. Ser. B. **28** (1987), 362–375.

18. M. R. S. Kulenović, G. Ladas, and Y. G. Sficas, *Oscillations of second - order linear delay differential equations*, Applicable Anal. **27** (1988), 109–123.

19. G. Ladas, E. C. Partheniadis, and Y. G. Sficas, *Necessary and sufficient conditions for oscillations of second - order neutral equations*, J. Math. Anal. Appl. **138** (1989), 214–231.

20. G. Ladas, E. C. Partheniadis, and Y. G. Sficas, *Oscillations of second-order neutral equations*, Can. J.. Math. **XL** (1988), 1301–1314.

21. G. Ladas, Y. G. Sficas, and I. P. Stavroulakis, *Necessary and sufficient conditions for oscillations of higher order delay differential equations*, Trans. Amer. Math. Soc. **285** (1984), 81–90.

22. G. Ladas and I. P. Stavroulakis, *Oscillations of differential equations of mixed type*, J. Math. Phys. Sciences **18** (1984), 245–262.

23. E. C. Partheniadis, *On bounded oscillations of neutral differential equations*, Applicable Anal. **29** (1988), 63–90.

24. S. W. Schultz, *Necessary and sufficient conditions for the oscillations of bounded solutions*, Applicable Anal. **30** (1988), 47–63.

25. Y. G. Sficas and I. P. Stavroulakis, *Necessary and sufficient conditions for oscillations of neutral differential equations*, J. Math. Anal. Appl. **123** (1987), 494–507.

26. M. Slemrod and E. F. Infante, *Asymptotic stability criteria for linear systems of difference - differential equations of neutral type and their discrete analogues*, J. Math. Anal. Appl. **38** (1972), 399–415.

27. W. Snow, *Existence, uniqueness, and stability for nonlinear differential - difference equation in the neutral case*, N. Y. U. Courant Inst. Math. Sci. Rep. IMM-NYU (February 1965).

28. Z. C. Wang, *A necessary and sufficient condition for the oscillation of higher - order neutral equations*, Tôhuku Math. J. **41** (1989), 575–588.

CENTRE OF MATHEMATICS, TECHNICAL UNIVERSITY, 7017, ROUSSE, BULGARIA

DEPARTMENT OF MATHEMATICS, UNIVERSITY OF IOANNINA, 451 10 IOANNINA, GREECE

DEPARTMENT OF MATHEMATICS, UNIVERSITY OF IOANNINA, 451 10 IOANNINA, GREECE

Contemporary Mathematics
Volume **129**, 1992

A Wave Equation Viewed as an
Ordinary Differential Equation

T.A. BURTON, JOZSEF TERJÉKI AND BO ZHANG

ABSTRACT. In this paper we look at six classical problems associated with the ordinary differential equation

$$u'' + f(t)g(u) = 0, \ ug(u) > 0 \text{ if } u \neq 0, \ f(t) > 0$$

and show that these have parallels for

$$u_{tt} = f(t)g(u_x)_x, \ u(t,0) = u(t,1) = 0.$$

The problems concern oscillation, continuation of solutions, decay of solutions, limit circle, and limiting behavior.

1. Introduction

Through dynamical system theory, many properties of evolution equations are found to be parallel to those of special ordinary differential equations. The theory of inertial manifolds (cf. [**11**]) establishes deep theoretical connections between infinite dimensional and finite dimensional dynamical systems in terms of limit sets which are exponentially asymptotically stable. Central to so much of the application of this theory is the use of energy methods or the equivalent use of Liapunov functions.

This work takes a close look at six very well known classical problems associated with the ordinary differential equation

$$(1) \qquad u'' + f(t)g(u) = 0, \quad ug(u) > 0 \text{ if } u \neq 0, \quad f(t) \geq 0,$$

1991 *Mathematics Subject Classification*. Primary 35B05; 35B40.

This work was done while the second author visited Southern Illinois University at Carbondale. He thanks SIU for the kind hospitality and the support. This work was also supported by the Hungarian National Foundation for Scientific Research with grant number 6032/6319.

and shows that these problems have parallels for the equation

(2) $u_{tt} = f(t)g(u_x)_x, \qquad u(t,0) = u(t,1) = 0,$

both in terms of results and methods of solution. These problems concern oscillation, continuation of solutions, decay of solutions, limit circle considerations, and limiting behavior of solutions.

The study actually began in [1] where it was noted that there were striking similarities between the classical Liénard equation

(L) $u'' + f(u)u' + g(u) = 0, \quad f(u) > 0, \quad ug(u) > 0$ if $u \neq 0,$

and several forms of the damped wave equation such as

(W) $u_{tt} = g(u_x)_x - f(u)u_t, \qquad u(t,0) = u(t,1) = 0.$

In particular:

   (i) Each of (L) and (W) has a natural Liapunov function with derivative which is negative semi-definite.
   (ii) Each of (L) and (W) has a Liénard transformation, the transformed form of which has a natural Liapunov function whose derivative is negative semidefinite.
   (iii) A combination of the Liapunov function in (i) and (ii) produces a Liapunov function whose derivative is negative definite.
   (iv) The forms of the Liapunov functions for (L) and (W) are very similar, as are the consequences derivable from them.

Here, we continue that type of study, selecting a Liapunov function for (1), converting it to a Liapunov function for (2), and deducing parallel results for oscillation, continuation, and other qualitative behavior of solutions.

2. Oscillation

Wintner [14] considered the linear equation

(3) $u'' + f(t)u = 0$

and generalized the following idea. Suppose that $f : [0,\infty) \to [0,\infty)$ is continuous and $\int_0^\infty f(t)dt = \infty$; then every solution oscillates. He proved this by assuming that a solution $u(t)$ has no zero past some t_0 and formed a Chetayev type Liapunov function

$$V(t) = u'(t)/u(t) \text{ for } t > t_0.$$

Then

$$V'(t) = [uu'' - (u')^2]/u^2 = -f(t) - V^2(t),$$

a Riccati equation having a solution reaching negative infinity in finite time $t > t_0$.

To extend the result to (2) we must first decide how to define oscillation for (2). Recall that a solution $u(t)$ of (1) is oscillatory if there is a sequence $\{t_n\} \uparrow \infty$

with $u(t_n) = 0$, $u(t) \not\equiv 0$. But a reading of classical oscillation papers reveals that for $f(t) \geq 0$ the property of most interest was the equivalent fact that $u''(t) = -f(t)g(u(t))$ oscillated. If we take that as a definition, then Wintner's proof works for (2).

In fact, such arguments showing oscillations have been equally effective for delay equations and instead of (2) we deal here with

$$(2^*) \qquad \begin{cases} u_{tt} = f(t)g(u_x(t-h,x))_x \\ u(t,0) = u(t,1) = 0 \end{cases}$$

where h is a nonnegative constant. It may be noted that if $h > 0$ then (2^*) can be solved by the method of steps, but it requires very smooth initial functions.

DEFINITION 1. A solution of (2^*) is oscillatory if there are sequences $\{t_n\} \uparrow \infty$ and $\{x_n\} \subset (0,1)$ such that $g(u_x(t_n, x_n))_x$ and $g(u_x(t_{n+1}, x_{n+1}))_x$ have opposite sign.

The reader may consider a vibrating string and conclude that Def. 1 is what we would intuitively mean by the string vibrating.

THEOREM 1. Suppose that for each $t_1 \geq 0$, the only solution of (2^*) satisfying $u(t_1 + \theta, x) = u_t(t_1 + \theta, x) \equiv 0$ for $-h \leq \theta \leq 0$ is the zero solution. Assume that $f(t) \geq 0$, that $\int_0^\infty f(s)ds = \infty$, and that $g'(r) \geq g_0 > 0$. Let $u(t,x)$ satisfy (2^*) on $[0,\infty)$ and be nonoscillatory. Then $u(t,x)$ is zero.

PROOF. Taking into account the boundary conditions, we follow Wintner and write

$$V(t) = \int_0^1 u_t(t,x)dx \Big/ \int_0^1 u(t-h,x)dx$$

for an assumed nonoscillatory solution u. This means that there is a $t_0 \geq 0$ such that u_{xx} has one sign for $t \geq t_0$. Suppose, to be definite, that $u_{xx}(t,x) \leq 0$ for $t \geq t_0$. Since $u(t,0) = u(t,1) = 0$ we will suppose that $u(t,x) \geq 0$ on $[t_0, \infty)$. From (2^*) we have $u_{tt}(t,x) \leq 0$ and so $u_t(t,x)$ is decreasing on $[t_0+h, \infty)$ for each fixed $x \in [0,1]$. Hence, $u_t(t,x) \geq 0$ for all $t \geq t_0+h$ and $x \in [0,1]$. If $u(t_1, x_1) = 0$ for some $t_1 > t_0 + h$ and $x_1 \in (0,1)$ then $u_{xx}(t_1, x) \leq 0$ and $u(t_1, x) \geq 0$ imply that $u(t_1, x) = 0$ for all $x \in [0,1]$. Consequently, $u(t,x)$ has a minimum at $t = t_1$ for all fixed x, so $u_t(t_1, x) = 0$ for all $x \in [0,1]$. Therefore we conclude that either $u(t,x) > 0$ for all $t > t_0$ and $x \in (0,1)$ or $u(t,x) \equiv 0$, $u_t(t,x) \equiv 0$ for all $x \in [0,1]$ and all large t. By the assumed uniqueness, $u(t,x) \equiv 0$.

Now assume that $\int_0^1 u(t,x)dx > 0$ for $t \geq t_0$ so $V(t)$ is defined and $V(t) \geq 0$

for $t \geq t_0 + h$. We then have

$$V'(t) = \left[\int_0^1 u(t-h,x)dx \int_0^1 f(t)g(u_x(t-h,x))_x dx \right.$$

$$\left. - \int_0^1 u_t(t,x)dx \int_0^1 u_t(t-h,x)dx \right] \Big/ \left[\int_0^1 u(t-h,x)dx \right]^2$$

$$= \left[f(t) \int_0^1 g'(u_x(t-h,x))u_{xx}(t-h,x)dx \Big/ \int_0^1 u(t-h,x)dx \right]$$

$$- \left\{ \int_0^1 u_t(t,x)dx \int_0^1 u_t(t-h,x)dx \Big/ \left[\int_0^1 u(t-h,x)dx \right]^2 \right\}.$$

Now $g'(r) \geq g_0 > 0$ and $u(t,x) \geq 0$, $u_{xx} \leq 0$, so for fixed $t \geq t_0 + h$ we have

$$- \int_0^1 g'(u_x(t-h,x))u_{xx}(t-h,x)dx \Big/ \int_0^1 u(t-h,x)dx$$

$$\geq g_0 \int_0^1 |u_{xx}(t-h,x)|dx \Big/ \int_0^1 u(t-h,x)dx$$

$$\geq g_0 \int_0^1 |u_{xx}(t-h,x)|dx/u(c)$$

where $u(c) = \sup_{0 \leq x \leq 1} u(t-h,x) > 0$ (t is fixed). But $u(t,0) = u(t,1) = 0$ so there is an $\xi \in (0,1)$ with $u_x(t-h,\xi) = 0$. This means that

$$\int_0^1 |u_{xx}(t-h,x)|dx \geq \sup_{0 \leq x \leq 1} |u_x(t-h,x)| \geq \int_0^1 |u_x(t-h,x)|dx$$

$$\geq \sup_{0 \leq x \leq 1} |u(t-h,x)| = u(c).$$

Hence,

$$g_0 \int_0^1 |u_{xx}(t-h,x)|dx/u(c) \geq g_0.$$

Moreover, since $u_{tt}(t,x) \leq 0$, $u_t(t,x) \geq 0$, for $t \geq t_0 + h$, we have

$$\int_0^1 u_t(t,x)dx \leq \int_0^1 u_t(t-h,x)dx \text{ for } t \geq t_0 + 2h$$

and

$$- \int_0^1 u_t(t,x)dx \int_0^1 u_t(t-h,x)dx \leq - \left(\int_0^1 u_t(t,x)dx \right)^2.$$

Thus,

$$V'(t) \leq -f(t)g_0 - V^2$$

a Riccati equation with $V(t_1) = -\infty$ for some $t_1 > t_0$. It then follows that there is a $t_1 > t_0 + 2h$ with $\int_0^1 u(t_1,x)dx = 0$ and so $u(t_1,x) \equiv 0$ on $[0,1]$, as required.

REMARK. Zlámal [15] generalized Wintner's theorem and showed that if

$$(*) \qquad \begin{cases} \text{there exists a function } w(t) > 0 \text{ such that} \\ \int_0^\infty w(t)f(t)dt = \infty \text{ and } \int_0^\infty (w'(t))^2 w^{-1}(t)dt < \infty \end{cases}$$

then solutions of (3) oscillate. Our theorem also remains valid if we assume (*) instead of $\int_0^\infty f(t)dt = \infty$. To see this we have to finish the proof in a different way: From the inequality $V'(t) \leq -f(t)g_0 - V^2$ we get for $T > t_2 \geq t_0 + 2h$ that

$$g_0 \int_{t_2}^T f(t)w(t)dt \leq -w(T)V(T) + w(t_2)V(t_2) + \int_{t_2}^T w'(t)V(t)dt$$

$$- \int_{t_2}^T w(t)V^2(t)dt \leq w(t_2)V(t_2)$$

$$+ \left(\int_{t_2}^\infty (w'(t))^2 w^{-1}(t)dt \right)^{1/2} \left(\int_{t_2}^T w(t)V^2(t)dt \right)^{1/2}$$

$$- \int_{t_2}^T w(t)V^2(t)dt$$

$$\leq w(t_2)V(t_2) + \int_{t_2}^\infty (w'(t))^2 w^{-1}(t)dt/4$$

$$- \left[\frac{1}{2} \left(\int_{t_2}^\infty (w'(t))^2 w^{-1}(t)dt \right)^{1/2} - \left(\int_{t_2}^T w(t)V^2(t)dt \right)^{1/2} \right]^2$$

$$\leq w(t_2)V(t_2) + \int_{t_2}^T (w'(t))^2 w^{-1}(t)dt/4 < \infty,$$

a contradiction as $T \to \infty$.

3. Continuation of Solutions

Frequently, in oscillation problems concerning (1), the function $f(t)$ is allowed to become negative some of the time. But then special care must be taken concerning the growth of g to be sure that a solution will not have finite escape time. In [2] it was shown that if $f(t_1) < 0$ for some $t_1 > 0$ and if

$$G(x) := \int_0^x g(s)ds,$$

then (1) has a solution not continuable to $t = \infty$ provided that either

(a) $\int_0^\infty [1 + G(x)]^{-1/2}dx < \infty$ or

(b) $\int_0^{-\infty} [1 + G(x)]^{-1/2}dx > -\infty$.

A partial converse was also obtained. Here, a similar result for (2) holds. It is to be noted that (1) can have a solution defined for $t \geq 0$ having $u(t) > 0$ and $u(t) \to \infty$; thus the conditions (a) and (b) are separate. The behavior of $g(u)$ for $u < 0$ is immaterial. But for (2); because of the boundary condition and the inequality $\int_0^1 \pi^2 u^2 dx \leq \int_0^1 u_x^2 dx$, if $u(t,x) \to +\infty$, then $g(u_x) \to \pm\infty$. Hence, the parallel result for (2) will involve g for both positive and negative values of its argument.

THEOREM 2. *Suppose that there is a* $t_1 > 0$ *with* $f(t_1) < 0$ *and suppose that there is a convex downward function* $\overline{g} : R^+ \to R^+$ *such that* $xg(x) \geq \overline{g}(x^2)$. *For*

$$\overline{G}(u) = \int_0^u \overline{g}(\xi)d\xi, \quad \text{if} \quad \int_0^\infty [1 + \overline{G}(x)]^{-1/2}dx < \infty,$$

then there are initial conditions for (2) such that any solution having those initial conditions can not be defined for all $t \geq t_1$.

PROOF. Since $f(t_1) < 0$ and $f(t)$ is continuous, there are positive constants δ, m, M such that $-M \leq f(t) \leq -m < 0$ if $t_1 \leq t \leq t_1 + \delta$. Let $u(t,x)$ be a solution of (2) and define $z(t) = \int_0^1 u^2(t,x)dx$ and $y(t) = 2\int_0^1 u(t,x)u_t(t,x)dx$. We then have the system of ordinary differential equations

(4)
$$\begin{cases} z' = y \\ y' = 2\int_0^1 u_t^2(t,x)dx - 2f(t)\int_0^1 u_x g(u_x)dx. \end{cases}$$

Denote by $(z(t), y(t))$ a solution of (4) satisfying $z(t_1) = 1$, $y(t_1) = y_1$ with y_1 large and to be determined later. So long as $(z(t), y(t))$ is defined on $[t_1, t_1 + \delta]$ we have both $y(t)$ and $z(t)$ monotonically increasing. From (4) we obtain

$$2yy' = 2y\int_0^1 u_t^2(t,x)dx - 2f(t)2y(t)\int_0^1 u_x g(u_x)dx$$
$$\geq 4my(t)\overline{g}\left(\int_0^1 u_x^2 dx\right)$$
$$\geq 4my(t)\overline{g}(z(t)),$$

using Jensen's inequality and then Wirtinger's inequality, so that

$$y^2(t) \geq y^2(t_1) + 4m\int_{t_1}^t z'(s)\overline{g}(z(s))ds$$
$$= y^2(t_1) + 4m\overline{G}(z(t)) - 4m\overline{G}(z(t_1))$$

and

$$z'(t) = y(t) \geq [y^2(t_1) - 4m\overline{G}(1) + 4m\overline{G}(z(t))]^{1/2}.$$

Divide both sides by the right-hand side and integrate from t_1 to t to obtain

$$\int_1^{z(t)} [y^2(t_1) - 4m\overline{G}(1) + 4m\overline{G}(z)]^{-1/2}dz \geq t - t_1.$$

That is,

$$\int_0^\infty [y^2(t_1) - 4m\overline{G}(1) + 4m\overline{G}(z)]^{-1/2}dz \geq t - t_1.$$

If $y^2(t_1) \geq 4m\overline{G}(1) + 4m$, then we have that

$$[y^2(t_1) - 4m\overline{G}(1) + 4m\overline{G}(z)]^{-1/2} \leq (4m)^{-1/2}(1 + \overline{G}(z))^{-1/2} \in L^1[0,\infty).$$

On the other hand, for fixed $z \in (0,\infty)$ we have

$$[y^2(t_1) - 4m\overline{G}(1) + 4m\overline{G}(z)]^{-1/2} \to 0 \quad \text{as} \quad y^2(t_1) \to \infty.$$

Therefore, by the Lebesgue dominated convergence theorem it follows that

$$\int_0^\infty [y^2(t_1) - 4m\overline{G}(1) + 4m\overline{G}(z)]^{-1/2}dz \to 0$$

as $y^2(t_1) \to \infty$. Consequently, we may take $y^2(t_1)$ so large that

$$\int_0^\infty [y^2(t_1) - 4m\overline{G}(1) + 4m\overline{G}(z)]^{-1/2}dz < \delta.$$

That is, $z(t) \to \infty$ before t reaches $t_1 + \delta$.

4. Instability

Section 3 deals with a drastic type of instability. But if $f(t) < 0$ for all $t \geq 0$, then a more gentle type of instability can occur.

As motivation we again consider equation (1) and suppose that $f(t) \leq -f_0 < 0$ on $[0, \infty)$. Then the classical theory of Chetayev (cf. [6; p. 27], for example) leads to the Liapunov function $V = uv$ for the system $\{u' = v, v' = -f(t)g(u)\}$ so that $V' = uv' + u'v = v^2 - f(t)g(u)u \geq v^2 + f_0 ug(u)$. Therefore, V vanishes on $u = 0$ and on $v = 0$, with $V' > 0$ on the set $uv > 0$. Thus, the zero solution is unstable.

We now give a very simple parallel for (2).

THEOREM 3. *If $f(t) \leq 0$ for $t \geq 0$ $ug(u) \geq 0$ for all $u \in R$, then the solution $u = 0$ is unstable.*

PROOF. Let $u(t, x)$ be a solution of (2) on $[0, \infty)$ with $u(0, x) \geq 0$, $u_t(0, x) \geq 0$, and $\int_0^1 u(0, x)u_t(0, x)dx > 0$. Define $\{u_t = v, v_t = f(t)g(u_x)_x\}$ and

$$V(t) = \int_0^1 u(t, x)v(t, x)dx$$

so that

$$V'(t) = \int_0^1 u_t v dx + \int_0^1 u v_t dx = \int_0^1 v^2 dx + \int_0^1 f(t)(g(u_x))_x u dx$$

$$= \int_0^1 v^2 dx - \int_0^1 f(t)g(u_x)u_x dx \geq \int_0^1 u_t^2 dx.$$

Suppose that $u = 0$ is stable. Then for a given $\epsilon > 0$ and $t_1 \geq 0$ there is a $\delta > 0$ such that $\int_0^1 u_t^2(t_1, x)dx < \delta^2$ and $\int_0^1 u^2(t_1, x)dx < \delta^2$ imply that any solution $u(t, x)$ satisfying those initial conditions will satisfy $\int_0^1 u^2(t, x)dx < \epsilon^2$ and $\int_0^1 u_t^2(t, x)dx < \epsilon^2$ for $t \geq t_1$. Now

$$V(t_1) \leq V(t) \leq \left(\int_0^1 u^2 dx \int_0^1 u_t^2 dx \right)^{1/2}$$

$$\leq \epsilon \left(\int_0^1 u_t^2 \right)^{1/2}.$$

Hence

$$(V(t_1)/\epsilon)^2 \leq \int_0^1 u_t^2(t, x)dx.$$

Then

$$V'(t) \geq (V(t_1)/\epsilon)^2$$

and so

$$\epsilon^2 \geq V(t) \geq V(t_1) + (V(t_1)/\epsilon)^2(t - t_1)$$

for all $t \geq t_1$ is a contradiction.

Because of the special form of this equation, the result is actually stronger than its ODE counterpart using the Chetayev theorem. We now give a simple generalization of Chetayev's theorem to abstract equations.

Consider the ordinary differential equation

(5) $$u'(t) = F(t, u(t)), \quad F(t, 0) = 0,$$

in a Banach space X with norm $|\cdot|_X$.

THEOREM 4. *Let A be an open subset of X with $O \in \partial A$ and let $B > 0$. Suppose that $V : \overline{A} \to R^+$, that $V(u)$ is bounded on $\{u \in A : |u|_X \leq B\}$, that $V(u) > 0$ for $u \in A$ and $|u|_X \leq B$, and that $V(u) = 0$ for $u \in \partial A$ and $|u|_X \leq B$. In addition, suppose that $V'_{(5)}(u(t)) \geq \alpha(t)W(u(t))$ for $u \in A$ and $|u|_X \leq B$ where $\alpha(t) \geq 0$, $\int_0^\infty \alpha(t)dt = \infty$, and $W(u) \geq 0$ for $u \in A$. Moreover, suppose that for any $\mu > 0$ there exists $\tilde{\mu} > 0$ such that $[u \in A, V(u) \geq \mu]$ imply that $W(u) \geq \tilde{\mu}$. Then the zero solution of (5) is unstable.*

PROOF. If the theorem is false, then for $\epsilon = B/2$ there is a $\delta > 0$ such that $|u_0|_X < \delta$ and $t > 0$ imply that $|u(t, 0, u_0)|_X < \epsilon$, where $u(t, 0, u_0)$ is a solution satisfying $u(0, 0, u_0) = u_0$; we also let $u(t, 0, u_0) = u(t)$. Choose $u_0 \in A$, $|u_0|_X = \delta/2$. Then $V(u_0) > 0$ and so long as $u(t, 0, u_0) \in A$ we have $V'(u(t, 0, u_0)) > 0$ so that

(6) $$V(u(t, 0, u_0)) \geq V(u_0) > 0.$$

This means that $u(t) \in \{\xi \in A : |\xi|_X \leq B\}$ and there is a $\tilde{\mu} > 0$ such that $W(u(t)) \geq \tilde{\mu}$ by (6). This yields

$$V'_{(5)}(u(t)) \geq \alpha(t)W(u(t)) \geq \alpha(t)\tilde{\mu}$$

for all $t > 0$. An integration yields a contradiction to V being bounded on A whenever $|u|_X \leq B$. This completes the proof.

The reader may verify the conditions of Theorem 4 for (2), $X = H_0^1 \times H^0$, $A = \{(u, v) \in X | \int_0^1 uv \, dx > 0\}$, $B = 1$, $\alpha(t) \equiv 1$, $W(u, v) = \int_0^1 v^2 dx$ and $V(u(t), v(t)) = \int_0^1 u(t)v(t)dx$. Jensen's inequality is used in this exercise.

5. Limiting Behavior

In 1893 Kneser [11] considered (3) with $f(t) \leq 0$ and gave conditions to ensure the "Kneser condition" that every solution $u(t)$ satisfies $u(t) \to 0$ or $|u(t)| \to \infty$. In 1962 Utz [13], motivated by Kneser's work, considered

$$(7) \qquad u'' = f(t)u^{2n-1}, \quad n \text{ a positive integer}$$

and proved the following result.

THEOREM (UTZ). *Let $f(t) > 0$ and continuous on $[0,\infty)$ and suppose that for each u_0, u_0' there is a unique solution on $[t_0,\infty)$ for each $t_0 \geq 0$. Then (7) has a solution $u(t) \not\equiv 0$ such that $u(t) \to 0$ and $u'(t) \to 0$, both monotonically, as $t \to \infty$.*

In view of our Theorem 2 and the continuation assumption, this result is valid only for $n = 1$; that is, (7) must be linear. Moreover, more must be added to the conditions on $f(t)$ to obtain the "Kneser condition" since $u(t) = 1 + e^{-t}$ is a solution of

$$u'' = [1/(1 + e^t)]u$$

and it tends to 1 as $t \to \infty$. In fact, equation (3), in the case $f(t) \leq 0$, has the Kneser property if and only if $\int_0^\infty sf(s)ds = \infty$ ([7; p. 103] and [10; Lemma 1]). This assertion is valid for the nonlinear case too as can be seen in the same way as in [7] when things are defined as follows. Let $h : [0,\infty) \times R \to R$ be continuous and locally Lipschitz in the second variable, $h(t,u)u > 0$ for $u \neq 0$, and suppose in addition that $h(t,u)$ is monotone increasing with respect to u for fixed t. If $\int_0^\infty th(t,c)dt < \infty$ for some $c > 0$, then $u'' = h(t,u)$ has a solution $u(t)$ such that $u(t) > 0$, $u'(t) < 0$, $u'(t) \to 0$, $u(t) \to \alpha$ as $t \to \infty$.

This will motivate the next result for (2) in that we, therefore, see that more is needed on $f(t)$.

THEOREM 5. *Let $f(t) \leq 0$, $g(u)/u \geq \alpha$ if $u \neq 0$ for some $\alpha > 0$, and let $\int_0^\infty tf(t)dt = -\infty$. If $u(t,x)$ is a solution of (2) on $[0,\infty)$, then either*
(a) $\int_0^1 u^2(t,x)dx \to 0$ as $t \to \infty$ or
(b) $\int_0^1 u^2(t,x)dx \to \infty$ as $t \to \infty$.

PROOF. Let $u(t) = u(t,x)$ be a solution of (2) on $[0,\infty)$. Then

$$(d/dt) \int_0^1 u^2(t,x)dx = \int_0^1 2uu_t dx$$

and

$$(d^2/dt^2) \int_0^1 u^2(t,x)dx = 2 \int_0^1 [u_t^2 + uu_{tt}]dx$$

$$= 2 \int_0^1 [u_t^2 - 2f(t)u_x g(u_x)]dx \geq 0$$

after use of (2) and an integration by parts. This implies that either

$$\lim_{t \to \infty} \int_0^1 u^2 dx = \infty$$

or (since the quantity is nonnegative)

$$\lim_{t \to \infty} \int_0^1 u^2 dx = c,$$

where c is a nonnegative constant. We claim that $c = 0$ in the latter case.

Suppose that $c > 0$. Then there is a $t_1 > 0$ such that $\int_0^1 u^2 dx \geq c/2$ on $[t_1, \infty)$.

(i) If there is a $t_2 \geq 0$ such that

$$(d/dt) \int_0^1 u^2 dx \Big|_{t=t_2} = 2 \int_0^1 u(t_2, x) u_t(t_2, x) dx =: \beta > 0,$$

then it follows readily that

$$\int_0^1 u^2 dx \geq \beta(t - t_2) \to \infty \text{ as } t \to \infty$$

(since the derivative is an increasing function), a contradiction.

(ii) If $(d/dt) \int_0^1 u^2(t, x) dx \leq 0$ on $[0, \infty)$, then

$$(d/dt) \int_0^1 u^2(t, x) dx = (d/dt) \int_0^1 u^2(t, x) dx \Big|_{t=0}$$
$$+ 2 \int_0^t \left(\int_0^1 u_t^2(s, x) dx - f(s) \int_0^1 u_x(s, x) g(u_x(s, x)) dx \right) ds$$

(since $\int_0^t F''(s) ds = F'(t) - F'(0)$). As the left side is nonpositive, this implies that

$$\int_0^\infty \left(\int_0^1 u_t^2(s, x) - f(s) \int_0^1 u_x(s, x) g(u_x(s, x)) dx \right) ds < \infty.$$

Clearly,

$$(d/dt) \int_0^1 u^2(t, x) dx =: F'(t) \to 0 \text{ as } t \to \infty;$$

for $F'(t) \leq 0$, $F''(t) \geq 0$, so if $F'(t) \leq -d < 0$ for all $t \geq t_2$, then $F(t) - F(t_2) \leq -d(t - t_2)$ yielding $F(t) \to -\infty$, a contradiction to $F(t) = \int_0^1 u^2 dx \geq 0$. Thus, we have

$$-2 \int_0^1 u u_t dx = 2 \int_t^\infty \left(\int_0^1 u_t^2(s, x) dx - f(s) \int_0^1 u_x(s, x) g(u_x(s, x)) dx \right) ds.$$

Let $\phi(x) = u(0,x)$, $\psi(x) = u_t(0,x)$, and then integrate the last expression from 0 to t and obtain

$$\int_0^1 \phi^2(x)dx = \int_0^1 u^2(t,x)dx$$

$$+ 2\int_0^t \int_w^\infty \left(\int_0^1 u_t^2(s,x)dx - f(s)\int_0^1 u_x(s,x)g(u_x(s,x))\right) dx\, ds\, dw$$

$$= \int_0^1 u^2(t,x)dx + 2\int_0^t \int_w^\infty \int_0^1 u_t^2(s,x)dx\, ds\, dw$$

$$- 2\int_0^t sf(s)\int_0^1 u_x(s,x)g(u_x(s,x))dx\, ds$$

$$- 2t\int_t^\infty f(s)\int_0^1 u_x(s,x)g(u_x(s,x))dx\, ds$$

$$\geq -2\alpha \int_0^t sf(s)\int_0^1 u_x^2(s,x)dx\, ds.$$

But $c/2 \leq \int_0^1 u^2 dx$ on $[t_1,\infty)$ so $c/2 \leq \int_0^1 u^2 dx \leq \int_0^1 u_x^2 dx$, together with $\int_0^\infty sf(s)ds = -\infty$ now yields $\int_0^1 \phi^2(x)dx = \infty$, a contradiction. This completes the proof.

6. Decay of Solutions and Limit Circle

Another classical problem is concerned with giving conditions on $f(t)$ in (3) to ensure that all solutions tend to zero. The literature is vast, but one may loosely state that it is sufficient to ask that $f(t) \to \infty$ monotonically and that $f'(t)/f^{3/2}(t)$ be bounded (cf. [3]). (It is not sufficient that $f(t) \to \infty$, as may be seen in [9].) But if one asks a bit more, then a trivial proof is available [4]. It goes as follows.

First, define a Liouville transformation $s = \int_0^t \sqrt{f(v)}dv$, $u(t) = w(s)$ and map (3) into $\ddot{w}(s) + [f'(t)/2f^{3/2}(t)]\dot{w}(s) + w(s) = 0$, where $\cdot = d/ds$. Let $\mu(s) = f'(t)/4f^{3/2}(t)$ and then define a system

$$\begin{cases} \dot{w} = z - \mu(s)w \\ \dot{z} = -w - \mu\dot{w} + \dot{\mu}w. \end{cases}$$

Define a Liapunov function
$$V(s) = w^2 + z^2$$

and obtain

$$\dot{V}(s) \leq -2\mu[w^2 + z^2] + |\mu^2 + \dot{\mu}|(z^2 + w^2)$$
$$= [-2\mu + |\mu^2 + \dot{\mu}|]V(s)$$

so that if

$$\int_0^\infty [-2\mu(s) + |\mu^2(s) + \dot{\mu}(s)|]ds = -\infty,$$

then every solution tends to zero.

Precisely the same sort of thing works for (2) and it also leads to a limit circle result. Preparatory to proving that theorem, recall that $\int_0^1 u_x^2 dx \geq |u|_\infty^2 \geq \int_0^1 u^2 dx$ when $u(t,0) = u(t,1) = 0$. Thus, when $rg(r) \geq \alpha r^2$ we will have $G(x) = \int_0^x g(s)ds$ and $\int_0^1 G(u_x)dx \geq (\alpha/2)\int_0^1 u_x^2 dx \geq (\alpha/2)|u|_\infty^2$.

THEOREM 6. *Let $g'(r) \geq 0$ for all r, $f(t) > 0$, $wg(w) \geq \alpha w^2$ for some $\alpha > 0$, $\alpha\pi^2 \leq 1$. Suppose that for $s = \int_0^t \sqrt{f(v)}dv$, for a > 0 and large, and for $\mu(s) = f'(t)/4f^{3/2}(t)$ we have*

$$(8) \qquad \int_0^\infty [-2\mu(s) + (|\dot\mu(s) + \mu^2(s)|/\alpha\pi^2)]ds = -\infty$$

and $2\mu(s) \geq |\dot\mu(s) + \mu^2(s)|/\alpha\pi^2$ for t sufficiently large. Then any solution $u(t,x)$ of (2) defined on $[0,\infty)$ satisfies

$$\int_0^1 G(u_x(t,x))dx \to 0 \text{ as } t \to \infty.$$

PROOF. First, the Liouville transformation

$$(9) \qquad s = \int_0^t \sqrt{f(v)}dv \text{ and } w(s,x) = u(t,x)$$

yields

$$u_t = w_s(ds/dt) = w_s\sqrt{f(t)}$$

and

$$u_{tt} = w_{st}\sqrt{f(t)} + w_s\left(f'(t)/2\sqrt{f(t)}\right)$$

so

$$u_{tt} = w_{ss}f(t) + w_s\left(f'(t)/2\sqrt{f(t)}\right).$$

Thus, (2) becomes

$$w_{ss} = g(w_x)_x - [f'(t)/2f^{3/2}(t)]w_s, \quad w(s,0) = w(s,1) = 0.$$

And this is equivalent to the system

$$\begin{cases} w_s = z - [f'(t)/4f^{3/2}(t)]w \\ z_s = g(w_x)_x - [f'(t)/4f^{3/2}(t)]w_s + (d/ds)[f'(t)/4f^{3/2}(t)]w. \end{cases}$$

This can be written as

$$(10) \qquad \begin{cases} w_s = z - \mu(s)w \\ z_s = g(w_x)_x - \mu(s)w_s + \dot\mu w. \end{cases}$$

With $G(r) = \int_0^r g(s)ds$, define a Liapunov function

$$V(s) = \int_0^1 [2G(w_x) + z^2]dx$$

and obtain the derivative of V along a solution as

$$\dot{V} = \int_0^1 (2g(w_x)w_{xs} + 2zz_s)dx$$

$$= \int_0^1 \{-2g(w_x)_x w_s + 2z[g(w_x)_x - \mu w_s + \dot{\mu}w]\}dx$$

(by the induced boundary conditions: $w_s(s,0) = w_s(s,1) = 0$)

$$= \int_0^1 \{-2g(w_x)_x[z - \mu w] + 2zg(w_x)_x - 2\mu zw_s + 2\dot{\mu}zw\}dx$$

$$= \int_0^1 [2\mu g(w_x)_x w - 2\mu zw_s + 2\dot{\mu}zw]dx$$

$$= \int_0^1 [-2\mu g(w_x)w_x - 2\mu z(-\mu w + z) + 2\dot{\mu}zw]dx$$

$$= \int_0^1 \{-2\mu[g(w_x)w_x + z^2] + 2\mu^2 zw + 2\dot{\mu}zw\}dx.$$

Now

$$\int_0^1 w^2 dx \le (1/\pi^2)\int_0^1 w_x^2 dx \le (1/\alpha\pi^2)\int_0^1 g(w_x)w_x dx$$

and since $G(w_x) \le g(w_x)w_x$ we have

$$\dot{V} \le \int_0^1 \{-2\mu[g(w_x)w_x + z^2] + |\mu^2 + \dot{\mu}|(z^2 + w^2)\}dx$$

$$\le \int_0^1 \{-2\mu[g(w_x)w_x + z^2] + |\mu^2 + \dot{\mu}|(z^2 + g(w_x)w_x/\alpha\pi^2)\}dx$$

$$\le (1/2)\int_0^1 \{-2\mu + (|\mu^2 + \dot{\mu}|/\alpha\pi^2)\}\{2G(w_x) + z^2\}dx$$

or

(11) $$\dot{V} \le \{-2\mu + (|\mu^2 + \dot{\mu}|/\alpha\pi^2)\}V/2$$

for t sufficiently large since $2\mu \ge |\mu^2 + \dot{\mu}|/\alpha\pi^2$ for t sufficiently large. The conclusion follows from this.

Note that the integral in the theorem, when changed to the variable t, is

$$\int_0^\infty [-2\{f'(t)/4f^{3/2}(t)\} + |\{(f'(t))^2/16\alpha\pi^2 f^3(t)\}$$

$$+ \{2f^{3/2}(t)f''(t) - 3(f'(t))^2\sqrt{f(t)}\}/8\alpha\pi^2 f^3(t)|]\sqrt{f(t)}dt$$

$$= \int_0^\infty [\{-f'(t)/2f(t)\} + |\{(f'(t))^2/16\alpha\pi^2 f^{5/2}(t)\}$$

$$+ \{2f(t)f''(t) - 3(f'(t))^2\}/8\alpha\pi^2 f^2(t)|]dt.$$

EXAMPLE 1. Let $f(t) = e^t$ and $\alpha\pi^2 > 1/4$ so that

$$\int_0^\infty [-2\mu(s) + |\dot\mu(s) + \mu^2(s)|/\alpha\pi^2]ds$$

$$= \int_0^\infty [-(1/2) + |(1/16\alpha\pi^2 e^{t/2}) - (1/8\alpha\pi^2)|]dt$$

$$= -\infty.$$

EXAMPLE 2. Let $f(t) = \ln(1 + t)$ so that

$$\int_1^\infty [-2\mu(s) + |\dot\mu(s) + \mu^2(s)|/\alpha\pi^2]ds$$

$$= \int_1^\infty [-(1/2(1+t)\ln(1+t)) + |\{1/16\alpha\pi^2(1+t)^2(\ln(1+t))^{5/2}\}$$

$$- [(2\ln(1+t)+3)/8\alpha\pi^2(1+t)^2(\ln(1+t))^2|]dt$$

$$= -\infty.$$

EXAMPLE 3. Let $f(t) = (1+t)^\beta$, $\beta > 0$. Then

$$\int_0^\infty [-2\mu(s) + |\dot\mu(s) + \mu^2(s)|/\alpha\pi^2]ds$$

$$= \int_0^\infty [-(\beta/2(1+t)) + |(\beta^2/16\alpha\pi^2(1+t)^2(1+t))^{\beta/2})$$

$$- (\beta(2+\beta)/8\alpha\pi^2(1+t)^2|]dt = -\infty.$$

From (11) it is very easy to obtain a result on the classical question of limit point-limit circle. If all solutions of (3) are in $L^2[0, \infty)$, then (3) is said to be in the limit circle case, otherwise it is in the limit point case. The terminology is explained, for example, in Coddington and Levinson [7; pp. 225-6]. The literature on the problem is vast and the reader is referred to Devinatz [8].

DEFINITION. Equation (2) is in the limit circle case if every solution $u(t, x)$ defined on $[0, \infty)$ satisfies $\int_0^\infty \int_0^1 u_x g(u_x)dxdt < \infty$.

The next result is an exact counterpart of [5] for (2).

THEOREM 7. *Let the conditions of Theorem 6 hold, let $G(r) \geq \beta rg(r)$ for some $\beta > 0$, and let*

(12)
$$\int_0^\infty \{(1/\sqrt{f(t)}) \exp(1/8\alpha\pi^2) \int_0^t |[f'(x))^2/2f^{5/2}(x)]$$

$$+ [2f(x)f''(x) - 3(f'(x))^2]/f^2(x)|dx\}dt < \infty.$$

Then (2) is in the limit circle case.

PROOF. We have

$$\beta \int_0^1 u_x g(u_x)dx \leq \int_0^1 [G(u_x) + z^2]dx \leq V(s)$$

and (11). The result now follows by integration of the bound on V obtained from integration of (11).

The next result extends [3] for (1) to (2). One may note that $f(t) = (1+t)^\beta$, $\beta > 0$, satisfies all conditions of this theorem.

THEOREM 8. *Suppose that $f'(t) \geq 0$, $f(t) > 0$, $f'(t)/f^{3/2}(t) \leq \gamma$ for some $\gamma > 0$, and there is a nonnegative decreasing function $\mu(t)$ such that $f'(t) \geq \mu(t)f(t)$ and $\int_0^\infty \mu(t)dt = \infty$. Let $ug(u) > 0$ if $u \neq 0$, $ug(u) \geq \alpha G(u)$ for some $\alpha > 0$ where $G(u) = \int_0^u g(s)ds$. If $u(t) = u(t,x)$ is a solution of (2) on $[0,\infty)$ with $\int_0^1 u_{xx}^2(t,x)dx \leq M$ for some $M > 0$ and all $t \geq 0$, then*

$$\int_0^1 G(u_x(t,x))dx + (1/f(t)) \int_0^1 u_t^2(t,x)dx \to 0 \text{ as } t \to \infty.$$

PROOF. Let

$$V(t) = 2 \int_0^1 G(u_x(t,x))dx + (1/f(t)) \int_0^1 u_t^2(t,x)dx.$$

Then by using the induced boundary conditions we obtain

$$V'(t) = -[f'(t)/f^2(t)] \int_0^1 u_t^2(t,x)dx.$$

Let

$$y(t) = \left(\int_0^1 u_t^2(t,x)dx \right)^{1/2} / \sqrt{f(t)};$$

then

(13) $$\begin{cases} V(t) = 2\int_0^1 G(u_x(t,x))dx + y^2(t), \\ V'(t) = -[f'(t)/f(t)]y^2(t). \end{cases}$$

Now

(14) $$\liminf_{t \to \infty} y(t) = 0;$$

for if $y^2(t) \geq \delta > 0$ on some interval $[t_1, \infty)$, then

$$V'(t) \leq -[f'(t)/f(t)]\delta^2 \text{ on } [t_1, \infty)$$

and a contradiction results from the properties of f. Suppose that

$$\limsup_{t \to \infty} y(t) = \lambda > 0;$$

then $\lim_{t \to \infty} V(t) = c > 0$, c constant. Let $\delta = \min\{1, \lambda/2, 3c\alpha/8(1 + \alpha + \gamma\sqrt{M})\}$; then there are sequences $\{t_n\}$, $\{t_n'\}$ having the following properties: $t_n < t_n' \leq t_{n+1}$, $y(t_n) = y(t_n') = \delta/2$, $y(t) > \delta/2$ on (t_n, t_n') with $\max_{s \in [t_n, t_n']} y(s) > \delta$, while $y(t) \leq \delta$ on $[t_n', t_{n+1}]$. To see that such sequences exist, let t_0 be defined such that $t_0 > 0$, $y(t_0) < \delta/2$ and consider the open set $\{t > t_0, y(t) > \delta/2\}$. It follows that $\{t > t_0, y(t) > \delta/2\}$ is a union of countable disjoint open intervals

H_i $(i = 1, 2, \ldots)$; that is, $\{t > t_0, y(t) > \delta/2\} = \bigcup\limits_{j=1}^{\infty} H_j$. Define $H = \{H_j:$ there exists $t_j^* \in H_j$ such that $y(t_j^*) > \delta\}$. Since $y'(t)$ is continuous, and consequently bounded on any finite interval, we may assume that $H = \bigcup\limits_{j=1}^{\infty} (a_j, b_j)$, with $a_j < b_j \leq a_{j+1}$, $j = 1, 2, \ldots$. Then $y(t) > \delta/2$ on (a_j, b_j), $\max\limits_{a_j \leq t \leq b_j} y(t) > \delta$, $y(a_j) = y(b_j) = \delta/2$, and $y(t) \leq \delta$ on $[b_j, a_{j+1}]$, $j = 1, 2, \ldots$. So $t_n = a_n$, $t'_n = b_n$ are the required sequences.

We shall show that

$$(15) \qquad \int_{t'_n}^{t_{n+1}} \sqrt{f(t)}\,dt \leq k \int_{t_n}^{t'_n} \sqrt{f(t)}\,dt, \quad n = 1, 2, \ldots$$

where $k > 0$ is a fixed constant. To that end we first note that

$$|(d/dt)y^2(t)| = \left| -[f'(t)/f^2(t)] \int_0^1 u_t^2\,dx + 2\int_0^1 u_t(g(u_x))_x\,dx \right|$$

$$\leqq [f'(t)/f(t)]y^2(t) + (2/\sqrt{f(t)})\left(\int_0^1 u_t^2\,dx\right)^{1/2}\sqrt{f(t)}$$

$$\times \left(\int_0^1 (g(u_x))_x^2\,dx\right)^{1/2}$$

and that

$$|(d/dt)y(t)| = [f'(t)/2f^{3/2}(t)]\sqrt{f(t)}y(t) + \sqrt{f(t)}\left(\int_0^1 (g(u_x))_x^2\,dx\right)^{1/2}$$

if $y(t) \neq 0$.

Now $[f'(t)/f^{3/2}(t)]$ and $y(t)$ are bounded, while $\left(\int_0^1 (g(u_x))_x^2\,dx\right)^{1/2}$ is bounded since $\int_0^1 u_{xx}^2\,dx$ is bounded. Hence, there exists a constant $k_1 > 0$ such that $|y'(t)| \leq k_1\sqrt{f(t)}$. This implies that

$$(16) \qquad (\delta/2) \leq \int_{t_n}^{t'_n} |y'(s)|\,ds \leq k_1 \int_{t_n}^{t'_n} \sqrt{f(s)}\,ds.$$

On the other hand,

$$(d/dt)\int_0^1 uu_t\,dx = \int_0^1 u_t^2\,dx - f(t)\int_0^1 u_x g(u_x)\,dx$$

$$= f(t)y^2(t) - f(t)\int_0^1 u_x g(u_x)\,dx.$$

Thus,

$$\alpha f(t)\int_0^1 G(u_x(t,x))\,dx \leq f(t)\int_0^1 u_x g(u_x)\,dx$$

$$\leq f(t)y^2(t) - (d/dt)\int_0^1 uu_t\,dx.$$

As $V(t) \to c > 0$ when $t \to \infty$, without loss of generality we may assume that $V(t) \geq 3c/4$ for $t \geq t_1$. Then

$$2 \int_{t'_n}^{t_{n+1}} \sqrt{f(s)} \int_0^1 G(u_x(s,x)) dx \, ds \geq \int_{t'_n}^{t_{n+1}} [(3c/4) - y^2(s)] \sqrt{f(s)} ds$$

$$\geq \int_{t'_n}^{t_{n+1}} \sqrt{f(s)} [(3c/4) - \delta^2] ds.$$

Hence,

$$\alpha \int_{t'_n}^{t_{n+1}} \sqrt{f(s)} ds [(3c/8) - (\delta^2/2)] \leq \alpha \int_{t'_n}^{t_{n+1}} \sqrt{f(s)} \int_0^1 G(u_x(s,x)) dx$$

$$\leq \int_{t'_n}^{t_{n+1}} \sqrt{f(s)} y^2(s) ds - \int_{t'_n}^{t_{n+1}} \{[(d/dt) \int_0^1 u u_t dx)] / \sqrt{f(s)}\} ds$$

and so

$$[(3c\alpha/8) - (1 + (\alpha/2))\delta^2)] \int_{t'_n}^{t_{n+1}} \sqrt{f(s)} ds$$

$$\leq - \int_{t'_n}^{t_{n+1}} (1/\sqrt{f(s)})[(d/dt) \int_0^1 u(s,x) u_t(s,x) dx] ds$$

$$= (1/\sqrt{f(t'_n)}) \int_0^1 u(t'_n) u_t(t'_n) dx - (1/\sqrt{f(t_{n+1})}) \int_0^1 u(t_{n+1},x) u_t(t_{n+1},x) dx$$

$$- 1/2 \int_{t'_n}^{t_{n+1}} [f'(s)/f^{3/2}(s)] \int_0^1 u(s,x) u_t(s,x) dx \, ds$$

$$\leq \left(\int_0^1 u^2(t'_n,x) dx \right)^{1/2} y(t'_n) + \left(\int_0^1 u^2(t_{n+1},x) dx \right)^{1/2} y(t_{n+1})$$

$$+ \gamma \sqrt{M} \delta \int_{t'_n}^{t_{n+1}} \sqrt{f(s)} ds$$

where $\gamma > 0$ is defined in the theorem. By the definition of δ and the boundedness of u and $y(t)$, we have

(17) $$\int_{t'_n}^{t_{n+1}} \sqrt{f(s)} ds \leq \beta, \text{ for some } \beta > 0, \quad n = 1, 2, \dots .$$

By (16) and (17), (15) follows.

Since $f'(t) \geq 0$, we have

(18) $$t_{n+1} - t'_n \leq k(t'_n - t_n).$$

Let $t > t_n$; then

$$V(t) \leq V(t_1) - \int_{t_1}^{t_n} [f'(s)/f(s)]y^2(s)ds$$

$$\leq V(t_1) - (\delta^2/4) \sum_{j=1}^{n} \int_{t_j}^{t_j'} [f'(s)/f(s)]ds$$

$$\leq V(t_1) - (\delta^2/4) \sum_{j=1}^{n} \int_{t_j}^{t_j'} \mu(s)ds$$

$$\leq V(t_1) - (\delta^2/8) \sum_{j=1}^{n} \left(\int_{t_j}^{t_j'} \mu(s)ds + 1/k \int_{t_j'}^{t_{j+1}} \mu(s)ds \right)$$

$$\leq V(t_1) - (\delta^2/8) \min\{1, 1/k\} \int_{t_1}^{t_{n+1}} \mu(s)ds \to \infty \text{ as } n \to \infty,$$

a contradiction. This implies that $y(t) \to 0$ as $t \to \infty$.

Also,

$$(d/dt) \int_0^1 uu_t dx = \int_0^1 u_t^2 dx - f(t) \int_0^1 u_x g(u_x)dx$$

so that

$$(19) \qquad \int_0^1 u_x g(u_x)dx = y^2(t) - (1/f(t))(d/dt) \int_0^1 uu_t dx.$$

Using the facts that $\int_0^1 u^2 dx$ and $[f'(t)/f^{3/2}(t)]$ are bounded, that $f(t) \to \infty$ as $t \to \infty$, and that $y(t) \to 0$ as $t \to \infty$, it follows that

$$\liminf_{t \to \infty} \int_0^1 u_x(t, x)g(u_x(t, x))dx = 0.$$

In fact, suppose that there exist $c > 0$, $t_1 > 0$ such that

$$\int_0^1 u_x(t, x)g(u_x(t, x))dx \geq c > 0$$

on $[t_1, \infty)$. We may assume that $|y(t)| \leq \min\{\sqrt{c/2}, c/4\gamma M\}$ for $t \geq t_1$. From (19) we have

$$c \leq (c/2) - (1/f(t))[(d/dt) \int_0^1 uu_t dx].$$

Thus

$$
c/2(t - t_1) \leq - \int_{t_1}^{t} \{(1/f(s))(d/ds) \int_{0}^{1} u(s,x)u_t(s,x)dx\}ds
$$

$$
= (1/f(t_1)) \int_{0}^{1} u(t_1,x)u_t(t_1,x)dx - (1/f(t)) \int_{0}^{1} u(t,x)u_t(t,x)dx
$$

$$
- \int_{t_1}^{t} \{(f'(s)/f^2(s)) \int_{0}^{1} u(s,x)u_t(s,x)dx\}ds
$$

$$
\leq \{My(t_1)/\sqrt{f(t_1)}\} + \big(My(t)/\sqrt{f(t)}\big)
$$

$$
+ \int_{t_1}^{t} \{(f'(s)/f^{3/2}(s))\Big(\int_{0}^{1} u^2(s,x)dx\Big)^{1/2}\Big(\int_{0}^{1} u_t^2(s,x)dx\Big)^{1/2}/\sqrt{f(s)}\}ds
$$

$$
\leq \big(My(t_1)/\sqrt{f(t_1)}\big) + \big(My(t)/\sqrt{f(t)}\big) + \gamma M \int_{t_1}^{t} y(s)ds
$$

$$
\leq \big(My(t_1)/\sqrt{f(t_1)}\big) + \big(My(t)/\sqrt{f(t)}\big) + [c/4(t - t_1)].
$$

This yields

$$
\big(My(t_1)/\sqrt{f(t_1)}\big) + \big(My(t)/\sqrt{f(t)}\big) \geq c/4(t - t_1)
$$

which tends to infinity, a contradiction.

As the

$$
\liminf_{t \to \infty} \int_{0}^{1} u_x(t,x)g(u_x(t,x))dx = 0
$$

we argue that

$$
\lim_{t \to \infty} \int_{0}^{1} G(u_x(t,x))dx = 0.
$$

Since

$$
y^2(t) + 2 \int_{0}^{1} G(u_x(t,x))dx \to c
$$

as $t \to \infty$, we conclude that $c = 0$. This completes the proof.

REFERENCES

1. T. A. Burton, *The nonlinear wave equation as a Liénard equation*, Funkcial. Ekvac. (to appear).

2. T. A. Burton and R. Grimmer, *On continuability of solutions of second order differential equations*, Proc. Amer. Math. Soc. **29** (1971), 277–283.

3. T. A. Burton and R. Grimmer, *On the asymptotic behavior of solutions of $x'' + a(t)f(x) = 0$*, Proc. Camb. Phil. Soc. **70** (1971), 77–88.

4. T. A. Burton and R. Grimmer, *On the asymptotic behavior of solutions of $x'' + a(t)f(x) = e(t)$*, Pacific J. Math. **41** (1972), 43–55.

5. T. A. Burton and W. T. Patula, *Limit circle results for second order equations*, Monatsh. Math. **81** (1976), 185–194.

6. N. G. Chetayev, *The Stability of Motion*, Pergamon Press, New York, 1961.

7. E. A. Coddington and N. Levinson, *Theory of Ordinary Differential Equations*, McGraw-Hill, New York, 1955.

8. A. Devinatz, *The deficiency index problem for ordinary self-adjoint differential operators*, Bull. Amer. Math. Soc. **79** (1973), 1109–1127.

9. A. S. Galbraith, E. J. McShane, and G. B. Parrish, *On the solutions of linear second order differential equations*, Proc. Nat. Acad. Sci. U.S.A. **53** (1965), 247–249.

10. S. P. Hastings, *Boundary value problems in one differential equation with a discontinuity*, J. Differential Equations **1** (1965), 346–369.

11. A. Kneser, *Untersuchungen über die reelen Nullstelles der Integrale linearer Differentialgleichungen*, Math. Ann. **42** (1893), 409–435.

12. B. Nicolaenko, C. Foias, and R. Temmam, *The Connection between Infinite Dimensional and Finite Dimensional Dynamical Systems*, Amer. Math. Soc., Providence, R.I., 1989.

13. W. R. Utz, *Properties of solutions of $u'' + g(t)u^{2n-1} = 0$*, Monatsh. Math. **66** (1962), 55–60.

14. A. Wintner, *A criterion of oscillatory stability*, Quart. Appl. Math. **5** (1947), 232–236.

15. M. Zlámal, *Oscillation criterion*, Časopis Pěst. Mat. Fys. **75** (1950), 213–218.

DEPARTMENT OF MATHEMATICS, SOUTHERN ILLINOIS UNIVERSITY, CARBONDALE, ILLINOIS 62901-4408

E-mail: GE0641@SIUCVMB.BITNET

BOLYAI INSTITUTE, ARADI VÉRTANÚK TERE 1, SZEGED, HUNGARY

DEPARTMENT OF MATHEMATICS, SOUTHERN ILLINOIS UNIVERSITY, CARBONDALE, ILLINOIS 62901-4408 (ON LEAVE FROM NORTHEAST NORMAL UNIVERSITY, CHANGCHUN, JILIN, PRC)

Contemporary Mathematics
Volume **129**, 1992

The Oscillation and Exponential Decay Rate of Solutions of Differential Delay Equations

Y. CAO

ABSTRACT. This paper generalize the discrete Lyapunov function to state-dependent delay differential equation. The discrete Lyapunov function is a measure of oscillation of solutions on intervals whose length equals the time delay at the state being zero. It establishes a relation among the oscillation, the exponential decay rate and the first order estimation of solutions which go to zero as t goes to positive infinity. Also it proves that the faster a solution oscillates, the faster it decays.

0. Introduction

In a previous paper [**3**], we established some relations between the oscillation and the exponential decay rate of solutions of differential equations with a constant delay by introducing a discrete Lyapunov function. This paper will generalize the results in [**3**] to solutions of differential equations with a state-dependent delay

$$(0.1) \qquad \dot{x}(t) = f(x(t), x(t - \mu), t), \quad t > 0,$$

where $\mu = \mu(x(t)) > 0$ depends on the current state $x(t)$. First of all, we generalized the discrete Lyapunov function of [**3**] to (0.1) (see Definitions 2.1 and 2.2). Through the definitions, one can see that the discrete Lyapunov function $V(x, t)$ of a solution x of (0.1) is a measure of oscillation of x on the unit interval $[t - \tau_0, t]$, where $\tau_0 = \mu(0) > 0$. Under the assumption of negative feedback or positive feedback (see (2.2)) on the delay term, it is proved in Theorem 2.1 that the oscillation of each solution will not speed up as the time elapses. Using the properties of the discrete Lyapunov function, we establish the relations among the oscillation (i.e. the discrete Lyapunov function), the exponential decay rate and the first order estimation of solutions which go to zero as $t \to +\infty$ (see Theorem 4.2). In particular, we prove that the faster a solution oscillates, the

1991 *Mathematics Subject Classification.* 34K15.

faster it decays. As a consequence, for each equation, we can find a non-negative number n_0 such that a solution will not decay to zero if it oscillates slower than n_0-times in some unit interval. The last result can be used to prove that all solutions oscillate for some equations, see Arino [1] for example.

There have been many papers discussing the oscillation of solutions of differential delay equations, where the oscillation is defined one way or another (see, for example, [1, 2, 8, 9, 10]). Our definition of the discrete Lyapunov function gives us a way to see how fast a solution oscillates. For example, for a solution which goes to zero as $t \to +\infty$, our results tell that the oscillation of the solution on $[t - \tau_0, t]$ for any $t > 0$ is not faster than on $[-\tau_0, 0]$, and is not slower than the n_0 mentioned in last paragraph. Using our result that a solution oscillates infinitely on unit intervals if and only if it decays faster than any exponential rate, it is possible to generalize the Morse-Decomposition of [12] to equation (0.1).

1. Constant Delay and Non-Constant Delay

Consider the differential delay equation

$$(1.1) \qquad \dot{x}(t) = f(x(t), x(t - \mu), t), \quad t > t_0,$$

where $\mu = \mu(x(t), t)$ depends on the current state $x(t)$ and the time t, and $f(x, y, t)$ is a function from $\mathbb{R}^n \times \mathbb{R}^n \times [t_0, +\infty)$ to $\mathbb{R}^n$. Assume that there exists a constant $\tau > 0$ such that

$$(1.2) \qquad 0 < \mu(x, t) \le \tau \quad \text{for all} \ \ x \in \mathbb{R}^n \ \ \text{and} \ \ t \in [t_0, +\infty).$$

Let $C^0 = C^0([-\tau, 0], \mathbb{R}^n)$ be the Banach space of Lipschitz continuous functions from $[-\tau, 0]$ to $\mathbb{R}^n$ with the norm:

$$(1.3) \qquad \|\varphi\| = \max_{\theta \in [-\tau, 0]} |\varphi(\theta)| + \sup\left\{ \frac{|\varphi(\theta) - \varphi(\theta')|}{|\theta - \theta'|} : \theta, \theta' \in [-\tau, 0], \theta \ne \theta' \right\}.$$

For any φ in C^0, similar to ODE, if $f(x, y, t)$ and $\mu(z, t)$ are continuous functions and are Lipschitzian in (x, y) and z respectively, then one can show that, given any initial condition

$$(1.4) \qquad x(\theta + t_0) = \varphi(\theta) \quad \text{for} \ \ \varphi \ \ \text{in} \ \ C^0 \ \ \text{and} \ \ \theta \in [-\tau, 0],$$

there exists a unique solution of (1.1) on $[t_0 - \tau, t_1]$ for some $t_1 > t_0$.

Example of Non-uniqueness. Consider the state-dependent delay differential equation with an inital value

$$(1.5) \qquad \begin{cases} \dot{x}(t) = x(t - \tau(x(t))) & t > 0 \\ x(\theta) = \varphi(\theta) & \theta \in [-1, 0], \end{cases}$$

where $\tau(z)$ is any C^∞-function satisfying $\frac{1}{4} \le \tau(z) \le 1$ for all $|z| > \frac{1}{4}$ and $\tau(z) = \frac{1}{2} - z$ for $|z| \le \frac{1}{4}$, and φ is any continuous function on $[-1, 0]$ satisfying

$$(1.6) \qquad \varphi(\theta) = \begin{cases} -1 + |\theta + \frac{1}{2}|^{\frac{1}{2}}, & \theta \ \text{near} \ -\frac{1}{2} \\ 0 & \theta \ \text{in} \ [-1, 0] \ \text{and near} \ 0. \end{cases}$$

For t near zero, the equation can be simplified by

(1.7) $\begin{cases} \dot{x}(t) = \varphi(t - \tau(x(t))) = -1 + |t + x(t)|^{\frac{1}{2}} & 0 < t << 1 \\ x(0) = 0. \end{cases}$

If $y(t) = t + x(t)$, then

(1.8) $\begin{cases} \dot{y}(t) = |y(t)|^{\frac{1}{2}} & 0 < t << 1 \\ y(0) = 0. \end{cases}$

This equation has two solutions $y(t) \equiv 0$ and $y(t) = \frac{1}{4}t^2$ for $t \geq 0$ sufficiently small. Thus $x(t) = -t$ and $x(t) = \frac{1}{4}t^2 - t$ are two different solutions for $t \geq 0$ sufficiently small. Note that the non-uniquness of solutions is caused by non-Lipschitzian of the initial value φ at $\theta = -\frac{1}{2}$.

We have already obtained some results on the oscillation and exponential decay rate of solutions for the differential equation with constant delay in [3]. In this section, we will make a change of time such that the equations with a non-constant delay become equations with a constant delay. Consider the time-dependent delay differential equations

(1.9) $$\dot{x}(t) = f(x(t), x(t - \mu(t)), t), \quad t > t_0,$$

where $f(x, y, t)$ is a continuous function from $\mathbb{R}^n \times \mathbb{R}^n \times [t_0, +\infty)$ to $\mathbb{R}^n$. Assume

(H1) $\mu(t) > 0$ for all $t \in [t_0, +\infty)$ and $\liminf\limits_{t \to +\infty} \mu(t) = \tau_* > 0$.

(H2) $\mu(t)$ is C^1 in $t \in [t_0, +\infty)$, $|\mu'(t)| < 1$ for all $t \in [t_0, +\infty)$ and there exists a constant $\alpha > 1$ such that $\lim\limits_{t \to +\infty} \mu'(t)t^\alpha = 0$.

THEOREM 1.1. *Suppose (H1) and (H2) hold. Then there exists an invertible time change*

(1.10) $$t = \sigma(s), \quad y(s) = x(\sigma(s)) \quad for \ [-1, +\infty)$$

such that Eq. (1.9) becomes

(1.11) $$\frac{dy(s)}{ds} = f(y(s), y(s - 1), \sigma(s))\sigma'(s), \quad s > 0,$$

where σ is a C^1-solution of the difference equation

(1.12) $$\sigma(s) - \mu(\sigma(s)) = \sigma(s - 1)$$

and satisfies $\sigma(0) = t_0$, $\lim\limits_{t \to +\infty} \sigma(t) = +\infty$ and $\sigma'(s)$ is bounded for s in $[-1, +\infty)$.

PROOF. By the time change (1.10), Eq. (1.9) becomes

(1.13) $$\frac{dy(s)}{ds} = f(x(\sigma(s)), x(\sigma(s) - \mu(\sigma(s)), \sigma(s))\sigma'(s).$$

We want to find a C^1-function σ such that

$$(1.14) \qquad \sigma(s) - \mu(\sigma(s)) = \sigma(s-1)$$

for all $s \in \mathbb{R}^+$. If $h(t) = t - \mu(t)$ and $\tau_0 = \mu(t_0) > 0$, then the equation $h(t) = \rho$ can be solved uniquely by a C^1-function $t = H(\rho)$ for $\rho \in [t_0 - \tau_0, +\infty)$ and $t \in [t_0, +\infty)$. From (H2), it is easy to see that $H'(\rho) > 0$ for all $\rho \in [t_0 - \tau_0, +\infty)$. We choose an arbitrary C^1-function $\sigma_0(\theta)$ ($\theta \in [-1, 0]$) satisfying

$$(1.15) \qquad \sigma_0(-1) = t_0 - \tau_0, \quad \sigma_0(0) = t_0, \quad \sigma_0'(0) = H'(t_0 - \tau_0)\sigma_0'(-1)$$

and

$$\sigma_0'(\theta) > 0 \qquad \text{for} \ \ \theta \in [-1, 0].$$

In this way, for $n = 1, 2, \cdots$, and $\theta \in [-1, 0]$, if we define

$$(1.16) \qquad \sigma(\theta + n) = H^n(\sigma_0(\theta)),$$

where H^n is the nth iteration of H, then $\sigma(s)$ is a C^1-function for $s \in [-1, +\infty)$ and satisfies (1.14) for all $s \in [0, +\infty)$. Since

$$(1.17) \qquad \frac{d}{d\theta}\sigma(\theta + n) = \left\{ \prod_{i=1}^{n} H'(H^{n-i}(\sigma_0(\theta))) \right\} \sigma_0'(\theta),$$

we know that $\sigma'(s) > 0$ for all $s \in [-1, +\infty)$. From $\sigma'(s) > 0$ and $\mu(t) > 0$, (1.14) implies $\lim_{s \to +\infty} \sigma(s) = +\infty$. The only thing we need to show is the boundness of $\sigma'(s)$. From (H1), we know that there exists $M_0 > t_0$ such that $\mu(t) \geq \frac{\tau_*}{2}$ for all $t \geq M_0$. Therefore $H(\rho) \geq \rho + \frac{\tau_*}{2}$ if $t = H(\rho) \geq M_0$. Thus, letting $n_0 > 0$ such that $\sigma(s) \geq M_0$ for all $s \geq n_0 - 1$, we have

$$(1.18) \qquad \sigma(\theta + n) = H^n(\sigma_0(\theta)) = H^{n-n_0}(\sigma(\theta + n_0))$$
$$\geq \sigma(\theta + n_0) + (n - n_0)\frac{\tau_*}{2}$$

for all $n > n_0$. On the other hand,

$$H'(\rho) = \frac{1}{1 - \mu'(t)} = [1 - \mu'(H(\rho))]^{-1},$$

where $\mu'(t)t^\alpha \to 0$ as $t \to +\infty$ according to (H2). Thus

$$H'(H^k(\sigma_0(\theta))) = [1 - \mu'(H^k(\sigma_0(\theta))]^{-1} = (1 - a_k(\theta))^{-1},$$

where it follows from (1.18) that $a_k(\theta)(\frac{\tau_*}{2}k)^\alpha \to 0$ uniformly for $\theta \in [-1, 0]$ as $k \to +\infty$ for $\alpha > 1$ in (H2). Therefore, the right-hand side of (1.17) converges to a bounded continuous function $\psi(\theta)$ as $n \to +\infty$. We have proved the boundness of $\sigma'(s)$. $\square$

2. Discrete Lyapunov Function for a Scalar Equation

From now on, we will study a state-dependent time delay differential equation

$$(2.1) \qquad \dot{x}(t) = f(x(t), x(t-\mu), t) \qquad t > 0,$$

where $f(0,0,t) = 0$ for all $t > 0$, $\mu = \mu(x(t)) > 0$ depends on the current state $x(t)$, and $f(x,y,t)$ and $\mu = \mu(z)$ are locally Lipschitzian in (x,y) and z, respectively. Assume (1.2) is still true for some $\tau > 0$ and denote $\tau_0 = \mu(0)$. In discussing (2.1), we always assume that there is an $M > 0$ (finite or infinite) such that

$$(2.2) \qquad f(0,y,t) \neq 0 \quad \text{for } 0 < |y| \leq M \text{ and all } t \in \mathbb{R}^+.$$

This assumption, together with the additional assumption that $f_y'(0,0,t) \neq 0$ for all $t \in \mathbb{R}^+$, means that the delay term in (2.1) is positive or negative feedback in a neighborhood of the origin. Generalizing Definitions 1.1 and 1.2 in [3], we give the following defintions.

DEFINITION 2.1. Let $\tau_0 = \mu(0) > 0$. For $\psi \in C^0$, define $\# : C^0 \to Z \cup +\infty$ by $\#(\psi) = $ the number of zeros of ψ on $[-\tau_0, 0]$, not counting multiplicities. Here, if $\psi(\theta) \equiv 0$ on $[-\tau_0, 0]$, then we define $\#(\psi) = +\infty$; if $\psi(\theta) \equiv 0$ on $[a,b] \subset [-\tau_0, 0]$ and $[a,b] \neq [-\tau_0, 0]$, then we count this interval as a single zero.

DEFINITION 2.2. Let x be a solution of (2.1) on $[-\tau, +\infty)$. For $t \geq 0$, define

$$\sigma(x,t) = \begin{cases} +\infty & \text{if } x(s) \neq 0 \text{ for all } s \geq t \\ \inf\{s \geq t : x(s) = 0\} & otherwise \end{cases}$$

and

$$V(x,t) = \begin{cases} 0 & \text{if } \sigma(x,t) = +\infty \\ \#(x_{\sigma(x,t)}) & otherwise. \end{cases}$$

Note that the function $\sigma(x,t)$ is the time shift of x to the first zero of x which is not less than t. Different from our original definiton in [3], we define $V(x,t) = 0$ instead of $V(x,t) = 1$, when $\sigma(x,t) = +\infty$. This modification is made to distinguish non-oscillating solutions from slowly oscillating solutions. This modification will not affect the essential properties of the discrete Lyapunov function $V(x,t)$ obtained in [3]. We should point out that, under the assumption (2.2), a solution x of (2.1) on $[-\tau, +\infty)$ is equal to zero for all $t \geq 0$ if and only if $x(\theta) = 0$ for all θ on $[-\tau_0, 0]$. Our next theorem says that $V(x,t)$ defined in Definition 2.2 is a Lyapunov function.

THEOREM 2.1. *Suppose x is a solution of (2.1) for $t \in [-\tau, +\infty)$ and $|x(t)| \leq M$ for $t \in [-\tau_0, +\infty)$, where M is the same as in (2.2) and $\tau_0 = \mu(0)$. Then $V(x,t)$ is nonincreasing in t. Moreover, if $x(t_0) = \dot{x}(t_0) = 0$ for some $t_0 > 0$, then*

$$V(x,t) \leq V(x,t_0) - 1 \qquad \text{for all } t \geq t_0 + \tau_0.$$

Furthermore, if $\lim\limits_{t \to +\infty} V(x,t) < +\infty$, then there exist a $t_1 > 0$ such that all zeros of $x(t)$ in $[t_1, +\infty)$ are simple.

PROOF. The proof of this theorem is almost the same as Theorem 1.5, Lemma 1.6 and Corollary 1.7 in [3]. We will prove a lemma similar to Lemma 1.3 in [3] and omit the other proofs.

LEMMA 2.2. *Suppose x is a solution of (2.1) for $t \in [-\tau, +\infty)$ and $|x(t)| \le M$ for $t \in [-\tau_0, +\infty)$, where M is given by (2.2) and $\tau_0 = \mu(0)$. If there exist $t_1 > t_0 \ge 0$ such that $x(t_1) = x(t_0) = 0$ and $x(t)$ is not identically zero for t in (t_0, t_1), then $x(t)$ has at least one zero between $t_1 - \tau_0$ and $t_0 - \tau_0$. More precisely, there are $s_i \in (t_0 - \tau_0, t_1 - \tau_0), i = 1, 2, 3$, with $s_1 < s_2 < s_3$ such that $x(s_2) = 0, x(s_1) \ne 0$ and $x(s_3) \ne 0$.*

PROOF. Without loss of generality, we assume that $x(t) \ne 0$ for all t in (t_0, t_1), since we can always replace (t_0, t_1) by a subinterval (t_0', t_1') such that $x(t_0') = x(t_1') = 0$ and $x(t) \ne 0$ for all t in (t_0', t_1'). We will prove the lemma for the case $x(t) > 0$ for all t in (t_0, t_1). The treatment of the other situation is similar. If $h(t) = \ln|x(t)|, t \in (t_0, t_1)$, then $h(t)$ is a C^1-function and $h(t) \to -\infty$ as $t \to t_1^+$ and $t \to t_1^-$. Therefore, there exist t_n^0 and t_0' in $(t_0, t_1), n = 1, 2, \cdots$, such that $t_n^0 \to t_0$ and $t_n' \to t_1$ as $n \to +\infty$ and

$$h'(t_n^0) = \dot{x}(t_n^0)/x(t_n^0) = M_n^0 \to +\infty,$$
(2.3)
$$h'(t_n') = \dot{x}(t_n')/x(t_n') = M_n' \to -\infty, \quad as \ n \to +\infty.$$

Let $M_0 > 0$ be a Lipschitz constant such that

(2.4) $|f(x, y, t) - f(0, y, t)| \le M_0|x|$ for $|x|, |y| \le M, t \in [t_0, t_1]$.

It follows from (2.1) and (2.3) that

$$f(0, x(t_n^0 - \mu_n^0), t_n^0) \ge \dot{x}(t_n^0) - M_0 x(t_n)$$
(2.5)
$$\ge (M_n^0 - M_0)x(t_n^0) > 0 \quad \text{for } n \text{ large enough}$$

where $\mu_n^0 = \mu(x(t_n^0)) \to \tau_0$ as $n \to +\infty$; and

$$f(0, x(t_n' - \mu_n'), t_n') = \dot{x}(t_n') + M_0 x(t_n')$$
(2.6)
$$\le (M_n' + M_0)x(t_n') < 0 \quad \text{for } n \text{ large enough},$$

where $\mu_n' = \mu(x(t_n')) \to \tau_0$ as $n \to +\infty$. If both $x(t_0 - \tau_0)$ and $x(t_1 - \tau_0)$ are not zero, then (2.5) and (2.6) imply that there exist $s_i \in (t_0 - \tau_0, t_1 - \tau_0), i = 1, 2, 3$, with $s_1 < s_2 < s_3$ such that $x(s_2) = 0, x(s_1) \ne 0$ and $x(s_3) \ne 0$ because of (2.2). If $x(t_0 - \tau_0) = 0$, then (2.1) implies $\dot{x}(t_0) = 0$. Since μ is Lipschitzian, $\mu(x) - \mu(0) = \alpha(x)x$ for some $\alpha(x)$ being bounded for $|x| \le M$. Therefore

$$t_n^0 - \mu_n^0 = t_0 - \tau_0 + (t_n^0 - t_0) + \alpha(x(t_0))\dot{x}(t_0)(t_n^0 - t_0) + o(|t_n^0 - t_0|)$$
$$= t_0 - \tau_0 + (t_n^0 - t_0) + o(|t_n^0 - t_0|) \quad \text{as } n \to +\infty,$$

that is, $t_n^0 - \mu_n^0$ is in $(t_0 - \tau_0, t_1 - \tau_0)$ for n large enough. Similarly, if $x(t_1 - \tau_0) = 0$, then $t_n' - \mu_n'$ is in $(t_0 - \tau_0, t_1 - \tau_0)$. These insure the existence of s_i in $(t_0 - \tau_0, t_1 - \tau_0), i = 1, 2, 3$. The proof is completed.   □

From this lemma, one can see that when $\sigma(x, t)$ moves from t_0 to t_1 for $x(t)$ to gain a zero, $x(t)$ has already lost at least one zero in the interval $(t_0 - \tau_0, t_1 - \tau_0)$. This explains why $V(x, t)$ is nonincreasing in t. When $x(t_0) = \dot{x}(t_0) = 0$ and

$x(t_1) = 0$ for some $t_1 \in (t_0, t_0 + \tau_0]$, $x(t)$ will loose one zero in the interval $(t_0 - \tau_0, t_1 - \tau_0)$ and one zero at $t = t_0 - \tau_0$ as $\sigma(x, t)$ moves from t_0 to t_1. When $x(t_0) = \dot{x}(t_0) = 0$ and $x(t) \neq 0$ for any $t \in (t_0, t_0 + \tau_0]$, $V(x, t) \leq 1$ for $t \geq t_0 + \tau_0$ and $V(x, t_0) \geq 2$ since $x(t_0 - \tau_0) = 0$ from (2.1) and (2.2). Thus $V(x, t)$ will decrease by at least one when $t \geq t_0 + \tau_0$ if the zero point t_0 of x is not simple. Consequently, if $\lim_{t \to +\infty} V(x, t) < +\infty$, then zeros of $x(t)$ are simple for all large t.

3. Exponential Decay Rate

In this section, we will set up a connection between the oscillation and the exponential decay rate for solutions of (2.1).

DEFINITION 3.1. Suppose x is a solution of (2.1) for $t \in [-\tau, +\infty)$ and $\lim_{t \to +\infty} x(t) = 0$. Define the exponential decay rate $\overline{\alpha}$ of x by

$$\overline{\alpha}(x) = \inf \left\{ \alpha \leq 0 : \lim_{t \to +\infty} x(t) e^{-\alpha t} = 0 \right\}.$$

Usually, when $\overline{\alpha}(x) = -\infty$, the solution x is called a small soltuion.

Consider the differential equation with a constant delay

$$(3.1) \qquad \dot{x}(t) = f(x(t), x(t-1), t),$$

where $f(x, y, t)$ is continuously differentiable in x and y, and $f(0, 0, t) = 0$ for all $t \in \mathbb{R}^+$. Denote

$$A(\xi, \eta, t) = \frac{\partial}{\partial \xi} f(\xi, \eta, t), \quad B(\xi, \eta, t) = \frac{\partial}{\partial \eta} f(\xi, \eta, t).$$

(H3)   Assume $f(0, 0, t) \equiv 0$ for $t \geq 0$ and there are some positive constants $\delta_0, \overline{A}, \overline{B}$ and $\overline{b}$ such that $\overline{b} \leq |B(\xi, \eta, t)| \leq \overline{B}$ and $|A(\xi, \eta, t)| \leq \overline{A}$ for all $|\xi|, |\eta| \leq \delta_0$ and $t \in \mathbb{R}^+$.

Before discussing the exponential decay rate for solutions of (2.1), we present a result for (3.1) obtained in our previous paper [3].

LEMMA 3.1 ([3, Theorem 2.8]). *Assume (H3) is satisfied. Suppose x is a solution of (3.1) for $t \in [-1, +\infty)$ with $\limsup_{t \to +\infty} |x(t)| < \delta_0$, where δ_0 is as in (H3). Then x is a small solution if and only if $\lim_{t \to +\infty} V(x, t) = +\infty$.*

The statement here is a little different from that of Theorem 2.8 in [3], where we assume $x(\theta) \not\equiv 0$ for $\theta \in [-1, 0]$ and use a super-exponential solution instead of a small solution. By a super-exponential solution we mean a small solution which is not identically zero for all large t. One can see that, under (H3), a solution x of (3.1) for $t \in [-1, +\infty)$ is identically zero on some unit interval if and only if $x(\theta)$ is zero for all θ in $[-1, 0]$. Therefore, this change of statement does not affect the truth of the lemma. The reason to make this change is as follows: if f is analytic in (ξ, η) around $(0, 0)$ and is indepdent of t, it was proved in [4] recently that there is no super-exponential solution of (3.1), provided $\frac{\partial f}{\partial \eta}(0, 0) \neq 0$. We are going to show a result for solutions of (2.1) similar to Lemma 3.1. Our next two lemmas are similar to Lemma 2.6 and Theorem 2.7 of [3]. For completeness we will give detailed proofs.

LEMMA 3.2. *Assume f in (2.1) satisfies (H3), and assume μ is a C^1-function satisfying $0 < \mu(x) \leq \tau$ for some constant τ and all $|x| \leq \delta_0$, where δ_0 is in (H3). Let $\tau_0 = \mu(0) > 0$. Suppose x is a solution of (2.1) for $t \in [-\tau, +\infty)$ with $\limsup\limits_{t \to +\infty} |x(t)| < \delta_0$. If $\lim\limits_{x \to +\infty} V(x, t) = +\infty$, then there exists a constant $\tilde{\theta} \in [-\tau_0, +\infty)$ such that $x(\tilde{\theta} + n\tau_0) = 0, n = 1, 2, \cdots$.*

PROOF. By the assumption, there exists $t_0 \geq 0$ such that $|x(t)| \leq \delta_0$ for all $t \geq t_0 - \tau$. From (H3) it is clear that (2.2) is satisfied for some $M \geq \delta_0$. Since $V(x, t)$ is nonincreasing in $t \geq t_0$, $\lim\limits_{t \to +\infty} V(x, t) = +\infty$ implies $V(x, t) = +\infty$ for all $t \geq t_0$. Therefore, if

$$B_k = \{\theta \in [-\tau_0, 0] : x(\theta + k\tau_0 + t_0) = 0\}$$

and

$$B_k^0 = \{\theta \in B_k : \theta \text{ is a limit point of } B_k\},$$

then $B_k^0 \subset B_k$ is a non-empty and closed set for $k = 0, 1, 2, \cdots$. We want to show $B_{k+1}^0 \subset B_k^0$. Assume $\theta_0 \in B_{k+1}^0$. If there is an interval $[a, b]$ which contains θ_0 and is in B_{k+1}^0, then (2.1) becomes

$$0 = f(0, x(t - \tau_0), t), \quad \text{for } t \in [a + (k+1)\tau_0 + t_0, b + (k+1)\tau_0 + t_0],$$

thus $x(t) = 0$ for $t \in [a + k\tau_0 + t_0, b + k\tau_0 + t_0]$ by (2.2), i.e. $\theta_0 \in [a, b] \subset B_k^0$. Now, assume there exist $\{\theta_n\} \subset B_{k+1}^0$ with $\theta_n \neq \theta_0$ for $n = 1, 2, \cdots$, such that $\theta_n \to \theta_0$ as $n \to +\infty$ and $x(t + (k+1)\tau_0 + t_0)$ is not identically zero for t between θ_n and $\theta_{n+1}(n = 1, 2, \cdots)$. From Lemma 2.2, we can find θ_n' between θ_n and θ_{n+1} such that $x(\theta_n' + k\tau_0 + t_0) = 0 \ (n = 1, 2, \cdots)$. Therefore $\theta_0 = \lim\limits_{n \to \infty} \theta_n'$ is in B_k^0. We have shown $B_{k+1}^0 \supset B_k^0, k = 0, 1, 2, \cdots$. Thus $\bigcap\limits_{k=0}^{\infty} B_k^0$ is non-empty. Choosing any $\theta^* \in \bigcap\limits_{k=1}^{\infty} B_k^0$ and letting $\tilde{\theta} = \theta^* + t_0$, we complete the proof.

LEMMA 3.3. *Assume the same conditions as in Lemma 3.2. If $\lim\limits_{t \to +\infty} V(x, t) = +\infty$, then x is a small solution.*

PROOF. From Lemma 3.2, we know that there is $\tilde{\theta} \in [-\tau_0, +\infty)$ such that $x(\tilde{\theta} + n\tau_0) = 0$ for $n = 0, 1, 2, \cdots$. By a time-shift, we can assume $\tilde{\theta} = 0$ and $|x(t)| \leq \delta_0$ for all $x \in [-\tau, +\infty)$. Let L_0 and L_1 be two positive constants that

$$|f(x, y, t)| \leq L_0 \quad \text{and} \quad |\mu(x) - \tau_0| \leq L_1|x|$$

for all $|x|, |y|, \leq \delta_0$ and $t \in \mathbb{R}^+$, where $\tau_0 = \mu(0)$. In this way, we have

$$|x(t - \mu(x(t)))| \leq |x(t - \tau_0)| + L_0 L_1 |x(t)| \quad \text{for } t \geq 0,$$

since $|\dot{x}(t)| = |f(x(t), x(t - \mu), t)| \leq L_0$. Thus,

$$|x(\theta + n\tau_0)| = \int_{-\tau_0}^{\theta} |f(x(\rho + n\tau_0), x(\rho + n\tau_0 - \mu), \rho + n\tau_0)| d\rho$$

$$\leq \int_{-\tau_0}^{\theta} \left\{ \overline{A}|x(\rho + n\tau_0)| + \overline{B}|x(\rho + n\tau_0 - \mu)| \right\} d\rho$$

$$\leq \int_{-\tau_0}^{\theta} \left\{ (\overline{A} + L_0 L_1 \overline{B})|x(\rho + n\tau_0)| + \overline{B}|x(\rho + (n-1)\tau_0)| \right\} d\rho,$$

where $\mu = \mu(x(\rho + n\tau_0))$. Letting $L_2 = \overline{A} + L_0 L_1 \overline{B}$ and using the Generalized Gronwall inequality, we have

$$|x(\theta + n\tau_0)| \le \int_{-\tau_0}^{\theta} \overline{B}|x(\rho + (n-1)\tau_0)|d\rho +$$

$$\int_{-\tau_0}^{\theta} \int_{-\tau_0}^{\rho} L_2 \overline{B}|x(s + (n-1)\tau_0)|ds\, e^{\overline{A}(\tau_0 + \rho)}d\rho.$$

Thus

$$|x(\theta + n\tau_0)| \le M_0 \int_{-\tau_0}^{\theta} |x(\rho + (n-1)\tau_0|d\rho, \quad \theta \in [-\tau_0, 0], n = 1, 2, \cdots,$$

where $M_0 = \overline{B}[1 + L_2 e^{\overline{A}\tau_0}]$. These inequalities imply, step by step, that

$$|x(\theta + n\rho_0)| \le \{[M_0(\theta + \tau_0)]^n / n!\} \max_{s \in [-\tau_0, 0]} |x(s)|$$

for $n = 1, 2, \cdots$. Thus x is a small solutions.   □

Now, we are ready to show the following result.

THEOREM 3.4. *Assume f in (2.1) satisfies (H3), and assume μ is a C^1-function satisfying $0 < \mu(x) \le \tau$ for some constant τ and all $|x| \le \delta_0$, where δ_0 is as in (H3). Suppose x is a solution of (2.1) for $t \in [-\tau, +\infty)$ with $\limsup_{t \to +\infty} |x(t)| < \delta_0$. Then x is a small solution if and only if $\lim_{t \to +\infty} V(x, t) = +\infty$.*

PROOF. If $\lim_{t \to +\infty} V(x, t) = +\infty$, then from Lemma 3.3 we know that x is a small solution. On the other hand, if $\tilde{x}$ is a small solution of (2.1), then $\tilde{x}$ satisfies the equaiton

$$\dot{x}(t) = f(x(t), x(t - \tilde{\mu}(t)), t),$$

where $\tilde{\mu}(t) = \mu(\tilde{x}(t))$. Since $\tilde{x}$ is a small solution, there exists $t_0 > 0$ such that $\tilde{\mu}$ satisfies (H1) and (H2). Let $\tau_0 = \tilde{\mu}(t_0)$. By Theorem 1.1, there exists an invertible time change $t = \sigma(s)$ from $[-1, +\infty)$ onto $[t_0 - \tau_0, +\infty)$ with $\sigma'(s)$ being bounded for all $s \in [-1, +\infty)$ such that, if $y(s) = x(\sigma(s))$, then

$$\dot{y}(s) = f(y(s), y(s - 1), \sigma(s))\sigma'(s) \quad \text{for} \quad s > 0.$$

By the boundness of $\sigma'(s)$ we know y is a small solution of the equation above. Therefore, from Lemma 3.1 we know that $\lim_{s \to \infty} V(y, s) = +\infty$. Since $V(x, t)$ is non-increasing in t, $\lim_{s \to \infty} V(y, s) = +\infty$ implies $\lim_{t \to +\infty} V(x, t) = +\infty$. The proof is completed.   □

This theorem tells us that, if a solution x of (2.1) oscillates finitely on the initial interval $[-\tau, 0]$, then x can decay only at a finite exponential rate. This property will be used for the first order estimate of solutions of (2.1) which go to zero as $t \to +\infty$ in the next section.

4. Oscillation and Exponential Decay Rate

The first order estimation of a solution of nonlinear ordinary differential equations is well known. Consider the ordinary differential equation $\dot{y}(t) = g(y(t), t)$, where $g(y, t)$ is C^1 in y and periodic in t. Suppose that $y = 0$ is a hyperbolic equilibrium of the equation. If a nonzero solution $y(t)$ decays to zero, then $y(t)$ has a finite exponential decay rate and can be written as $y(t) = y_1(t) + y_2(t)$, where $y_1(t)$ is a generalized eigensolution of the linearization around $y = 0$ and $y_2(t)$ is a higher order term as $t \to +\infty$. Generally speaking, this property does not hold for periodic differential delay equations because some of the solutions of the equations may be small solutions which decay faster than any eigensolution. In the last section, we presented conditions which ensure that a solution is not a small solution, therefore, we are in position to give some results about the first order estimation of solutions of state-dependent delay differential equations. Consider the scalar linear differential equation

$$(4.1) \qquad \dot{y}(t) = a(t)y(t) + b(t)y(t-1),$$

where $a(t)$ and $b(t)$ are periodic functions with period one. The following lemma is somewhat well known and can be found in [**3, Lemma 3.1**].

LEMMA 4.1. *Suppse $b_0 = \int_0^1 b(t)dt \neq 0, a_0 = \int_0^1 a(t)dt$. Then $\mu = e^\lambda$ is a characteristic multiplier of (4.1) if and only if $\lambda \in \mathbb{C}$ satisfies the following equation:*

$$(4.2) \qquad \lambda = a_0 + b_0 e^{-\lambda}.$$

Moreover, if all solutions of (4.2) are denoted by Σ, then Σ can be listed as
$$(4.3)$$
$$\Sigma = \{\alpha_1 \pm i\beta_1, \alpha_3 \pm i\beta_3, \cdots, \alpha_{2k+1} \pm i\beta_{2k+1}, \cdots\} \quad when \ -e^{a_0-1} < b_0 < 0,$$

or

$$(4.4) \quad \Sigma = \{\tilde{\alpha}_0, \alpha_0, \alpha_3 \pm i\beta_3, \cdots, \alpha_{2k+1} \pm i\beta_{2k+1}, \cdots\} \quad when \ b_0 < -e^{a_0-1},$$

or

$$(4.5) \qquad \Sigma = \{\alpha_0, \alpha_2 \pm i\beta_2, \cdots, \alpha_{2k} \pm i\beta_{2k}, \cdots\} \quad when \ b_0 > 0,$$

where α_n and β_n satisfy the following: $\beta_n \in ((n-1)\pi, n\pi)$ for $n = 1, 2, \cdots$, $\alpha_k < \alpha_n$ for $k > n$ in each one of (4.3) - (4.5), $\tilde{\alpha}_0 \geq \alpha_0$ in (4.4), and finally $\lim_{n\to\infty} \alpha_n = -\infty$.

Furthermore, the generalized eigensolutions corresponding to $\lambda_n = \alpha_n \pm i\beta_n$ can be expressed in one of the following forms

$$(4.6) \qquad y(t) = Ce^{\alpha_n t}p_{\lambda_n}(t)\cos\left(\int_0^t \beta_n b(s)/b_0 \, ds + \theta\right) \quad when \ \beta_n \neq 0$$

$$(4.7) \qquad y(t) = Ce^{\alpha_n t}p_{\lambda_n}(t) \quad when \ \lambda_n \ is \ real \ and \ simple$$

(4.8) $y(t) = (C_1 t + C_2) e^{\alpha_n t} p_{\lambda_n}(t)$ when λ_n is real and of multiplicity two,

where $p_{\lambda_n}(t) = \exp \int_0^t [(a(s) - a_0) + (\alpha_n - a)(b(s) - b_0)/b_0] ds$ is a periodic function of period one, and C_1, C_2, C and θ are constants.

Consider the nonlinear periodic differential equation

(4.9) $\dot{x}(t) = f(x(t), x(t - \mu), t),$

where $f(\xi, \eta, t)$ is C^1 in (ξ, η) and is a periodic function in t with periodic τ_0, $\mu = \mu(x(t))$ is the state-dependent time delay satisfying $\mu(0) = \tau_0$ and $0 < \mu(z) \leq \tau$ for some fixed τ and for all $z \in \mathbb{R}$. Similar to (H2) in [3], we need the following assumption in order to discuss the first order estimation of solutions of (4.9) which go to zero as $t \to +\infty$:

(H4) Let $g(\xi, \eta, t) = f(\xi, \eta, t) - (a(t)\xi + b(t)\eta)$, where $a(t) = \frac{\partial f}{\partial \xi}(0, 0, t)$ and $b(t) = \frac{\partial f}{\partial \eta}(0, 0, t)$. Assume $|b(t)| > 0$ for all $t \in \mathbb{R}$ and $g(\xi, \eta, t) = 0(|\xi| + |\eta|)^{1+\sigma}$ uniformly with respect to $t \in \mathbb{R}^+$ as $(\xi, \eta) \to 0$ for some constant $\sigma > 0$. Also, assume that $\mu = \mu(z)$ is continuously differentiable for z near zero.

THEOREM 4.2. *Suppose (H4) is satisfied. If x is a solution of (4.9) for $t \in \mathbb{R}^+$ with the exponential decay rate $\overline{\alpha} = \overline{\alpha}(x) < 0$ then the folowing statements are equivalent:*

   (i)   *The exponential decay rate $\overline{\alpha}(x) > -\infty$.*
   (ii)   *The limit of the Lyapunov function $\lim\limits_{t \to +\infty} V(x, t) < +\infty$.*
   (iii)   *The solution has the first order estimation $x(t) = y(t) + h(t)$, where y is given by (4.6) - (4.8) with $\alpha_n < 0$ (or $\tilde{\alpha}_n < 0, n = 0$) and h is a higher order term than y, that is, $\overline{\alpha}(h) < \overline{\alpha}(y)$.*

Moreover, $\overline{\alpha}(x) = \alpha_n$ and $\lim\limits_{t \to +\infty} V(x, t) = n$.

PROOF. We omit the proof because it is the same as [**3, Theorem 3.4**].

COROLLARY 4.3. *If $x = 0$ is a hyperbolic equilibrium, and the assumption $\overline{\alpha}(x) < 0$ in Theorem 4.2 is replaced by the condition that $x(t) \to 0$ as $t \to +\infty$, then the same results as in Theorem 4.2 are true.*

PROOF. If $x = 0$ is hyperbolic, then $x \to 0$ as $t \to +\infty$ implies $\overline{\alpha}(x) < 0$. Therefore, the same results as in Theorem 4.2 are true.

COROLLARY 4.4. *Suppose (H4) is satisfied. Let x and $\tilde{x}$ be two solutions of (4.9) with $\alpha(x), \alpha(\tilde{x}) < 0$. Then $\alpha(x) < \alpha(\tilde{x})$ if and only if $\lim\limits_{t \to +\infty} V(x, t) > \lim\limits_{t \to +\infty} V(\tilde{x}, t)$.*

PROOF. It is obvious from Theorem 4.2 and Lemma 4.1.

COROLLARY 4.5. *Suppse (H4) is satisfied. Assume $x = 0$ is a hyperbolic equilibrium of (4.9). There exists a non-negative constant n_0 such that a solution x of (4.9) goes to zero as $t \to +\infty$ only if $\lim\limits_{t \to +\infty} V(x, t) \geq n_0$.*

PROOF. If we choose n_0 be the smallest index of $\alpha_n < 0$ in the appropriate one of (4.2), (4.3) and (4.4) (depending on a_0 and b_0), then the result follows from Corollary 4.3.

From Corollary 4.4, we learn that the faster a soltuion decays, the faster it oscillates. From Corolalry 4.5, we learn that, for a given equaiton (4.9), there exists a non-negative number n_0 such that a solution does not go to zero as $t \to +\infty$ unless it oscillates faster than n_0. In applications, when (2.2) is satisfied for $M = +\infty$ and $V(x, t_0) < n_0$ for a solution x and $t_0 > 0$, we can claim that x does not go to zero as $t \to +\infty$ by the non-increasing property of the Lyapunov function.

References

1. O. Arino, *A note on "Discrete Lyapunov Function $\cdots$ "*, J. Differential Equations (to appear).

2. O. Arino and P. Ségnier, *Existence of oscillating solutions for certain differential equations with delay*, Lect. Notes Math. **730** (1979), 46–64.

3. Y. Cao, *The discete Lyapunov function for scalar differential delay equation*, J. Differential Equations **87** (1990), 365–390.

4. Y. Cao, *Non-Existence of small solutions for scalar differential delay equation*, J. Math. Anal. Appl. (to appear).

5. Yu. I. Domshlak, *Comparison theorem of Sturm type for first and second order differential equations with alternating devating of argument*, Ukrain. Math. Zh. **34** (1982), 158–163.

6. J. K. Hale, *Theory of Functional Differential Equations*, Springer-Verlag, New York, 1977.

7. K. G. Kartsatos and T. K. Canturia, *On the oscillating and monotone solutions of first order differential equations with a deviating argument*, Differensial'nye Uraveniya **18** (1982), 1463–1465.

8. G. Ladas, *Sharp conditions for oscillation caused by delays*, Appl. Anal. **9** (1979), 93–98.

9. G. Ladas, Y. G. Sficas and I. P. Stavroulakis, *Necessary and sufficient conditions for oscillations*, Amer. Math. Monthly **90** (1983), 637–640.

10. G. S. Ladas, *Oscillations caused by retarded perturbations of first order linear ordinary equations*, Atti Accad. Naz. Lincie. Rend. Cl. Sci. Fis. Math. Matur. (8) **63** (1977), 351–359.

11. W. Mahfoud, *Comparison theorems for delay differential equations*, Pacific J. Math. **83** (1979), 187–197.

12. J. Mallet-Paret, *Morese decomposition for delay-differentail equations*, J. Differential Equations **62** (1986), 266–274.

13. V. A. Staikos and Ch. G. Philos, *Basic comparison results for the oscillatory and asymptotic behavior of differential equations with deviating argument*, Bull. Inst. Math. Accad. Sinica **9** (1981), 417–445.

Department of Mathematics, University of Georgia, Athens, Georgia 30602

E-mail: cao @ math.uga.edu

Contemporary Mathematics
Volume **129**, 1992

Transition Layers for a Singularly Perturbed 2-Dimensional System of Differential-Difference Equations

SHUI-NEE CHOW[1] AND WENZHANG HUANG

ABSTRACT. A two-dimensional system of singularly perturbed differential-difference equations of the form

$$\epsilon \dot{x}(t) = -x(t) + g(y(t), \lambda)$$
$$y(t) = f(x(t-1), y(t-1), \lambda),$$

is studied with an emphasis on the existence of heteroclinic orbits for some transition layer equations. These transition layers correspond to the square-wave-like periodic solutions for the singularly perturbed systems. A technique is introduced to overcome the discontinuity from the singur perturbation by defining an inverse of a singularly perturbed linear operator.

1. Introduction

A system of singularly perturded differential-difference equations of the form

$$(1.1)_\epsilon \qquad \begin{aligned} \epsilon \dot{x}(t) &= -Ax(t) + g(y(t), \lambda) \\ y(t) &= f(x(t-1), y(t-1), \lambda), \end{aligned}$$

has occured in many physical and biological problems as a mathematicsl model [**1, 8, 9, 13, 14, 18**]. A special case of $(1.1)_\epsilon$ is the scalar delay differential equation

$$(1.2) \qquad \epsilon \dot{x}(t) = -x(t) + f(x(t-1)),$$

which has been studied by many authors (see, for example, Chow and Mallet-Paret [**5**], Chow and Green [**2**], Chow, Lin and Mallet-Paret [**3**], Chow and

---

1991 *Mathematics Subject Classification*. 34E05, 34E10, 34K10, 34K15, 34K25, 34K30.
[1]RESEARCH PARTIALLY SUPPORTED BY NSF DMS 8401719 AND DARPA 70NANB8H0860

Walther [6], Mallet-Paret and Nussbaum [14, 15], Mackey and Glass [8], Lasota
[17], Gibbs [8], Ikeda [12, 13], Sharkovsky and Ivanov [16]). It is interesting
to see from these and other works that many different qualitative behaviors
have been observed for equations (1.2). There are still many interesting open
questions.

In the higher dimensional case, very little is known about the relations, for
sufficiently small ϵ, between $(1.1)_\epsilon$ and the following difference equation obtained
by setting $\epsilon = 0$:

$$(1.3) \qquad \begin{aligned} x(t) &= A^{-1}g(y(t), \lambda) \\ y(t) &= f(A^{-1}g(y(t-1), y(t-1), \lambda). \end{aligned}$$

One of the problems concerns the existence of the so-called square wave periodc
solution. Specifically, suppose $F(y, \lambda) = f(A^{-1}g(y, \lambda), y, \lambda)$ has a pair of period
doubling points $p_1(\lambda)$ and $p_2(\lambda)$, that is

$$F(p_1(\lambda), \lambda) = p_2(\lambda), \quad F(p_2(\lambda), \lambda) = p_1(\lambda).$$

Thus (1.3) possesses a 2-periodic solution which is a step function. Hence, we
anticipate that for sufficiently small ϵ, $(1.1)_\epsilon$ possesses a square wave like pe-
riodic solution $(x(t), y(t))$ of period $2 + 2r(\epsilon, \lambda)\epsilon$ which alternates near $P_1 = (A^{-1}g(p_1, \lambda), p_1)$ and $P_2 = (A^{-1}g(p_2, \lambda), p_2)$, with the length of time interval of
passing from P_1 to P_2 and P_2 to P_1 being $O(\epsilon)$. A formal approach introduced
in [6] suggests a transformation

$$\begin{aligned} x_1(t) &= x(-\epsilon t), & x_2(t) &= x(-\epsilon t + 1 + r\epsilon), \\ y_1(t) &= y(-\epsilon t), & y_2(t) &= y(-\epsilon t + 1 + r\epsilon). \end{aligned}$$

Let $\epsilon \to 0$, then $x_i(t)$, $y_i(t)$, $i = 1, 2$ satisfy the transition layer equation

$$(1.4) \qquad \begin{aligned} \dot{x}_i(t) &= Ax_i(t) - g(y_i(t), \lambda) \quad i = 1, 2 \\ y_1(t) &= f(x_2(t-r), y_2(t-r), \lambda) \\ y_2(t) &= f(x_1(t-r), y_1(t-r), \lambda), \end{aligned}$$

with boundary condition

$$(1.5) \qquad \begin{aligned} (x_1(t), y_1(t), x_2(t), y_2(t)) &\longrightarrow (P_1, P_2) & \text{as } t \to +\infty, \\ (x_1(t), y_1(t), x_2(t), y_2(t)) &\longrightarrow (P_2, P_1) & \text{as } t \to -\infty. \end{aligned}$$

Note that in equation (1.4), r is an unknown constant. This leads to the question
of existence of a heteroclinic orbit of (1.4) joining (P_1, P_2) and (P_2, P_1).

In the scalar case, the existence of a heteroclinic orbit has been studied by
Chow and Mallet-Paret [6], Mallet-Paret and Nussbaum [15, 16], Chow, Lin and
Mallet-Paret [5]. The purpose of this paper is to study the problem for our two
dimensional system by a new approach. Under certain weak conditions, we show
that if a pair of period doubling points of $F(y, \lambda)$ bifurcates from $\lambda = 0$ (treat
λ as parameter), then a heteroclinic orbit joining this pair of period doubling

points will also bifurcate from $\lambda = 0$. We remark that this is also true for the higher dimensional systems.

2. A Main Result for Two Dimensional System

Throughout this paper we assume $A = 1$, and $g : \mathbb{R} \times \mathbb{R} \to \mathbb{R}$, $f : \mathbb{R}^2 \times \mathbb{R} \to \mathbb{R}$ are C^4 functions satisfying the following hypotheses:

(H)$_1$ $g(0, \lambda) \equiv 0$, $f(0, 0, \lambda) \equiv 0$, $\lambda \in \mathbb{R}$.

(H)$_2$ $F(y, \lambda) = f(g(y, \lambda), y, \lambda)$ has the form

$$F(y, \lambda) = -(1 + \lambda)y + a(\lambda)y^2 + b(\lambda)y^3 + O(y^4)$$

(H)$_3$ $R_0 = a^2(0) + b(0) \neq 0$.

(H)$_4$ $r_0 = -f_x(0, 0, 0)g_y(0, 0) \neq 0$, where $a(\lambda)$, $b(\lambda)$ are continuously differentiable.

With hypothesis (H)$_2$ and (H)$_3$ we have a result concerning the existence of period 2 points of F.

LEMMA 2.1. *If $R_0 > 0$ (< 0), then there exist $\lambda_1 > 0$ and neighborhoods $O^+, O^- \subset \mathbb{R}$ of $\gamma_+^* = \dfrac{1}{\sqrt{|R_0|}}$ and $\gamma_-^* = -\dfrac{1}{\sqrt{|R_0|}}$, respectively, such that for each $\lambda \in (0, \lambda_1)$ ($\lambda \in (-\lambda_1, 0)$), there is a unique pair of points $\gamma_+(\lambda) \in O^+$, $\gamma_-(\lambda) \in O^-$ such that*

$$p_1(\lambda) = \sqrt{|\lambda|}\gamma_+(\lambda) \quad and \quad p_2(\lambda) = \sqrt{|\lambda|}\gamma_-(\lambda)$$

are a pair of period 2 points of $F(\cdot, \lambda)$.

For the proof of this lemma we refer the readers to [11; Theorem 3.21]. We now state our main theorem.

THEOREM 2.2. *Assume hypotheses (H)$_1$–(H)$_4$ are satisfied and let $p_1(\lambda)$, $p_2(\lambda)$ be defined as in Lemma 2.1. If*

$$(2r_0 - r_0^2)R_0 > 0,$$

then a heteroclinic orbit of the transition layer equations

$$\dot{x}_i(t) = x_i(t) - g(y_i(t), \lambda) \quad i = 1, 2$$

(2.1)$_\lambda$
$$y_1(t) = f(x_2(t - r), y_2(t - r), \lambda)$$

$$y_2(t) = f(x_1(t - r), y_1(t - r), \lambda)$$

(2.2)$_\lambda$
$$(x_1(t), y_1(t), x_2(t), y_2(t)) \longrightarrow (P_1(\lambda), P_2(\lambda)) \quad as\ t \to +\infty,$$

$$(x_1(t), y_1(t), x_2(t), y_2(t)) \longrightarrow (P_2(\lambda), P_1(\lambda)) \quad as\ t \to -\infty,$$

$$P_i(\lambda) = (g(p_i(\lambda), \lambda), p_i(\lambda)), \quad i = 1, 2$$

associated to a square wave periodic solution will bifurcate from $\lambda = 0$. Specifically, if $R_0 > 0$, $2r_0 - r_0^2 > 0$ (resp. $R_0 < 0, 2r_0 - r_0^2 < 0$), then there is $\lambda_0 > 0$ such that for each $\lambda \in (0, \lambda_0)$ (resp. $\lambda \in (-\lambda_0, 0)$)

(i) *there is an $r(\lambda)$ such that equations $(2.1)_\lambda$ - $(2.2)_\lambda$ has a solution, provided $r = r(\lambda)$,*

(ii) $r(\lambda) \to r_0 = -f_x(0,0,0)g_y(0,0)$.

The proof of this theorem will be given in Section 7.

3. Functional Equations

Our first strategy is to reduce $(2.1)_\lambda$ - $(2.2)_\lambda$ to a functonal equation in a Banach space. In fact if $(2.1)_\lambda$ has a heteroclinic orbit $(x_1(t), y_1(t), x_2(t), y_2(t))$ connecting (P_1, P_2) and (P_2, P_1), then by the variation of constants formula we have

$$x_i(t) = e^{t-\sigma} x_i(\sigma) - \int_\sigma^t e^{t-\tau} g(y_i(\tau), \lambda) d\tau, \quad i = 1, 2.$$

By setting $\tau = t + s$ and letting $\sigma \to \infty$ we have

$$(3.1) \qquad x_i(t) = \int_0^\infty e^{-s} g(y_i(t+s), \lambda) ds, \quad i = 1, 2.$$

Substituting (3.1) into the remaining two equations in $(2.1)_\lambda$ gives

$$(3.2) \qquad y_1(t+r) = f\left(\int_0^\infty e^{-s} g(y_2(t+s), \lambda) ds, y_2(t), \lambda \right),$$

$$(3.3) \qquad y_2(t+r) = f\left(\int_0^\infty e^{-s} g(y_1(t+s), \lambda) ds, y_1(t), \lambda \right).$$

Furthermore, substituting (3.2) into (3.3) gives

$$(3.4) \qquad y_2(t+2r) = \hat{\mathcal{F}}(\hat{\mathcal{F}}(y_2, \lambda), \lambda)(t) \quad t \in \mathbb{R}$$

and

$$(3.5) \qquad \begin{array}{ll} y_2(t) \longrightarrow p_2(\lambda) & \text{as } t \to +\infty \\ y_2(t) \longrightarrow) p_1(\lambda) & \text{as } t \to -\infty. \end{array}$$

where

$$\hat{\mathcal{F}}(y, \lambda)(t) = f\left(\int_0^\infty e^{-s} g(y(t+s), \lambda) ds, y(t), \lambda \right), \quad t \in \mathbb{R}.$$

Conversely, if (3.4) - (3.5) have a solution $y_2(t)$, defining $x_i(t)$ by (3.1) and (3.2) gives a heteroclinic orbit of $(2.1)_\lambda$ - $(2.2)_\lambda$.

Finally, for $\lambda \neq 0$, let $\mu = \sqrt{|\lambda|}$, $w(t) = \dfrac{1}{\mu} y_2\left(\dfrac{r}{\mu} t\right)$, $t \in \mathbb{R}$ and let

$$\mathcal{F} : C(\mathbb{R}) \times \mathbb{R} \times \mathbb{R} \times \mathbb{R} \to C(\mathbb{R})$$

be defined by

$$(3.6) \qquad \mathcal{F}(w, r, \epsilon, \lambda)(t) = f\left(\int_0^\infty e^{-s} g\left(w\left(t + \frac{\epsilon s}{r}\right), \lambda \right) ds, w(t), \lambda \right)$$

Then $y_2(t))$ satisfy (3.4) and (3.5) if and only if w satisfies the equation

$(3.7)_\mu \qquad \mu w(t + 2\mu) = [\mathcal{F} \circ \mathcal{F}(\mu w, r, \mu, \mathrm{sgn}(\lambda)\mu^2)](t), \quad t \in \mathbb{R}$

and

$(3.8)_\mu \qquad\qquad\qquad \lim_{t \to \pm\infty} w(t) = \gamma_\pm(\mathrm{sgn}(\lambda)\mu^2).$

Here

$$[\mathcal{F} \circ \mathcal{F}(w, r, \epsilon, \lambda)](t) = [\mathcal{F}(\mathcal{F}(w, r, \epsilon, \lambda), r, \epsilon, \lambda)](t)$$

and γ_-, γ_+ are defined as in Lemma 2.1.

So instead of studying $(2.1)_\lambda$ and $(2.2)_\lambda$, we will study equation $(3.7)_\mu$ with boundary condition $(3.8)_\mu$.

4. Formal Computation

In this section we will give a formal computation to see how a heteroclinic orbit of $(3.7)_\mu$ and $(3.8)_\mu$ may bifurcate from $\mu = 0$. For this purpose we shall first derive a more explicit form for equation $(3.7)_\mu$ for the small μ. From now on we only consider $(3.7)_\mu$ under the assumption

$(4.1) \qquad\qquad\qquad \lambda > 0, \quad R_0 > 0, \quad 2r_0 - r_0^2 > 0.$

We can see later that all the methods introduced here are applicable to the case of $R_0 < 0$, $2r_0 - r_0^2 < 0$ and $\lambda < 0$.

Define a map

$$J : C(\mathbb{R}) \times \mathbb{R} \times \mathbb{R} \times \mathbb{R} \to C(\mathbb{R})$$

by

$(4.2) \qquad\qquad J(w, r, \epsilon, \lambda) = \mathcal{F} \circ \mathcal{F}(w, r, \epsilon, \lambda) - F \circ F(w(\cdot), \lambda),$

where $F \circ F(w(\cdot), \lambda) = F(F(w(\cdot), \lambda), \lambda)$. Thus we can rewrite equation $(3.7)_\mu$ as

$$\frac{1}{\mu^2} \left[D_w J(0, r, \mu, \mu^2)(w) - w(\cdot + 2\mu) + w \right]$$

$(4.3)_\mu \qquad\qquad = \frac{1}{\mu^2} \left[w - \frac{1}{\mu} F \circ F(\mu w, \mu^2) \right]$

$$- \frac{1}{\mu^3} \left[J(\mu w, r, \mu, \mu^2) - D_w J(0, r, \mu, \mu^2)(\mu w) \right],$$

where "D" denotes the differentiation operator. Now for $r > 0$ and $\epsilon \in \mathbb{R}$, we define a linear operator $U_{r,\epsilon} : C(\mathbb{R}) \to C(\mathbb{R})$ by

$(4.4) \qquad [U_{r,\epsilon} w](t) = \int_0^\infty r e^{-rs} w(t + \epsilon s)\, ds = \int_0^\infty e^{-s} w(t + \frac{\epsilon s}{r})\, ds.$

Then for $r > 0$ and $w \in C(\mathbb{R})$,

$$\mathcal{F}(w, r, \epsilon, \lambda)(t) = f\left([U_{r,\epsilon} g(w(\cdot), \lambda)](t), w(t), \lambda\right).$$

It follows that

(4.5) $[D_w \mathcal{F}(0, r, \mu, \lambda)(w)](t) = f_x(0, 0, \lambda)g_y(0, \lambda)[U_{r,\mu}w](t) + f_y(0, 0, \lambda)w(t).$

From $(H)_2$ we have

$$-1 - \lambda = D_y F(0, \lambda) = F_w(0, \lambda) = f_x(0, 0, \lambda)g_y(0, \lambda) + f_y(0, 0, \lambda),$$

so

(4.6) $f_y(0, 0, \lambda) = -(1 + \lambda + f_x(0, 0, \lambda)g_y(0, \lambda)).$

Using (4.5) and (4.6), we obtain that

$$D_w \mathcal{F} \circ \mathcal{F}(0, r, \mu, \lambda)(w)$$

(4.7)
$$= [D_w \mathcal{F}(0, r, \mu, \lambda)]^2 (w)$$
$$= b_1(\lambda) \left(U_{r,\mu}^2 w - U_{r,\mu}w \right) + b_2(\lambda)(U_{r,\mu}w - w) + (1 + \lambda)w$$

where

$$b_1(\lambda) = [f_x(0, 0, \lambda)g_y(0, \lambda)]^2,$$
$$b_2(\lambda) = -f_x(0, 0, \lambda)g_y(0, 0, \lambda)[2(1 + \lambda) + f_x(0, 0, \lambda)g_y(0, 0, \lambda)].$$

Therefore
(4.8)
$$D_w J(0, r, \mu, \lambda)w$$

$$= [D_w \mathcal{F}(0, r, \mu, \lambda)]^2 (w) - [D_w F(0, \lambda)]^2 w$$
$$= b_1(\lambda) \left(U_{r,\mu}^2 w - U_{r,\mu}w \right) + b_2(\lambda)(U_{r,\mu}w - w)$$
$$= b_1(0) \left(U_{r,\mu}^2 w - U_{r,\mu}w \right) + b_2(0)(U_{r,\mu}w - w) + \lambda\Theta(w, r, \mu) + O(\lambda^2 \|w\|)$$
$$= r_0^2 \left(U_{r,\mu}^2 w - U_{r,\mu}w \right) + r_0(2 - r_0)(U_{r,\mu}w - w) + \lambda\Theta(w, r, \mu) + O(\lambda^2 \|w\|),$$

where

(4.9) $\Theta(w, r, \mu) = \dot{b}_1(0) \left(U_{r,\mu}^2 w - U_{r,\mu}w \right) + \dot{b}_2(0)(U_{r,\mu}w - w).$

Define a linear operator $\mathcal{L}(r, \mu) : C(\mathbb{R}) \to C(\mathbb{R})$ by

(4.10)
$$\mathcal{L}(r, \mu)w = D_w J(0, r, \mu, 0)w$$
$$= r_0^2 \left(U_{r,\mu}^2 w - U_{r,\mu}w \right) + (2r_0 - r_0^2)(U_{r,\mu}w - w).$$

It follows from (4.8), (4.10) and the generalized Taylor's theorem that we can rewrite equation $(4.3)_\mu$ as

$$\frac{1}{\mu^2} \left[\mathcal{L}(r, \mu)w - w(\cdot + 2\mu) + w \right]$$

$(4.11)_\mu$
$$= \frac{1}{\mu^2} \left[w - \frac{1}{\mu} F \circ F(\mu w, \mu^2)) \right] - \frac{1}{2\mu} D_w^2 J(0, r, \mu, \mu^2)(w)^2$$
$$- \frac{1}{3!} D_w^3 J(0, r, \mu, \mu^2)(w)^3 - \Theta(w, r, \mu) + O(\mu).$$

We now make a formal computation in $(4.11)_\mu$ for small μ. Let $w \in C^3(\mathbb{R})$, by expanding

$$w(t + \delta) = w(t) + \delta \dot{w}(t) + \frac{\delta^2}{2} \ddot{w}(t) + O(\delta^3)$$

and setting $r = r_0 + h\mu$, the left hand side of $(4.11)_\mu$ becomes

$$\frac{1}{\mu^2} \left[(\mathcal{L}(r, \mu)w)(t) - w(t + 2\mu) + w(t) \right]$$

$$= \frac{1}{\mu^2} \left\{ r_0^2 \int_0^\infty re^{-rs} \int_0^\infty re^{-r\tau} (\mu \dot{w}(t)\tau + \frac{\mu^2}{2} \ddot{w}(t)(2s\tau + \tau^2)) d\tau ds \right.$$

$$+ (2r_0 - r_0^2) \int_0^\infty re^{-rs} (\mu \dot{w}(t)s + \frac{\mu^2}{2} \ddot{w}(t)s^2) ds$$

(4.12)

$$\left. - 2\mu \dot{w}(t) - 2\mu^2 \ddot{w}(t) + O(\mu^3) \right\}$$

$$= \frac{1}{\mu^2} \left\{ \frac{2\mu(r_0 - r)}{r} \dot{w}(t) + \mu^2 \frac{r_0^2 + 2r_0 - 2r^2}{r^2} \ddot{w}(t) + O(\mu^3) \right\}$$

$$= -\frac{2h}{r_0} \dot{w}(t) + \frac{2 - r_0}{r_0} \ddot{w}(t) + O(\mu).$$

For the right hand side of $(4.11)_\mu$, it follows from assumption $(H)_2$ that

$$(4.13) \qquad \frac{1}{\mu^2}[w - \frac{1}{\mu} F \circ F(\mu w, \mu^2))] = -2w + 2R_0 w^2 + O(\mu).$$

By the definition of $\mathcal{F}(w, r, \epsilon, \lambda)$ one is able to verify that

$$D_w^2 (\mathcal{F} \circ \mathcal{F}(0, r, \epsilon, \lambda))(w)^2 = \hat{a}(\lambda) U_{r,\epsilon} w^2 + \sum_{i+j=1}^2 a_{ij}(\lambda)(U_{r,\epsilon}^i w)(U_{r,\epsilon}^j w),$$

$$D_w^2 (F \circ F(0, \lambda))(w)^2 = a^*(\lambda) w^2,$$

where $U_{r,\epsilon}^i = [U_{r,\epsilon}]^i$. We note that

$$F \circ F(w, \lambda) \equiv \mathcal{F} \circ \mathcal{F}(w, r, 0, \lambda), \qquad \text{and} \qquad U^i(r, 0)w = w, \ i = 0, 1, 2.$$

Thus,

$$a^*(\lambda) \equiv \hat{a}(\lambda) + \sum_{i+j=1}^2 a_{ij}(\lambda).$$

This implies that

$$D_w^2 J(0, r, \mu, \mu^2)(w)^2$$

$$= D_w^2 (\mathcal{F} \circ \mathcal{F}(0, r, \mu, \mu^2))(w)^2 - D_w^2 (F \circ F(0, \mu^2))w^2$$

(4.14)

$$= \hat{a}(\mu^2)[U_{r,\epsilon} w^2 - w^2] + \sum_{i+j=1}^2 a_{ij}(\mu^2)[(U_{r,\epsilon}^i w)(U_{r,\epsilon}^j w) - w^2].$$

Observe that for $r > 0$,

$$
(4.15) \qquad
\begin{aligned}
[U_{r,\mu}w](t) &= \int_0^\infty re^{-rs}w(t+\mu s)\,ds \\
&= w(t) + \frac{\mu}{r}\dot{w}(t) + O(\mu^2\|w\|_{C^2}).
\end{aligned}
$$

By setting $r = r_0 + h\mu$, it follows from (4.14)-(4.15) that

$$
(4.16) \qquad \Theta(w, r_0 + h\mu, \mu) = O(\mu\|w\|_{C^1}),
$$

$$
(4.17) \qquad
\begin{aligned}
&D_w^2 J(0, r_0 + h\mu, \mu, \mu^2)(w)^2 \\
&= \hat{a}(\mu^2)\frac{2\mu}{r_0 + h\mu}w(t)\dot{w}(t) \\
&\quad + \sum_{i+j=1}^2 a_{ij}(\mu^2)\frac{(i+j)\mu}{r_0 + h\mu}w(t)\dot{w}(t) + O(\mu^2\|w\|_{C^2}) \\
&= \frac{\mu}{r_0}\left[2\hat{a}(0) + \sum_{i+j=1}^2 a_{ij}(0)(i+j)\right]w(t)\dot{w}(t) + O(\mu^2\|w\|_{C^2}).
\end{aligned}
$$

Furthermore, it is clear that if w is continuously differentiable, then

$$
D_w^3 J(0, r, \epsilon, \lambda)(w)^3
$$

is continuously differentiable with respect to ϵ and

$$
D_w^3 J(0, r, \epsilon, \lambda)(w)^3 = D_w^3 J(0, r, 0, \lambda)(w)^3 + O(\epsilon\|w\|_{C^1}) = O(\epsilon\|w\|_{C^1}),
$$

for $J(w, r, 0, \lambda) \equiv 0$. Therefore, we have

$$
(4.18) \qquad D_w^3 J(0, r_0 + h\mu, \mu, \mu^2)(w)^3 = O(\mu\|w\|_{C^1}).
$$

By (4.13) and (4.16)-(4.18) , we obtain the following estimate for the right hand side of $(4.11)_\mu$:

$$
-2w(t) + 2R_0 w^3(t) - R_1 w(t)\dot{w}(t) + O(\mu)
$$

where

$$
(4.19) \qquad R_1 = \frac{1}{r_0}\left[2\hat{a}(0) + \sum_{i+j=1}^2 a_{ij}(0)(i+j)\right].
$$

Therefore, by formally letting $\mu \searrow 0$ and $h \to 0$, we have the second order equation

$$
(4.20) \qquad \frac{2 - r_0}{r_0}\ddot{w}(t) + 2w(t) - 2R_0 w^3(t) + R_1 w(t)\dot{w}(t) = 0.
$$

Since $2r_0 - r_0^2 > 0$ and $R_0 > 0$, we have the existence of a heteroclinic orbit of (4.20) joining $\pm\sqrt{\dfrac{1}{R_0}}$. This shows why we expect heteroclinic orbits of $(4.11)_\mu$ to bifurcate from $\mu = 0$.

We are not able to use the implicit function theorem directly here because the linear operator $T(r, \mu)$ defined by the left hand side of $(4.11)_\mu$ is not invertble, has the same smoothness as w, and is singular as $\mu \to 0$.

To overcome this difficulty, one considers the inverse of $T(r, \mu) + \alpha I$, where α is a real number and I is the identity, such that $T(r, \mu) + \alpha I$ is invertible. So we define a linear operator $L(\mu) : C(\mathbb{R}) \to C(\mathbb{R})$ by

$$
(4.21) \qquad
\begin{aligned}
L(\mu)w &= \left[T(r_0, \mu) - \frac{2 - r_0}{r_0} I \right] w \\
&= \frac{1}{\mu^2} \left[\mathcal{L}(r_0, \mu)w - w(\cdot + 2\mu) + w - \frac{\mu^2(2 - r_0)}{r_0} w \right].
\end{aligned}
$$

We rewrite $(4.11)_\mu$ as

$$
(4.22)_\mu \qquad L(\mu)w + \frac{1}{\mu^2} [\mathcal{L}(r, \mu) - \mathcal{L}(r_0, \mu)]w = Z(w, r, \mu)
$$

with

$$
\begin{aligned}
Z(w, r, \mu) &= -\frac{2 - r_0}{r_0} w + \frac{1}{\mu^2} \left[w - \frac{1}{\mu} F \circ F(\mu w, \mu^2)) \right] - \frac{1}{2\mu} D_w^2 J(0, r, \mu, \mu^2)(w)^2 \\
&\quad - \frac{1}{3!} D_w^3 J(0, r, \mu, \mu^2)(w)^3 - \Theta(w, r, \mu) + O(\mu).
\end{aligned}
$$

In the next section, we will show that $L(\mu)$ is invertible for small positive μ and $L^{-1}(\mu)$ has a nice asymptotical behavior as $\mu \searrow 0$. Thus $(4.22)_\mu$ is equivalent to

$$
(4.23)_\mu \qquad w + \frac{1}{\mu^2} L^{-1}(\mu)([\mathcal{L}(r, \mu) - \mathcal{L}(r_0, \mu)]w) - L^{-1}(\mu)Z(w, r, \mu) = 0.
$$

5. Invertibility of $L(\mu)$

Before proceeding to show the invertibility of $L(\mu)$, we remark that a hetero-clinic orbit of $(2.1)_\lambda$ - $(2.2)_\lambda$ is actually a C^1 function. So for technical reasons we will consider equation $(4.22)_\mu$ in the space $C^1(\mathbb{R})$ instead of $C(\mathbb{R})$. This makes our main convergence result (Lemma 6.3) easier to prove. Therefore, from now on $L(\mu)$ will be considered as an operator from $C^1(\mathbb{R})$ to $C^1(\mathbb{R})$.

Let $\eta(\mu, \cdot)$ be a function of bounded variation on $\mathbb{R}$ such that

$$
[L(\mu)w](t) = \int_{\mathbb{R}} d_\theta \eta(\mu, \theta) w(t - \theta), \quad w \in C(\mathbb{R}), \quad t \in \mathbb{R}.
$$

For Re $z = 0$, let $\hat{\eta}(\mu, z)$ be the Laplace transform of $\eta(\mu, \cdot)$ defined as
(5.1)

$$
\begin{aligned}
\hat{\eta}(\mu, z) &= \int_{\mathbb{R}} d_\theta \eta(\mu, \theta) e^{-\theta z} = [L(\mu) e^{\cdot z}](0) \\
&= \frac{1}{\mu^2} \left\{ [\mathcal{L}(r_0, \mu) e^{\cdot z}](0) - e^{2\mu z} + 1 - \mu^2 \frac{2 - r_0}{r_0} \right\} \\
&= \frac{1}{\mu^2} \left\{ r_0^2 [\int_0^\infty r_0 e^{-(r_0 - \mu z)s} ds]^2 - r_0^2 \int_0^\infty r_0 e^{-(r_0 - \mu z)s} ds \right. \\
&\quad \left. + (2r_0 - r_0^2)(\int_0^\infty r_0 e^{-(r_0 - \mu z)s} ds - 1) - e^{2\mu z} + 1 - \mu^2 \frac{2 - r_0}{r_0} \right\} \\
&= \frac{1}{\mu^2} \left\{ r_0^2 \left(\frac{r_0}{r_0 - \mu z} - 1 \right)^2 + 2r_0 \left(\frac{r_0}{r_0 - \mu z} - 1 \right) + (1 - e^{2\mu z}) - \mu^2 \beta \right\} \\
&= \frac{1}{\mu^2 (r_0 - \mu z)^2} \Delta(\mu, z),
\end{aligned}
$$

where

$$
\begin{aligned}
\Delta(\mu, z) &= \mu^2 z^2 [(r_0 - 1)^2 - \frac{\mu^2(2 - r_0)}{r_0} - e^{2\mu z}] \\
&\quad + 2r_0 \mu z (r_0 - 1 + \frac{\mu^2(2 - r_0)}{r_0} + e^{2\mu z}) \\
&\quad + r_0^2 (1 - e^{2\mu z}) - \frac{\mu^2(2 - r_0)}{r_0} r_0^2.
\end{aligned}
$$

We have the following results.

LEMMA 5.1. *There exist $\mu_0 > 0$ and $\delta > 0$ such that*

(i) *For each $\mu \in (0, \mu_0)$, $\Delta(\mu, z) = 0$ has exactly two roots $z_+(\mu), z_-(\mu)$ in the strip $\{|Re\ z| \le \frac{\delta}{\mu}\}$. Both $z_+(\mu)$ and $z_-(\mu)$ are real, $z_+(\mu) > 0 > z_-(\mu)$ and*

$$
\lim_{\mu \searrow 0} z_\pm(\mu) \pm 1.
$$

(ii) *Let Ω be any bounded domain of the complex plane, then*

$$
\lim_{\mu \to 0} \frac{1}{\mu^2} \Delta(\mu, z) = (2r_0 - r_0^2)(z^2 - 1)
$$

uniformly for z in Ω.

We omit the proof of this lemma (see Chow and Huang [4] for the details). As a consequence of (i) of Lemma 5.1 we have:

LEMMA 5.2. *For each $\mu \in (0, \mu_0)$, the linear operator $L(\mu)$ defined by (4.21) is invertible.*

PROOF. Lemma 5.1 implies that for $\mu \in (0, \mu_0)$, the Laplace transform $\hat{\eta}(\mu, z)$ of $\eta(\mu, \cdot)$ does not vanish for Re $z = 0$. By applying the whole line Gelfand Theorem [[9, Theorem 4.4] we obtain Lemma 5.2.

6. Properties of $L^{-1}(\mu)$

We now discuss the asymptotic behavior of $L^{-1}(\mu)$ as $\mu \to 0$.

Let $S_1(\mu), S_2(\mu) : C^1(\mathbb{R}) \to C^1(\mathbb{R})$ be linear operators defined by

(6.1)
$$
\begin{aligned}
[S_1(\mu)w](t) = {} & \int_{-\infty}^{t} \left(\int_{(-\frac{\delta}{\mu})} \frac{\mu^2(r_0^2 - 2r_0\mu z)e^{z(t-s)}dz}{\Delta(\mu, z)} \right) w(s)ds \\
& + \int_{t}^{\infty} \left(\int_{(\frac{\delta}{\mu})} \frac{\mu^2(r_0^2 - 2r_0\mu z)e^{z(t-s)}dz}{\Delta(\mu, z)} \right) w(s)ds \\
& - \int_{-\infty}^{t^-} d_s \left(\int_{(-\frac{\delta}{\mu})} \frac{\mu^4 z e^{z(t-s)}dz}{\Delta(\mu, z)} \right) w(s) \\
& - \int_{t^+}^{\infty} d_s \left(\int_{(\frac{\delta}{\mu})} \frac{\mu^4 z e^{z(t-s)}dz}{\Delta(\mu, z)} \right) w(s), \quad t \in \mathbb{R}
\end{aligned}
$$

and

(6.2)
$$
\begin{aligned}
[S_2(\mu)w](t) = {} & \int_{-\infty}^{t} \left(\frac{1}{2\pi i} \oint_{\Gamma_-} \frac{\mu^2(r_0 - \mu z)^2 e^{z(t-s)}dz}{\Delta(\mu, z)} \right) w(s)ds \\
& - \int_{t}^{\infty} \left(\frac{1}{2\pi i} \oint_{\Gamma_+} \frac{\mu^2(r_0 - \mu z)^2 e^{z(t-s)}dz}{\Delta(\mu, z)} \right) w(s)ds, \quad t \in \mathbb{R},
\end{aligned}
$$

where

$$
\int_{(C)} = \frac{1}{2\pi i} \lim_{T \to \infty} \int_{C-iT}^{C+iT},
$$

$\Gamma_\pm$ are small circles centered at $z = \pm 1$ containing $z_\pm(\mu)$ respectively, and the integrals $\oint_{\Gamma_\pm}$ are taken in the counterclockwise sense.

By applying Lemma 5.1 we have the following:

THEOREM 6.1. *If $\mu_0 > 0$ is sufficiently small, then for each $\mu \in (0, \mu_0)$,*

$$
L^{-1}(\mu) = S_1(\mu) + S_2(\mu).
$$

LEMMA 6.2. *We have*

(6.3)
$$
\|S_1(\mu)\|_{\mathcal{L}(C^1, C^1)} = O(\mu^2) \to 0 \quad \text{as } \mu \searrow 0
$$

and

(6.4)
$$
\lim_{\mu \searrow 0} \|S_2(\mu) - S\|_{\mathcal{L}(C^1, C^1)} = 0,
$$

where the linear operator S is defined by

(6.5)
$$
[Sw](t) = -\frac{r_0^2}{2(2r_0 - r_0^2)} \left[\int_{-\infty}^{t} e^{-(t-s)}w(s)ds + \int_{t}^{\infty} e^{t-s}w(s)ds \right], \quad t \in \mathbb{R}.
$$

Lemma 6.2 says that $S_1(\mu)$ behaves like a regular perturbation or fast motion which vanishes quickly as $\mu \to 0$, and $S_2(\mu)$ has a nice limit since the linear operator S takes a C^0 function to a C^2 function.

LEMMA 6.3. *Let $\mu_0 > 0$ and $h_0 > 0$ be sufficiently small. Introduce a mapping $G : C^1(\mathbb{R}) \times [-h_0, h_0] \times (0, \mu_0] \to C^1(\mathbb{R})$ defined by the left hand side of equation* $(4.23)_\mu$:

(6.6)

$$G(w, h, \mu)$$
$$= w + \frac{1}{\mu^2} L^{-1}(\mu) \left\{ [\mathcal{L}(r_0 + h\mu, \mu) - \mathcal{L}(r_0, \mu)]w \right\} - Z(w, r_0 + h\mu, \mu), \quad \mu \neq 0.$$

Then G and $D_{(w,h)}G$ are continuous,

(6.7)
$$\lim_{\mu \to 0} G(w, h, \mu) = w + S \left(-\frac{2h}{r_0} \dot{w} + 2(w - R_0 w^3) + R_1 w \dot{w} + \frac{2 - r_0}{r_0} w \right)$$
$$\overset{\text{def}}{=} \widehat{G}(w, h)$$

and

(6.8)
$$\lim_{\mu \to 0} D_{(w,h)}G(w, h, \mu) = D_{(w,h)}\widehat{G}(w, h),$$

where the operator S is defined as in (6.5), R_1 is defined as in (4.19), and the limits in (6.6) and (6.7) are uniform for $(w, h) \in$ bounded subset of $C^1(\mathbb{R}) \times [-h_0, h_0]$.

The detailed proofs of can be found in [**4**].

7. Proof of Theorem 2.2

We will prove Theorem 2.2 in this section.

LEMMA 7.1. *There is a nonzero heteroclinic orbit w_0 of the equation $\widehat{G}(w, 0) = 0$ satisfying*

(7.1)
$$\lim_{t \to \pm\infty} w_0(t) = \pm\sqrt{\frac{1}{R_0}}.$$

PROOF. Note that the linear opeartor $S : C(\mathbb{R}) \to C^2(\mathbb{R})$ defined in (6.5) is invertible and

(7.2)
$$S^{-1}(w) = \frac{2 - r_0}{r_0} (\ddot{w} - w).$$

This implies that $\widehat{G}(w, 0) = 0$ if and only if $w \in C^2(\mathbb{R})$ and

(7.3)
$$\frac{2 - r_0}{r_0} \ddot{w} + 2[w - R_0 w^3] + R_1 w \dot{w} = 0.$$

Since (7.3) is Hamiltonian, there exists a unique heteroclinic orbit w_0 up to a time translation satisfying (7.1) provided $\frac{2-r_0}{r_0} > 0$ and $R_0 > 0$. This concludes the proof of Lemma 7.1.

Let $\mathcal{T} : C^2(\mathbb{R}) \to C(\mathbb{R})$ be defined by

$$\mathcal{T}v = \frac{2-r_0}{r_0}\ddot{v} + 2(1 - 3R_0w_0^2)v + R_1(w_0\dot{v} + \dot{w}_0v),$$

where w_0 is the unique heteroclinic orbit of (7.3) obtained by Lemma 7.1.

LEMMA 7.2. *For the operator $\mathcal{T}$ we have that*

$$\mathcal{N}(\mathcal{T}) = \ span\{\dot{w}_0\}, \qquad \mathcal{R}(\mathcal{T}) = \{v \in C(\mathbb{R}) \mid \int_{-\infty}^{\infty} \dot{w}_0(t)v(t)dt = 0\}$$

PROOF. Consider the equation

$$(7.4) \qquad\qquad\qquad \mathcal{T}v = 0.$$

Observe that $\dot{w}_0$ satisfies (7.4) and $\dot{w}_0(t) \to 0$ as $t \to \pm\infty$. Hence, the limiting equations of (7.4) as $t \to \pm\infty$ are given by

$$(7.5)_a \qquad\qquad \frac{2-r_0}{r_0}\ddot{v} - 4v + R_1\sqrt{\frac{1}{R_0}}\dot{v} = 0$$

and

$$(7.5)_b \qquad\qquad \frac{2-r_0}{r_0}\ddot{v} - 4v - R_1\sqrt{\frac{1}{R_0}}\dot{v} = 0.$$

The characteristic equations of $(7.5)_a$ and $(7.5)_b$ are

$$\frac{2-r_0}{r_0}\lambda^2 \pm R_1\sqrt{\frac{1}{R_0}}\lambda - 4 = 0.$$

Each of the above equations contains exactly one positive and one negative root. It follows from [5, p.9] that

$$\dim \mathcal{N}(\mathcal{T}) = \operatorname{codim} \mathcal{R}(\mathcal{T}) = 1.$$

The lemma now follows from the fact that $\mathcal{T}$ is formally self adjoint.

LEMMA 7.3. *Suppose $R_0 > 0$ and $2r_0 - r_0^2 > 0$. Then there are neighborhoods $O_{w_0} \times I_0 \subset C^1(\mathbb{R}) \times [-h_0, h_0]$ of $(w_0, 0)$ and $\mu_1 > 0$ such that for each $\mu \in (0, \mu_1)$, there exist unique $w(\mu, \cdot) \in O_{w_0}$ and $h(\mu) \in I_0$ satisfying*
(i) $\int_{-\infty}^{\infty} \dot{w}_0(s)[w(\mu, s) - w_0(s)]ds = 0$,
(ii) $G(w(\mu, \cdot), h(\mu), \mu) = 0$,
(iii) $(w(\mu, \cdot), h(\mu))$ is continuous with respect to μ and $(w(\mu, \cdot), h(\mu)) \to (w_0, 0)$ as $\mu \searrow 0$.

PROOF. Let $\mathcal{G} : C^1(\mathbb{R}) \times I_0 \times V_0 \to C^1(\mathbb{R}) \times \mathbb{R}$ be defined by

$$\mathcal{G}(w, h, \mu) = (G(w, h, \mu), \int_{-\infty}^{\infty} \dot{w}_0(s)[w(s) - w_0(s)]ds).$$

Then it follows from Lemma 6.3 and Lemma 7.1 that $\mathcal{G}$ and $D_{(w,h)}\mathcal{G}$ are continuous,

$$\lim_{\mu \to 0} \mathcal{G}(w_0, 0, \mu) = (\widehat{G}(w_0, 0), 0) = 0,$$

and

$$\lim_{\mu \to 0} D_{(w,h)}\mathcal{G}(w_0, 0, \mu) = \Lambda,$$

where $\Lambda : C^1(\mathbb{R}) \times R \to C^1(\mathbb{R}) \times \mathbb{R}$ is defined by

$$\Lambda(w, h) = \left(D_{w,h}\widehat{G}(w_0, 0)(w, h), \int_{-\infty}^{\infty} \dot{w}_0(s)w(s)ds\right).$$

We claim that Λ is one-to-one and onto. To see this, let $(v, \alpha) \in C^1(\mathbb{R}) \times \mathbb{R}$. Consider the equations

$$D_{(w,h)}\widehat{G}(w_0, 0)(w, h)$$

$$(7.6)_a \quad = w - \frac{2h}{r_0}S(\dot{w}_0) + S\left(2(1 - 3R_0w_0^2)w + \frac{2 - r_0}{r_0}\ w + R_1(w_0\dot{w} + \dot{w}_0w)\right)$$

$$= v,$$

$$(7.6)_b \qquad\qquad \int_{-\infty}^{\infty} \dot{w}_0(s)w(s)ds = \alpha.$$

Let $y = w - v$. By inverting S we have the following equivalent system of equations:

$$(7.7)_a \qquad Ty = \frac{2h}{r_0}\ \dot{w}_0 - 2(1 - 3R_0w_0^2)v + \frac{2 - r_0}{r_0}\ v + R_1(w_0\dot{v} + \dot{w}_0v),$$

$$(7.7)_b \qquad\qquad \int_{-\infty}^{\infty} \dot{w}_0(s)[y(s) + v(s)]ds = \alpha.$$

By Lemma 7.2, $(7.7)_a$-$(7.7)_b$ has a unique solution (w, h) for each $(v, \alpha) \in C^1(\mathbb{R}) \times \mathbb{R}$. Hence Λ is one-to-one and onto. Lemma 7.3 follows from the implicit function theorem.

PROOF OF THEOREM 2.2. Let $w(\mu, \cdot) \in C^1(\mathbb{R})$ and $h(\mu)$ be defined as in Lemma 7.3. By the definition of $G(w, h, \mu)$, one sees that $w(\mu, \cdot)$ is a solution of $(6.2)_\mu$ with $r = r_0 + h(\mu)\mu$. Thus, it is sufficient to show that for sufficiently small $\mu > 0$, the function $w(\mu, \cdot)$ satisfies

$$\lim_{t \to \pm\infty} w(\mu, t) = \gamma_{\pm}(\mu^2),$$

where $\gamma_{\pm}(\lambda)$ are defined as in Lemma 2.1.

Let $\mu_1 > 0$ be as in Lemma 7.3. Let $\Phi : C^1(\mathbb{R}) \times [0, \mu_1) \to C^1(\mathbb{R})$ be defined by

$$\Phi(w, \mu) = G(w, h(\mu), \mu).$$

Since $\mu\gamma_{+}(\mu^2)$ and $\mu\gamma_{-}(\mu^2)$ is a pair of the period doubling points of $F(\cdot, \mu^2)$, $\mu\gamma_{\pm}(\mu^2)$ are fixed points of $F \circ F(\cdot, \mu^2)$. Therefore, $\gamma_{+}(\mu^2)$ and $\gamma_{-}(\mu^2)$ are equilibria of $(6.2)_\mu$. Consequently,

$$(7.8) \qquad\qquad \Phi(\gamma_{\pm}(\mu^2), \mu) = 0,$$

where $\gamma_\pm(\mu^2)$ are constant functions in $C^1(\mathbb{R})$. By Lemma 6.3 we have

$$K \stackrel{\text{def}}{=} \lim_{\mu \to 0} D_w \Phi(\gamma_+(\mu), \mu) = D_w \widehat{G}\left(\sqrt{\frac{1}{R_0}}, 0\right).$$

So for $w \in C^1(\mathbb{R})$ one has

$$Kw = w + S\left(2\left(1 - 3R_0 \left(\sqrt{\frac{1}{R_0}}\right)^2\right)w + \frac{2 - r_0}{r_0}w + R_1\sqrt{\frac{1}{R_0}}\,\dot{w}\right)$$

$$= w + S\left(-4w + \frac{2 - r_0}{r_0}\,w + R_1\sqrt{\frac{1}{R_0}}\,\dot{w}\right).$$

By inverting S it is now not difficult to see that $K : C^1(\mathbb{R}) \to C^1(\mathbb{R})$ is invertible. Therefore it follows from the implicit function theorem and (7.8) that, if μ_1 is small enough, there is a neighborhood $V_+ \subset C^1(\mathbb{R})$ of $\sqrt{\frac{1}{R_0}}$ such that for each $\mu \in (0, \mu_1)$, the only solution of $\Phi(\cdot, \mu) = 0$ in V_+ is the constant function $w \equiv \gamma_+(\mu^2)$. Now for each $\mu \in (0, \mu_1)$, let

(7.9) $$\omega_\mu = \bigcap_{s \geq 0} \overline{\bigcup_{t \geq s} w_t(\mu)}, \quad \alpha_\mu = \bigcap_{s \leq 0} \overline{\bigcup_{t \leq s} w_t(\mu)},$$

where $w_t(\mu) \in C([-\mu, 0], \mathbb{R})$ is defined by

$$w_t(\mu)(\theta) = w(\mu, t + \theta), \quad \theta \in [-\mu, 0], \quad t \in \mathbb{R}.$$

Since

$$\lim_{t \to \pm\infty} w_0(t) = \pm\sqrt{\frac{1}{R_0}}$$

and

$$\lim_{\mu \searrow 0} w(\mu, \cdot) = w_0,$$

we have

$$\lim_{\mu \searrow 0} \text{dist}\left(\frac{1}{\sqrt{R_0}}, \omega_\mu\right) = 0, \quad \lim_{\mu \searrow 0} \text{dist}\left(-\frac{1}{\sqrt{R_0}}, \alpha_\mu\right) = 0.$$

Here we identify $\sqrt{\frac{1}{R_0}}$ as a constant function in $C^1([-\mu, 0], \mathbb{R})$. Hence, if μ_1 is small, for each $\mu \in (0, \mu_1)$

$$\widehat{\omega}_\mu \stackrel{\text{def}}{=} \{u \in C^1(\mathbb{R}), u_t \in \omega_\mu, t \in \mathbb{R}\} \subset V_+.$$

Observe further that the solution of $\Phi(w, \mu) = 0$ corresponds to a bounded solution of delay equations with delay μ, and ω and α limit sets of bounded solution of delay equations are univariant. These imply that for each $y \in \omega_\mu$, there is $u(\mu, y) \in C^1(\mathbb{R})$ such that

$$u_0(\mu, y) = y, \quad u_t(\mu, y) \in \omega_\mu, \quad t \in \mathbb{R} \qquad \text{and} \qquad \Phi(u(\mu, y), \mu) = 0.$$

It follows that $u(\mu, y) \in \widehat{\omega}_\mu \subset V_+$ and, therefore, the uniqueness of the zero of $\Phi(\cdot, \mu)$ in V_+ implies $u(\mu, y) \equiv \gamma_+(\mu)$. In particular, we have $y \equiv \gamma_+(\mu)$. This yields that

$$\omega_\mu = \{\gamma_+(\mu^2)\}$$

and hence

$$\lim_{t \to +\infty} w(\mu, t) = \gamma_+(\mu^2).$$

With the same argument as above, we have

$$\lim_{t \to -\infty} w(\mu)(t) = \gamma_-(\mu^2).$$

The proof of Theorem 2.2 is completed.

REFERENCES

1. M. L. Berre, E. Ressayre, A. Tallet and H. M. Gibbs, *High-dimension chaotic attractors of a nonlinear ring cavity*, Phys. Rev. Lett. **56** (1986), 274–277.

2. S. N. Chow and D. Green, *Some results on singular delay-diferential equations*, Chaos, Fractal and Dynamics (P. Fisher and W. R. Smith, eds.), Lecture Notes in Pure and Applied Math., Vol. 98, Dekker, 1985, pp. 161–182.

3. S. N. Chow, J. K. Hale and W. Huang, *From sine waves to square waves in the singularly perturbed delay differential equations*, Proc. Royal Soc. Edinburgh (to appear)Preprint, Ga. Tech. CDSNS91-44 .

4. S. N. Chow and W. Huang, *Singular perturbation problems for a system of differential-difference equation* (to appear).

5. S. N. Chow, X. B. Lin and J. Mallet-Paret, *Transition layers for singularly perturbed delay differential equations with monotone nonlinearitys*, J. Dynamics Diff. Equations **1** (1989), 3–43.

6. S. N. Chow and J. Mallet-Paret, *Singularly perturbed delay-differential equations*, Coupled Nonlinear Oscillators (J. Chandra and A. Scott, eds.), North-Holland Math. Studies, Vol. 80, North-Holland, 1983.

7. S. N. Chow and H. O. Walther, *Characteristic multipliers and stability of periodic solutions of $\dot{x}(t) = g(x(t-1))$*, Trans Amer. Math. Soc. **307** (1988), 127–142.

8. H. M. Gibbs, F. A. Hopf, D. D. L. Kaplan and R. L. Shoemaker, *Observation of chaos in optical bistability*, Phys. Rev. Lett. **46** (1981), 474–477.

9. L. Glass and M. Mackey, *Oscillation and chaos in physiological control systems*, Science **197** (1977), 287-289.

10. G. Gripenberg, S. O. Londen and O. Staffans, *Volterra integral and functional equations* (to appear).

11. J. Guckenheimer and P. Holmes, *Nonlinear Oscillations, Dynamical Systems, and Bifurcations of Vector Fields*, Springer-Verlag, 1983.

12. J. K. Hale, *The Theory of Functional Differential Equations*, Springer-Verlag, 1977.

13. K. Ikeda, *Multiple-valued stationary state and its instability of the transmitted light by a ring cavity systems*, Opt. Commun. **30** (1979), 257–261.

14. K. Iketa, H. Daido and O. Akimoto, *Optical turbulence: Chaotic behavior of transmitted light from a ring cavity*, Phys. Rev. Lett. **45** (1980), 709–712.

15. J. Mallet-Paret and R. D. Nussbaum, *Global continuation and asymptotic behavior for periodic solutions of a differential-delay equation*, Ann. Mat. Pura Appl. (4) **145** (1986), 33–128.

16. J. Mallet-Paret and R. D. Nussbaum, *Global continuation and complicated trajectories for periodic solutions for a differential-delay equation*, Proc. Sympos. Pure Math. Vol. 45, Part 2, Amer. Math. Soc., 1986, pp. 155–167.

17. A. N. Shakovsky and A. F. Ivanov, *Oscillations in singularly perturbed delay equations*, Dynamics Reported (U. Kirchgraber and H. O. Walther, eds.) (to appear).

18. M. Wazewska-Czyzewska and A. Lasota, *Mathematical models of the red cell system*, Mat. Stos. **6** (1976), 25–40. (Polish)

CENTER FOR DYNAMICAL SYSTEMS AND NONLINEAR STUDIES, SCHOOL OF MATHEMATICS, GEORGIA INSTITUTE OF TECHNOLOGY, ATLANTA, GA 30332

CENTER FOR DYNAMICAL SYSTEMS AND NONLINEAR STUDIES, SCHOOL OF MATHEMATICS, GEORGIA INSTITUTE OF TECHNOLOGY, ATLANTA, GA 30332

Contemporary Mathematics
Volume **129**, 1992

Differential Difference Equations
in Analytic Number Theory

HAROLD G. DIAMOND AND H. HALBERSTAM

ABSTRACT. Differential difference equations have come to play a significant role in analytic number theory, particularly in problems related to sieves. Here we survey a number of occurrences of these equations, starting with one arising from the sieve of Eratosthenes and including surprising recent results showing maldistribution of primes in certain intervals.

0. Introduction

The objects of number theory, the integers, are the paradigm of discreteness. On the surface it is quite surprising that the methodology of analysis, which depends on continuity and smoothness, should play a decisive role in many number theoretic investigations. The problems that we discuss here, which usually involve counting the number of integers in a certain large set, display few recognizable patterns locally, but do possess statistical patterns in the large that one can exploit in some cases to show regular behavior and in other cases oscillation.

Our aim here is to show how differential difference equations (D.D.E.'s)—some with delayed and others with advanced arguments—arise in some problems of number theory. Our survey will include: an estimate of the number of uncancelled elements in an Eratosthenes–type sieve; a dual problem of counting integers devoid of large prime factors; a combinatorial sieve; recent oscillation results of Maier and others showing abnormal concentrations of primes in certain intervals; and a comparison of the relative size of the largest and second largest prime factors of an integer.

1991 *Mathematics Subject Classification*. Primary 11N25, 34K25.

1. Buschstab's Theorem on Uncancelled
Elements in the Sieve of Eratosthenes

In the classical method of isolating primes ascribed to Eratosthenes, the positive integers up to some point x are written down and the primes in this set are then identified by the following steps. Ignoring 1, which by convention is not prime, 2 is noted as being a prime, and then all its multiples (composite numbers!) are struck out of the list. We proceed inductively, identifying at each subsequent stage the first surviving element of the list as a prime and then deleting all of its multiples. The process can be halted when we have identified all primes up to $\sqrt{x}$, for all survivors in the range $(\sqrt{x}, x]$ are necessarily primes.

Suppose now that we quit the sieving procedure at some point y, which may differ from $\sqrt{x}$. For $n \geq 2$ let $p(n)$ denote the least prime factor of n, and let $p(1) = +\infty$. Define

$$\Phi(x, y) = \#\{n \in \mathbb{N} : n \leq x, \quad p(n) \geq y\},$$

i.e. $\Phi(x, y)$ counts the number of integers in $[1, x]$ having no prime factor smaller than y. If we could estimate $\Phi(x, \sqrt{x})$ accurately, then by the preceding discussion we would have a good approximation of $\pi(x)$, the number of primes between 1 and x. Indeed, since

$$\Phi(x, \sqrt{x}) = \pi(x) - \pi(\sqrt{x}) + 1,$$

we get

$$\pi(x) = \Phi(x, \sqrt{x}) + O(\sqrt{x}).$$

There is, alas, no simple method of estimating $\Phi(x, \sqrt{x})$ accurately.

For small values of y, good estimates of $\Phi(x, y)$ can be had rather easily. For $2 < y \leq 3$, $\Phi(x, y)$ counts the number of odd integers in $[1, x]$:

$$\Phi(x, y) \sim \frac{x}{2} = \left(1 - \frac{1}{2}\right) x.$$

(We say $f(x) \sim g(x)$ if $f(x)/g(x) \to 1$ as $x \to \infty$.) For $3 < y \leq 5$, $\Phi(x, y)$ counts the integers relatively prime to 2 or 3 in $[1, x]$:

$$\Phi(x, y) \sim \left(1 - \frac{1}{2}\right)\left(1 - \frac{1}{3}\right) x.$$

For any fixed y we have similarly $\Phi(x, y) \sim x V(y)$, where

$$V(y) := \prod_{p < y} \left(1 - \frac{1}{p}\right).$$

It is known after Mertens (see e.g. [19; Th. 429], [34; Th. 8.8]) that $V(y) \sim e^{-\gamma}/\log y$, where γ denotes Euler's constant.

If y tends to infinity with x, the situation becomes significantly more complicated, because the statistical independence of primes gets weaker as more primes are used in the sieve. Buchstab [6] gave the first successful treatment of $\Phi(x, y)$

for y tending to infinity as a power of x. He introduced a function ω, now named for him, which satisfies

(1.1)
$$\begin{cases} \omega(u) = 0, & u \leq 1 \\ \omega(u) = 1/u, & 1 < u \leq 2 \\ (u\omega(u))' = \omega(u-1), & u > 2 \end{cases}$$

and is continuous at $u = 2$. A form of Buchstab's theorem is as follows.

THEOREM 1.1. *Let U be any fixed number exceeding 1. Let $x^{1/U} \leq y \leq x/\log x$ and $y = x^{1/u}$. Then*

$$\Phi(x,y) \sim \omega(u)x/\log y$$

holds uniformly as $x \to \infty$.

We sketch a proof of Buchstab's result, assuming the truth of the Prime Number Theorem in the form $\pi(x) = x/\log x + O(x/\log^2 x)$ [39; **Th. 1**] and Mertens' formula for $V(y)$. Note that the theorem is also valid, but is trivial, for $0 < u \leq 1$, since $\Phi(x,y) \leq 2$ when $y \leq x$: all integers in the range $1 < n < x$ have been deleted.

Suppose now that $x^{\frac{1}{2}} \leq y \leq x/\log x$, so $1 < u \leq 2$. Following the discussion of Eratosthenes' sieve we have the formula

$$\Phi(x,y) = \pi(x) - \pi(y) + O(1) \sim x/logx.$$

On the other hand, the right side of Buchstab's formula also gives $x/\log x$, since $\omega(u) = (\log y)/\log x$ here. Thus the theorem holds on the initial range. This analysis shows why ω is discontinuous at 1 and why the formula for $\Phi(x,y)$ lacks uniformity for $y \to x^-$.

We continue our argument with the aid of the following identity of Buchstab.

LEMMA 1.2. *For $y < z$ we have*

$$\Phi(x,y) = \Phi(x,z) + \sum_{y \leq p < z} \Phi(x/p,p).$$

PROOF. Suppose that p is a prime and p' its successor. Then

$$\Phi(x,p) - \Phi(x,p') = \#\{np \leq x : p(np) \geq p\}$$
$$= \#\{n \leq x/p : p(n) \geq p\} = \Phi(x/p,p).$$

If we sum this relation over all primes p lying in the range $y \leq p < z$ we get the stated formula. $\square$

The "delay" in our D.D.E. for ω arises naturally from the terms $\Phi(x/p,p)$ in the lemma.

We continue the proof of Theorem 1.1 by an inductive argument. Suppose that N is some integer at least 2 and $y = x^{1/u}$, where $N < u \leq N+1$, and that we have established Theorem 1.1 for all $u \leq N$. The case $N = 2$ requires special attention, for in the initial interval we have established the theorem uniformly only for $x^{\frac{1}{2}} \le y \le x/\log x$, and not the entire range $1 \le u \le 2$.

For $x^{1/(N+1)} \le y = x^{1/u} \le p < x^{1/N}$ we have

$$N - 1 \le \frac{\log(x/p)}{\log p} = \frac{\log x}{\log p} - 1 \le u - 1 < N,$$

and thus for $N > 2$ we have uniformly

$$(1.3) \qquad \sum_{y \le p < x^{1/N}} \Phi\left(\frac{x}{p}, p\right) \sim \sum_{y \le p < x^{1/N}} \omega\left(\frac{\log x}{\log p} - 1\right) \frac{x}{p \log p}$$

by the induction hypothesis.

For $N = 2$ and $x^{\frac{1}{3}} \le y \le p < x^{\frac{1}{2}}$ we use the uniform estimate

$$\Phi\left(\frac{x}{p}, p\right) = \pi(x/p) - \pi(p) + O(1)$$

$$= \frac{x/p}{\log x/p} + O\left(\frac{x/p}{\log^2 x/p}\right) + O\left(\frac{p}{\log p}\right).$$

The sum of the two error terms over the interval $x^{\frac{1}{3}} \le p < x^{\frac{1}{2}}$ is of order $x/\log^2 x$, a term of lower order than the main contribution in (1.3). Thus (1.3) is valid also for $N = 2$.

By the inductive hypothesis and Lemma 1.2 we obtain

$$(1.4) \qquad \Phi(x, y) \sim \frac{N\omega(N)x}{\log x} + x \sum_{y \le p < x^{1/N}} \omega\left(\frac{\log x}{\log p} - 1\right) \frac{1}{p \log p}.$$

The ω function does not wobble violently for arguments exceeding 1, and by prime number theory

$$\sum_{p > z} \frac{1}{p \log p} - \int_z^\infty \frac{dt}{t \log^2 t} = O(\log^{-\nu} z)$$

for any fixed $\nu > 0$ as $z \to \infty$. It follows by an integration by parts that the sum in (1.4) is asymptotic to

$$x \int_y^{x^{1/N}} \omega\left(\frac{\log x}{\log t} - 1\right) \frac{dt}{t \log^2 t}.$$

A change of variable, $v = \log x / \log t$, brings this expression to the form

$$(x/\log x) \int_N^u \omega(v - 1)dv.$$

We now have

$$\Phi(x, y) \sim N\omega(N)\frac{x}{\log x} + \frac{x}{\log x} \int_N^u \omega(v - 1)dv.$$

If we apply the D.D.E. of ω to the integral, we get finally $\Phi(x, y) \sim u\omega(u)x/\log x$, which is equivalent to the assertion of the theorem. $\square$

We conclude this section by mentioning a related summatory function that will play a role in discussing sieves. Let $\Omega(n)$ be the total number of prime factors of n, including repetitions. For example $\Omega(12) = 3$. The arithmetic function $(-1)^{\Omega(n)}$ describes the parity of $\Omega(n)$, information that is useful in sieve theory. Introduce

$$\Phi_{-1}(x,y) = \sum_{n \in I}(-1)^{\Omega(n)},$$

where $I = \{n \in [1,x] : p(n) \geq y\}$. If we essentially repeat the argument establishing the asymptotic formula $\Phi(x,y) \sim xw(u)/\log y$ we can show the following result.

THEOREM 1.2. *Under the hypothesis of Theorem 1.1,*

$$\Phi_{-1}(x,y) \sim -xw^*(u)/\log y,$$

where w^* is a continuous function on $[1,\infty)$ satisfying

$$w^*(u) = 1/u, \quad 1 \leq u \leq 2,$$
$$(uw^*(u))' = -w^*(u-1), \quad u > 2.$$

Looking ahead to Section 4, in which we discuss the Dickman function ρ, there is a close connection between w^* and ρ: $w^*(u) = \rho(u-1)/u$.

2. Properties of Buchstab's w Function and Oscillation of $\pi(x)$

In this section we establish some properties of w, including its damped oscillatory behavior, which are used in the proof of Maier's theorem on irregularities in the distribution of primes numbers. We begin with two results that show that the oscillation of w is strongly damped.

PROPOSITION 2.1. $w(u) > 0$ *for all $u > 1$. Moreover, the following monotonicity relations hold for w on $(1,\infty)$:*

   (i) $uw(u) \uparrow$
   (ii) $uw(u)/(u-1) \downarrow$
   (iii) $uw(u)\exp(-1/(u-1))/(u-1) \uparrow$

PROOF. If we integrate the D.D.E. (1.1) we get

$$(2.1) \qquad uw(u) = 1 + \int_2^u w(v-1)dv, \quad u > 2.$$

Since $w(n)$ is initially positive, there cannot exist a first point $u_1 > 2$ at which $w(u_1) \leq 0$, for the right side of (2.1) is positive at such a u_1; thus $w(u) > 0$ on $(1,\infty)$.

We show the monotonicity relations (i), (ii), and (iii) successively for $1 < u < \infty$, $u \neq 2$, by differentiation. First, the positivity of w and (1.1) imply that $(uw(u))' > 0$. Next, we have

$$\left(\frac{uw(u)}{u-1}\right)' = \frac{(u-1)w(u-1) - uw(u)}{(u-1)^2} < 0.$$

Note that this formula is valid for $1 < u < 2$, for we have taken $\omega(v) = 0$ for $v \leq 1$. Finally, we have

$$\left\{ \frac{u\omega(u)}{u-1} \exp\left(\frac{-1}{u-1}\right) \right\}' = \frac{u-2}{(u-1)^2} \exp\left(-\frac{1}{u-1}\right) (f(u-1) - f(u)),$$

where $f(u) = u\omega(u)/(u-1)$, a decreasing function on $(1, \infty)$ by (ii); thus (iii) holds for $u > 2$. For $1 < u < 2$, $u\omega(u)$ is constant and $\frac{1}{u-1} \exp\left(\frac{-1}{u-1}\right) \uparrow$. $\quad\square$

Now we show that ω' tends rapidly to zero as u tends to ∞. This result along with Proposition 2.1 shows the rapid damping of ω. Our argument occurs in articles of de Bruijn [4] and Hua [27]; better estimates can be found, e.g., in [22] or [39].

PROPOSITION 2.2. $\omega'(u) \ll 1/\Gamma(u+1)$, where Γ denotes Euler's gamma function.

PROOF. From (1.1) we have

$$u\omega'(u) = -\omega(u) + \omega(u-1), \quad u > 2,$$

and hence

$$|\omega'(u)| \leq u^{-1} \max_{u-1 \leq t \leq u} |\omega'(t)|.$$

It follows that $\omega'(u)$ is bounded for $u > 2$. If we set $M(u) = \sup_{t \geq u} |\omega'(t)|$, we get the functional relation

$$M(u) \leq u^{-1} M(u-1), \quad u > 3.$$

The bound $M(u) \ll 1/\Gamma(u+1)$ is established inductively from this inequality.

COROLLARY 2.2. $\lim_{u \to \infty} \omega(u)$ exists.

One can show by a heuristic argument based on Theorem 1.1 and the relation $\Phi(x, y) \sim xV(y)$ for modest sized y that the limit of ω at ∞ is $e^{-\gamma}$. A rigorous argument can be given with the aid of the Laplace transform. We set

$$\hat{\omega}(s) = \int_1^\infty e^{-su}\omega(u)du, \quad \operatorname{Re} s > 0.$$

Familiar manipulations yield the differential equation

$$\hat{\omega}'(s) + (e^{-s}/s)\hat{\omega}(s) = -e^{-s}/s.$$

If we define $E_1(s) = \int_s^\infty e^{-t}t^{-1}dt$, we find that $\hat{\omega}$ satisfies the equation

$$(2.2) \qquad\qquad\qquad \hat{\omega}(s) = e^{E_1(s)} - 1.$$

PROPOSITION 2.3. $\lim_{u \to \infty} \omega(u) = e^{-\gamma}$.

PROOF. Let L denote the limit. By Abel's method we have $\hat{\omega}(s) \sim L/s$ as $s \to 0^+$. We rewrite (2.2) with the aid of the following identity for E_1 [1; 5.1.1]:

$$E_1(s) = -\log s - \gamma + Ein \ s,$$

where

$$Ein \ s = \int_0^s (1 - e^{-t})t^{-1}, \quad s \to 0 + .$$

Thus $L/s \sim e^{-\gamma}/s$, and $L = e^{-\gamma}$. $\square$

We study the oscillation of $\omega(u) - e^{-\gamma}$ with the aid of the adjoint function

$$(2.3) \qquad p(u) = \int_0^\infty e^{-xu - Ein \ x} dx, \quad u > 0.$$

This function was used for this purpose by de Bruijn [4]. Some easy manipulations yield the formula

$$(2.4) \qquad up'(u) = -p(u+1),$$

a D.D.E. with an advanced argument. It is immediate from (2.3) that $p(u) > 0$ for $u > 0$ and then from (2.3) or (2.4) that $p'(u) < 0$ and generally $sgn \ p^{(\nu)}(u) = (-1)^\nu$ for all $u > 0$. By another application of Abel's method, this time to (2.3),

$$p(u) \sim e^{-Ein \ 0}/u = 1/u \quad \text{as } u \to \infty.$$

PROPOSITION 2.4. For $u > 0$

$$(2.5) \qquad up(u) + \int_{u-1}^u p(t+1)dt = 1.$$

PROOF. The left–hand side of the formula has derivative zero. It follows that the left–hand side is some constant, and that constant must be 1, since $up(u) \to 1$ as $u \to \infty$. $\square$

We now consider the oscillatory behavior of ω. First, ω cannot be constant in any interval of length 1. Indeed, if $\omega(u) = c$, $a \le u \le a+1$ for some $a \ge 2$, then by (1.1)

$$0 = u\omega'(u) = -\omega(u) + \omega(u-1), \quad a \le u \le a+1,$$

and hence $\omega = c$ for $a - 1 \le u \le a$. By induction, we would obtain $\omega(u) = c$ in $1 < u \le 2$, in violation of (1.1).

Next, ω' cannot be of one sign (e.g. ≥ 0) on any subinterval of $(2, \infty)$ of length 1. Equation (1.1) yields

$$u\omega'(u) = -\{\omega(u) - \omega(u-1)\} = -\int_{u-1}^u \omega'(v)dv, \quad u > 2,$$

and if $\omega'(v) \ge 0$ for $u - 1 < v < u$ and $\omega' \not\equiv 0$ there, then the integral equation would give $\omega'(u) < 0$, violating the continuity of this function on $(2, \infty)$.

To see the oscillation of

$$W(u) := \omega(u) - e^{-\gamma},$$

we introduce an adjoint equation for ω:

(2.6) $$up(u)\omega(u) + \int_{u-1}^{u} \omega(t)p(t+1)dt = e^{-\gamma}, \quad u > 1.$$

The proof of this formula mirrors that of Proposition 2.4, except that as $u \to \infty$ $up(u)\omega(u) \to e^{-\gamma}$.

If we multiply (2.5) by $e^{-\gamma}$ and subtract it from (2.6) we get the homogeneous equation

$$up(u)W(u) = -\int_{u-1}^{u} W(t)p(t+1)dt, \quad u > 1.$$

Since p is everywhere positive, we obtain:

PROPOSITION 2.5 [7]. $W(u)$ has sign changes in each interval of length one in $(1, \infty)$.

3. Oscillation of $\pi(x)$

The Prime Number Theorem asserts that $\pi(x)$, the number of primes not exceeding x, is asymptotic to $x/\log x$. If I is an interval of the form $(x, x+y]$ and y is not too small relative to x, we may expect to have an asymptotic result

(3.1) $$\#\{p \in I : p \text{ prime}\} \sim y/\log x.$$

Considerable efforts have been made in finding small values of θ such that if $y = x^\theta$, then (3.1) holds. The present record is $\theta = 7/12$ [20].

If y is sufficiently small, then (3.1) cannot hold; for example $y = c \log x$, where $c > 0$ but c is not integral, cannot satisfy (3.1). It is not easy to go beyond this trivial result; for nearly 50 years the largest y for which (3.1) was known to fail was [36]

$$y = (\log x)(\log_2 x)(\log_4 x)/(\log_3 x)^2,$$

where the subscripts denote interated logs. Recently, H. Maier [33] and Hilde-brand–Maier [23] have shown that (3.1) can fail for rather larger values of y, such as $y = (\log x)^\lambda$ for any fixed $\lambda > 1$.

For $x > 1$ and $\lambda > 1$ let

$$f_\lambda(x) = \{\pi(x + (\log x)^\lambda) - \pi(x)\}/(\log x)^{\lambda-1}.$$

If the interval $(x, x + (\log x)^\lambda]$ contained the number of primes predicted by the Prime Number Theorem, then we would have $f_\lambda(x) \approx 1$. Maier showed that $\underline{\lim} f_\lambda(x)$ and $\overline{\lim} f_\lambda(x)$ can be estimated in terms of ω. Let

$$M(v) = \max_{u \geq v} \omega(u) > e^{-\gamma}, \quad m(v) = \min_{u \geq v} \omega(u) < e^{-\gamma}.$$

Both of these quantities are well defined; in fact Cheer–Goldston [7] show that $M(v)$ and $m(v)$ are to be found among values of $\omega(u)$ for $v \leq u \leq v + 2$.

Maier showed that, for any fixed $\lambda > 1$ we have

$$\varlimsup_{x \to \infty} f_\lambda(x) \geq e^\gamma M(\lambda) > 1,$$

$$\varliminf_{x \to \infty} f_\lambda(x) \leq e^\gamma m(\lambda) < 1.$$

As λ grows, $e^\gamma M(\lambda)$ and $e^\gamma m(\lambda)$ each converge to 1 at faster than exponential speed. Data of Cheer–Goldston show, for example, that for $\lambda \geq 4.22$, we have $e^\gamma M(\lambda) < 1.00012$.

4. Numbers without Large Prime Factors and Dickman's Function

For $n \in \mathbb{N}$, $n \geq 2$, let $P(n)$ denote the largest prime factor of n and set $P(1) = 1$. Define

$$\Psi(x, y) = \#\{n \in \mathbb{N} : n \leq x, \quad P(n) < y\},$$

the counting function of integers without large prime factors. There is a duality between $\Psi(x, y)$ and $\Phi(x, y)$ that we shall presently explain. In this section we also introduce the Dickman function ρ and give an estimate of $\Psi(x, y)$ in terms of ρ.

To see the connection between Φ and Ψ, we introduce arithmetic functions f_y and g_y by setting $f_y(1) = g_y(1) = 1$ and for $n > 1$,

$$f_y(n) = \begin{cases} 1, & p(n) \geq y \\ 0, & \text{else} \end{cases} \qquad g_y(n) = \begin{cases} 1, & P(n) < y \\ 0, & \text{otherwise.} \end{cases}$$

Thus f is the indicator function of 1 and integers with large prime factors, while g is the indicator function of 1 and numbers with small prime factors. Note that

$$\Phi(x, y) = \sum_{n \leq x} f_y(n), \quad \Psi(x, y) = \sum_{n \leq x} g_y(n).$$

Each integer $n \geq 1$ can be written uniquely as $n = st$, with $f_y(s) = 1 = g_y(t)$. Thus we have the convolution identity

$$f_y * g_y(n) := \sum_{st=n} f_y(s)g_y(t) = 1, \quad n \in \mathbb{N}.$$

We sum this formula over the range $1 \leq n \leq x$, and rewrite the resulting double sum as

$$\sum_{1 \leq n \leq x} \sum_{st=n} f_y(s)g_y(t) = \sum_{t \leq x} \left\{ \sum_{s \leq x/t} f_y(s) \right\} g_y(t)$$

or

$$\sum_{s \leq x} \left\{ \sum_{t \leq x/s} g_y(t) \right\} f_y(s).$$

We obtain a Stieltjes integral equation for Φ and Ψ:

PROPOSITION 4.1. *For x, $y \geq 1$,*

$$\int_{t=1^-}^{x} \Phi(x/t, y) d\Psi(t, y) = [x].$$

This equation can also be written with Φ and Ψ interchanged.

Like $\Phi(x, y)$, $\Psi(x, y)$ has an asymptotic formula involving a function of $u :=$ $(\log x)/\log y$. The function associated with Ψ is called ρ, the Dickman function [13], and it too satisfies a D.D.E. We shall sketch a proof of the formula for Ψ, following the general lines of the one for Φ but omitting details already discussed.

There are other approaches to the formula for Ψ. Analytic proofs are known [5], [25], [39] as well as another elementary argument that replaces a Buchstab-type identity by the following Chebyshev-type identity [22]. Let $\Lambda_y(n) = \log p$ if $n = p^\alpha$ for some prime $p < y$ and some $\alpha \in \mathbb{N}$ and $\Lambda_y(n) = 0$ otherwise. Then

$$\log n \, g_y(n) = g_y * \Lambda_y(n),$$

where $*$ is the multiplicative convolution introduced above.

For $y > x \geq 1$ we have $\Psi(x, y) = [x]$, since no numbers from $[1, x]$ are excluded, and $\Psi(x, x) = x + O(1)$. Thus, for $0 \leq u \leq 1$ we have

(4.1) $\Psi(x, y) \sim x\rho(u)$

where $\rho(v) = 1$ for $0 \leq v \leq 1$.

To continue our argument we use a Buchstab-type identity for Ψ.

PROPOSITION 4.2. *Let $z > y$. Then*

$$\Psi(x, z) - \Psi(x, y) = \sum_{y \leq p < z} \sum_{\alpha=1}^{\infty} \Psi(xp^{-\alpha}, p^\alpha).$$

The inner sum is zero, of course, if $p^\alpha > x$. Moreover, the higher prime powers contribute little to the inner sum, for $\Psi(s, t) \leq s$ trivially, and so

$$\sum_{\alpha \geq 2} \Psi(xp^{-\alpha}, p^\alpha) \leq \sum_{\alpha \geq 2} x/p^\alpha < 2xp^{-2}.$$

For p not too small, $2xp^{-2}$ is much smaller than $\Psi(x/p, p)$, and we ignore the contribution of the higher prime powers.

Proposition 4.2 is proved in much the same way as the Buchstab formula for Φ except that if p' is the successor prime of p, then we have

$$\Psi(x, p') - \Psi(x, p) = \Psi(x/p, p) + \Psi(x/p^2, p^2) + \cdots.$$

We return to the derivation of the asymptotic formula for $\Psi(x, y)$. Suppose that $\sqrt{x} \leq y < x$, i.e. $1 < u \leq 2$. Proposition 4.2 gives

$$\Psi(x, x) - \Psi(x, y) = \sum_{y \leq p < x} \sum_{\alpha \geq 1} \Psi\left(\frac{x}{p^\alpha}, p^\alpha\right).$$

Since $x/p \le x/y \le \sqrt{x}$ and $p^\alpha \ge p \ge \sqrt{x}$, we have $\Psi(x/p^\alpha, p^\alpha) = [x/p^\alpha]$, unless $y = \sqrt{x} = q$, a prime, in which case $\Psi(x/q, q) = [x/q] - 1$. Thus

$$\sum_{y \le p < x} \sum_{\alpha \ge 1} \Psi(x/p^\alpha, p^\alpha) \sim \sum_{y \le p < x} \sum_{\alpha \ge 1} \frac{x}{p^\alpha}$$

$$\sum_{y \le p < x} \frac{x}{p} \sim x \int_y^x \frac{dv}{v \log v} = x \log u.$$

The last approximation comes from the asymptotic prime number formula [19; Th. 427], [34; Th. 8.8])

$$(4.2) \qquad \sum_{p < t} \frac{1}{p} = \log \log t + c + o(1)$$

which is applicable if y is not too close to x; small changes are needed if y and x are very close.

It follows that in this range

$$\Psi(x, y) \sim x - x \log u = x \rho(u),$$

where

$$\rho(v) = 1 - \log v, \qquad 1 \le v \le 2.$$

Note that ρ is a continuous decreasing function on $[0, 2]$ and for $1 \le v \le 2$ satisfies

$$(4.3) \qquad \rho(v) = 1 - \int_1^v \rho(w - 1) dw/w.$$

We continue inductively, assuming that formula (4.1) is known to hold for $y \le x^{1/N}$ for some integer $N \ge 2$, where ρ is a positive continuous decreasing function on $[0, N]$ that satisfies (4.3) on $[1, N]$. Suppose that $x^{1/(N+1)} \le y < x^{1/N}$. We have

$$\Psi(x, x^{1/N}) - \Psi(x, y) \sim \sum_{y \le p < x^{1/N}} \Psi(x/p, p)$$

$$\sim \sum_{y \le p < x^{1/N}} \frac{x}{p} \rho\left(\frac{\log x}{\log p} - 1\right)$$

$$\sim \int_y^{x^{1/N}} \frac{x}{v} \rho\left(\frac{\log x}{\log v} - 1\right) \frac{dv}{\log v} = x \int_N^u \rho(w - 1) \frac{dw}{w}.$$

The last step arises from the change of variable $w = \log x / \log v$; the penultimate step is made with the aid of (4.2) and the assumed monotonicity of ρ on $[0, N]$. We find that

$$\Psi(x, y) \sim x\rho(N) - x \int_N^u \rho(v - 1) dv/v,$$

and thus (4.1) and (4.3) hold for $u \le N + 1$.

In the course of showing (4.1) we have defined Dickman's function on $[0, \infty)$: $\rho(v) = 1$ for $0 \le v \le 1$ and ρ satisfies equation (4.3) for $v > 1$. The latter is clearly equivalent to the D.D.E. $v\rho'(v) = -\rho(v-1)$, $v > 1$. It is convenient also to define $\rho(u) = 0$ for $u < 0$.

PROPOSITION 4.3. *The Dickman function possess the following properties:*

$$(4.4) \qquad\qquad v\rho(v) = \int_{v-1}^{v} \rho(w)dw, \qquad v \in \mathbb{R},$$

$$(4.5) \qquad\qquad \rho(v) > 0, \qquad v \geq 0,$$

$$(4.6) \qquad\qquad \rho'(v) < 0, \qquad v > 1.$$

PROOF. For $v \leq 1$, (4.4) holds by inspection. For $v > 1$, note that $v\rho(v) - \int_{v-1}^{v} \rho(w)dw$ is a constant, since the derivative is zero. The value of the constant is 0, as we see by letting $v \to 1+$.

Next, (4.5) holds for $0 \leq v \leq 1$ by definition and somewhat beyond $v = 1$ by continuity. If there were a first value a at which $\rho(a) \leq 0$, then by (4.4) we would have the inequality

$$\int_{a-1}^{a} \rho(w)dw \leq 0,$$

which is impossible, since $\rho(w) > 0$ for $a - 1 < w < a$.

Finally, the positivity of ρ and the D.D.E. $v\rho'(v) = -\rho(v-1)$, $v > 1$, imply that $\rho' < 0$ there. $\square$

The argument we used to estimate ω' shows also that $|\rho(v)| \leq 1/\Gamma(v+1)$. The D.D.E. for ρ then gives

$$|\rho'(v)| = \frac{\rho(v-1)}{v} \leq \frac{1}{v\Gamma(v)} = \frac{1}{\Gamma(v+1)}, \qquad v > 1.$$

We have shown the following theorem.

THEOREM 4.4. *Let $x > 1$, $y \geq 2$ and $u = (\log x)/(\log y)$. The relation*

$$(4.1) \qquad\qquad \Psi(x,y) \sim x\rho(u)$$

holds as $x \to \infty$.

Our argument gives a uniform estimate for u in any bounded range. It is known [22] that

$$\Psi(x,y) = x\rho(u)\left\{1 + O\left(\frac{\log(u+1)}{\log y}\right)\right\}$$

holds uniformly for $y \geq 2$ and $1 \leq u \leq \exp\{(\log y)^{3/5-\epsilon}\}$. Still other more accurate but more complicated approximations are known for $\psi(x,y)$ in terms of ρ (cf. [25]).

5. An Advanced Argument D.D.E.

In most examples known to date, when a D.D.E. describes a number theoretic situation, the equation is one having a delayed argument. D.D.E.'s with advanced arguments have occurred in such problems as adjoint functions. All the examples discussed in the other sections of this article have this character. Here, we consider briefly a special case of a result of F. Wheeler [40] in which a number theoretic phenomenon is modeled by a D.D.E. having an advanced argument.

We shall evaluate the probability that the second largest prime factor of an integer is less than some fixed power of the largest prime factor of the integer. As before, let $P(n)$ denote the largest prime factor of n, and now define $P_2(n)$ to be the second largest prime factor of n; in particular $P(1) = 1$ and $P_2(n) = P_1(n/P_1(n))$. We make the notion of probability precise by defining, for $u \geq 1$,

$$\text{Prob}(u) = \lim_{x \to \infty} \frac{1}{x} \#\{n \leq x : P_2(n) \leq P_1(n)^{1/u}\}.$$

Also, recall the function $p(u)$, defined by (2.3), which is the adjoint of Buchstab's function.

THEOREM 5.1. *For $u \geq 1$ we have* $\text{Prob}(u) = e^\gamma p(u)$.

Since Prob(1) is clearly 1, we must have $p(1) = e^{-\gamma}$. Also, the value of Prob(2), $e^\gamma p(2)$, is a number known as Golomb's constant $\lambda = .62432\ldots$ (see [32]).

Wheeler estimates $\sum_{n \in I}(\log P(n))^\alpha$ where

$$I = \{n \leq x : P_2(n) \leq P(n)^{1/u}\},$$

for fixed $u \geq 1$ and $\alpha \in \mathbb{R}$ (we took $\alpha = 0$). The approximation is given in terms of a family of functions whose representative for $\alpha = 0$ is $p(u)$. His argument is based upon uniform estimates of $\Psi(x, y)$.

6. A Combinatorial Sieve

The famous sieve of Eratosthenes, described in the discussion of $\Phi(x, y)$, can be extended to a much more general procedure: to isolate in any given integer sequence $\mathcal{A}$ those "almost primes" that have no prime factors which come from some assigned prime set $\mathcal{P}$ and, moreover, to estimate as accurately as possible $S(\mathcal{A}, \mathcal{P})$, the number of integers that have been so isolated, i.e. the number of elements of $\mathcal{A}$ having no prime factors from $\mathcal{P}$. In the Buchstab problem we have $\mathcal{A} = \{n \in \mathbb{N} : n \leq x\}$, $\mathcal{P} = \{p : p < y\}$ and $S(\mathcal{A}, \mathcal{P}) = \Phi(x, y)$.

The first to create an effective sifting procedure at some level of generality was V. Brun, but it took another 50 years—and many *ad hoc* applications of the same basic ideas–to arrive at a formulation of a general sieve problem and at estimates of corresponding generality that are reasonably sharp. Here we shall describe such a sieve procedure.

It is instructive to illustrate our discussion with another sieve example. Let $\mathcal{A} = \{n^2 + 1 : 1 \leq n \leq x\}$, and let $\mathcal{P}$ be a finite set of primes that may divide elements of $\mathcal{A}$. The set $\mathcal{P}$ need not contain any primes $p \equiv 3 \pmod{4}$, since such primes do not divide numbers of the form $n^2 + 1$ [19; Th. 82], [34; Th. 2.12]. The basic problem is to estimate $S(\mathcal{A}, \mathcal{P})$, the number of elements of $\mathcal{A}$ that have survived sifting by $\mathcal{P}$.

In principle, $S(\mathcal{A}, \mathcal{P})$ can be computed by the Inclusion–Exclusion Principle: if we set $P = \Pi\{p : p \in \mathcal{P}\}$ and

$$A_d = \#\{a \in \mathcal{A} : a \equiv 0 \pmod{d}\},$$

then

(6.1) $$S(\mathcal{A}, \mathcal{P}) = \sum_{d \mid P} \mu(d) A_d.$$

Here μ denotes the Möbius arithmetical function which satisfies $\mu(n) = (-1)^r$ if n is the product of r prime factors that are all distinct and $\mu(n) = 0$ if n is divisible by the square of any prime.

Formula (6.1) is of practical use only when $|\mathcal{P}|$ is relatively small (compared with x in our example). In most serious arithmetical applications the size of the appropriate set $\mathcal{P}$ is rather large. A further difficulty is that A_d can be determined only approximately, and since the sum on the right side of (6.1) contains $2^{|\mathcal{P}|}$ terms, even a bounded error in the estimate of A_d will entail an unacceptable accumulation of error terms in (6.1). In our illustration we would need to take $\mathcal{P}$ to be, say, $\{2\} \cup \{p : p \equiv 3 \pmod{4}, \ p \leq x^{1/4}\}$, and we would have

$$A_d = (x/d + O(1))\xi(d),$$

where $\xi(2) = 1$, $\xi(d) = 2^r$ if d or $d/2$ is a product of r distinct primes $p \equiv 1 \pmod{4}$, and $\xi(d) = 0$ if d is divisible by any prime $p \equiv 3 \pmod{4}$. Note that the integers d with $\xi(d) = 0$ do not occur in the sum on the right side of (6.1).

The central idea of sieve theory is to transmute (6.1) into a computationally effective form by TRUNCATING in some way the sum on the right side of the formula. The notion of truncation is to be understood in the broad sense of restricting the sum to a much smaller number of terms, possibly with "weights," which still yields estimates of $S(\mathcal{A}, \mathcal{P})$—perhaps only upper or lower estimates— that are correct at least in order of magnitude.

There are probably many potential applications of truncated versions of (6.1) in combinatorial analysis, but in most instances effective information about the counting numbers A_d is lacking at present. For two simple applications in probability theory we refer the reader to the inequalities of Bonferroni [14; §IV.5].

Only in number theory has it been possible to turn modified inclusion–exclusion principles into a powerful general method, and it is extraordinary, or so it seems to the present authors, that this method depends crucially on the solution of a boundary value problem for a pair of simultaneous differential delay equations, a problem that might easily have arisen in control theory.

To describe the problem requires a little additional preparation. We now take $\mathcal{P}$ to be an infinite set of primes, write, in place of P,

$$P(z) = \Pi\{p : p \in \mathcal{P}, p < z\},$$

and consider

$$S(\mathcal{A}, \mathcal{P}, z) = \#\{a \in \mathcal{A} : (a, P(z)) = 1\}$$

in place of $S(\mathcal{A}, \mathcal{P})$. Here (a, b) denotes the greatest common divisor of a and b. We introduce a convenient approximation X to $|\mathcal{A}|$ $(= A_1)$ and assume that there exists a non negative multiplicative arithmetic function $w(\cdot)$ so that the "remainders"

$$r_d := A_d - \frac{w(d)}{d}X$$

are small on average (perhaps in some "weighted" form) over the divisors d of $P(z)$ satisfying $d \leq y$; the two parameters z and y are basic to the formulation of the problem. (An arithmetic function $f \neq 0$ is called multiplicative if $f(mn) = f(m)f(n)$ whenever $(m, n) = 1$.)

Viewing this configuration heuristically, we may regard $w(p)/p$ as the "probability" that a prime p of $\mathcal{P}$ divides an element a of $\mathcal{A}$. It seems natural, therefore, to assume that

$$0 < w(p) < p, \qquad p \in \mathcal{P},$$

to put

$$w(p) = 0, \qquad p \notin \mathcal{P},$$

and to expect

$$X \prod_{p<z} \left(1 - \frac{w(p)}{p}\right)$$

to "measure" $S(\mathcal{A}, \mathcal{P}, z)$. Then the result showing the remainders r_d to be small on average over $d|p(z)$, $d \leq y$, is to be seen as a "quasi–independence" relation that shows the "events" ($a \in \mathcal{A}$ divisible by $p \in \mathcal{P}$) to be "nearly" independent. In practice, this relation often embodies the deepest available number–theoretic information about $\mathcal{A}$.

We need to introduce one last basic parameter of a sieve problem: its "dimension." Following Iwaniec [29] and writing

$$V(z) = \prod_{p<z}\left(1 - \frac{w(p)}{p}\right),$$

we assume that there exists a real number $\kappa \geq 0$ such that the one sided estimate

$$(6.2) \qquad V(t)/V(z) = \prod_{t \leq p < z}\left(1 - \frac{w(p)}{p}\right)^{-1} \leq \left(\frac{\log z}{\log t}\right)^{\kappa}\left\{1 + O\left(\frac{1}{\log t}\right)\right\}$$

holds uniformly for $2 \leq t < z$. Note that if (6.2) holds for some κ_1, then it holds for all $\kappa \geq \kappa_1$; in practice we choose κ as small as possible.

We refer to κ as the *dimension*. Clearly (6.2) implies that

$$\sum_{t \leq p < z} \frac{w(p)}{p} \leq \kappa \log \left(\frac{\log z}{\log t} \right) + O \left(\frac{1}{\log t} \right), \qquad 2 \leq t < z.$$

In the illustrative problem mentioned earlier, $w(p) = 2$ on half the primes and 0 on the other half; hence the dimension of that problem is 1, and the sieve method involved there is referred to as 'linear'.

The precision with which $S(\mathcal{A}, \mathcal{P}, z)$ can be estimated in terms of $XV(z)$ depends on the relative size of z and y, more precisely, on the size of

$$s = \frac{\log y}{\log z}.$$

One would expect the precision to be greatest when z is small compared with y, and indeed we do have, for this situation, a result of great interest and wide applicability that is known in sieve literature as a fundamental lemma.

FUNDAMENTAL LEMMA [15]. *Assume only that there exist constants $C \geq 1$ and $\kappa > 0$ such that (cf. (6.2))*

(6.3) $$V(t)/V(z) \leq C \left(\frac{\log z}{\log t} \right)^{\kappa}, \qquad 2 \leq t < z.$$

Then, for any given numbers $L \geq 2$, $z_0 \geq 2$, we have

$$-R^- \leq S(\mathcal{A}, \mathcal{P}, z_0) - XV(z_0)\{1 + O(e^{-L})\} \leq R^+,$$

where the O-constant depends at most on C and κ, and the remainder sums have the form

$$R^{\pm} = \sum_{\substack{m | P(z_0) \\ m < z_0^L}} \mu(m) \gamma_m^{\pm} r_m, \qquad \gamma_m^{\pm} = 0 \quad or \quad 1.$$

For example, with y large (as it is in practice), let $L = \log \log y$. Then, if

$$z_0 \leq \exp \left(\frac{\log y}{\log \log y} \right),$$

we have

$$|S(\mathcal{A}, \mathcal{P}, z_0) - XV(z_0)\{1 + O \left(\frac{1}{\log y} \right)\}| \leq \sum_{\substack{m | P(z_0) \\ m < y}} |r_m|.$$

We could have chosen for L, in place of $\log \log y$, any function $\psi(y)$ that tends monotonically to $+\infty$; thus the Fundamental Lemma shows $S(\mathcal{A}, \mathcal{P}, z)$ to have precisely the size predicted by probability provided only that $s = (\log y)/\log z \to +\infty$. For bounded s this need no longer be the case. We have, instead, the following result which, for the present expository purpose, is not stated with a sharp error term.

THEOREM 6.1 [28], [29], [31], [8], [11], [12]. *Suppose* $\kappa \geq 0$ *and that condition (6.2) holds. Then there exist functions* $F = F_\kappa$ *and* $f = f_k$ *and arithmetic coefficients* $c^\pm(\cdot)$ *such that for any numbers* y *and* z *satisfying* $y \geq z \geq 2$ *we have*

$$S(\mathcal{A}, \mathcal{P}, z) \leq XV(z)\left\{ F\left(\frac{\log y}{\log z}\right) + o(1) \right\} + \sum_{\substack{m|P(z)\\m<y}} c^+(m)r_m$$

and

$$S(\mathcal{A}, \mathcal{P}, z) \geq XV(z)\left\{ f\left(\frac{\log y}{\log z}\right) + o(1) \right\} - \sum_{\substack{m|P(z)\\m<y}} c^-(m)r_m.$$

The functions F *and* f *are determined by three kinds of specification: initial conditions, a pair of linked differential delay equations, and behavior at infinity. Thus, there exist positive numbers* $\alpha = \alpha_\kappa$ *and* $\beta = \beta_\kappa$ *such that* F *and* f *satisfy*

$$(s^\kappa F(s))' = \kappa s^{\kappa-1} f(s-1), \qquad s > \alpha,$$

$$(s^\kappa f(s))' = \kappa s^{\kappa-1} F(s-1), \qquad s > \beta,$$

and have the properties

$$F(s) = 1 + O(e^{-s}), \qquad f(s) + 1 + O(e^{-s}),$$

$$F(s) \text{ decreases monotonically to } 1 \text{ as } s \to \infty,$$

and

$$f(s) \text{ increases monotonically to } 1 \text{ as } s \to \infty.$$

The initial conditions, depending on κ, *are as follows:*

(i) *If* $0 \leq \kappa < \frac{1}{2}$, *then* $\alpha = 2$, $\beta = 1$ *and*

$$sF(s) = C_\kappa, \quad s \leq 2, \qquad sf(s) = c_\kappa, \quad s \leq 1,$$

where

$$c_\kappa = e^{\kappa\gamma} / \int_0^\infty e^{-t} t^{-\kappa} \cosh(\kappa E_1(t)) dt$$

and

$$C_k = \{c_k / \Gamma(1-\kappa)\} \int_0^\infty e^{-t} t^{-\kappa} \exp(\kappa E_1(t)) dt.$$

(ii) *If* $\kappa = \frac{1}{2}$, *then too* $\alpha = 2$, $\beta = 1$ *and*

$$s^{1/2} F(s) = 2e^{\gamma/2} \pi^{-1/2}, \quad s \leq 2; \qquad f(s) = 0, \quad s \leq 1.$$

(iii) *If* $\frac{1}{2} < \kappa \leq 1$, *then* $\alpha = 2 + \rho_\kappa$, $\beta = 1 + \rho_\kappa$, *where* ρ_κ *is the zero of the transcendental function*

$$q_\kappa(s) := \frac{\Gamma(2\kappa)}{2\pi i} \int_C z^{-2\kappa} e^{sz+\kappa Ein(-z)} dz.$$

Here, Γ is Euler's gamma function and $\mathcal{C}$ a path from $-\infty$ back to $-\infty$ which surrounds the negative real axis in the positive sense. Then

$$s^\kappa F(s) = C_k, \qquad s \le 2 + \rho_\kappa,$$
$$f(s) = 0, \qquad s \le 1 + \rho_\kappa,$$

where

$$C_\kappa = 2\rho_\kappa^{\kappa-1}/p_\kappa(\rho_\kappa), \qquad p_\kappa(s) = \int_0^\infty \exp(-st - \kappa Ein\, t)\,dt.$$

For $\kappa = 1$, $q(s) = -1$, so $\rho_1 = 1$, $\beta = 2$, and $\alpha = 3$. Also $p_1(s) = p(s)$, defined by (2.3), and so $C_1 = 2/p(1) = 2e^\gamma$, by the remark following Theorem 5.1.

(iv) *If $\kappa > 1$, then $\alpha > \beta > 2$ are zeros of a pair of transcendental equations to be described below, and*

$$F(s) = 1/\sigma(s), \quad s \le \alpha; \quad f(s) = 0, \quad s \le \beta,$$

where $\sigma(s) = \sigma_\kappa(s)$ is the continuous solution of the differential delay system

$$s^{-\kappa}\sigma(s) = (2e^\gamma)^{-\kappa}/\Gamma(\kappa+1), \qquad s \le 2$$
$$(s^{-\kappa}\sigma(s))' = -\kappa s^{-\kappa-1}\sigma(s-2), \qquad s > 2.$$

This theorem is known to be best possible only for $\kappa = \frac{1}{2}$ and $\kappa = 1$; and the results for $\kappa > 1$, while the hardest to achieve at present, are apparently not optimal. Work of Selberg [**Sel**] with very large κ supports the conjecture $\beta_\kappa \sim 2\kappa$ as $\kappa \to \infty$, whereas $\beta_\kappa \approx 2.41\ldots\kappa$ in the result stated above. We shall outline the argument for $\kappa > 1$, which represents current work. Before we do so, we offer several remarks on the cases $\kappa = \frac{1}{2}$ and $\kappa = 1$.

When $\kappa = \frac{1}{2}$, Iwaniec [**28**] was able to derive, by means of the sieve, accurate information about the distribution of numbers that are representable as sums of two squares. Such numbers correspond closely to sifting by the set $\mathcal{P}$ of primes $\equiv -1$ (mod 4), that is, by essentially half the totality of all primes; Iwaniec showed that here truncating the sifting set $\mathcal{P}$ does not prevent one, in the end, from isolating the desired integer sequence.

For $\kappa = 1$, which at the present time is the most important sieve, it is interesting to note that

$$F_1(s) = e^\gamma(\omega(s) + \omega^*(s)), \qquad f_1(s) = e^\gamma(\omega(s) - \omega^*(s)), \qquad s > 1,$$

where ω is Buchstab's function and ω^* the Dickman–like function defined in §1. One might say that we are sifting not only by small primes, as in the case of Φ, but also by (relatively) large ones, as in the case of Ψ, and that the structure of F_1, f_1 reflects the hybrid character of the procedure.

The upper estimate

$$F_1\left(\frac{\log x}{\log z}\right) X \prod_{p<z}\left(1-\frac{1}{p}\right)$$

is actually the asymptotic estimate of $S(\mathcal{A}^+, \mathcal{P}, z)$ when

$$\mathcal{A}^+ = \{n : 1 \le n \le x, 2 \nmid \Omega(n)\}, \quad \mathcal{P} \text{ is the set of all primes;}$$

the lower estimate

$$f_1\left(\frac{\log x}{\log z}\right) X \prod_{p<z}\left(1-\frac{1}{p}\right)$$

is the asymptotic estimate of $S(\mathcal{A}^-, \mathcal{P}, z)$ when

$$\mathcal{A}^- = \{n : 1 \le n \le x, \quad 2 | \Omega(n)\}, \quad \mathcal{P} \text{ is the set of all primes.}$$

We verify the formulas for $S(\mathcal{A}^\pm, \mathcal{P}, z)$ readily with the aid of the asymptotic estimates developed for Φ and Φ_{-1}. We have

$$S(\mathcal{A}^+, \mathcal{P}, z) = \#\{n \le x : p(n) \ge z, \quad \Omega(n) \text{ odd}\}$$
$$= \frac{1}{2}\{\Phi(x,z) - \Phi_{-1}(x,z)\}$$
$$\sim (x/2)\left\{\omega\left(\frac{\log x}{\log z}\right) + \omega^*\left(\frac{\log x}{\log z}\right)\right\} e^\gamma \prod_{p<z}\left(1-\frac{1}{p}\right).$$

It is known from prime number theory that

$$\sum_{n\le x}(-1)^{\Omega(n)} = o(x),$$

i.e. $X \sim |\mathcal{A}^+| \sim |A^-| \sim x/2$. Thus

$$S(\mathcal{A}^+, \mathcal{P}, z) \sim F_1(\log x/\log z)X \prod_{p<z}\left(1-\frac{1}{p}\right).$$

The asymptotic formula for $S(\mathcal{A}^-, \mathcal{P}, z)$ follows at once.

We conclude with a brief discussion of part (iv) of Theorem 6.1 above. The truncation of the Inclusion–Exclusion Principle underlying (iv) was established in [8] and is too elaborate to include here; suffice it to say that it is a hybrid of Selberg's sieve (which provides the initial conditions), the fundamental Lemma (which determines behavior at infinity) and an adaptation of the Buchstab-Rosser-Iwaniec sieve, and that insofar as (iv) is an imperfect result, the imperfection derives from this combinatorial foundation. We shall describe here how one proves that there exist number pairs α, β for which the D. D. boundary value problem has solutions F, f with the requisite properties.

One introduces the functions $P(s) = F(s) + f(s)$, $Q(s) = F(s) - f(s)$ and reformulates the problem in terms of P and Q. The functions P and Q have adjoint functions p_κ and q_κ respectively, introduced earlier in this section, and combine with them to yield the "inner products" (cf. [29] and [30] above)

$$sp(s)P(s) + \kappa \int_{s-1}^{s} P(t)p(t+1)dt = 2, \quad s \geq \alpha,$$

and

$$sq(s)Q(s) + \kappa \int_{s-1}^{s} Q(t)q(t+1)dt = 0, \quad s \geq \alpha.$$

(The constants 2 and 0 on the right derive from the conditions at infinity.) Upon analyzing these relations at $s = \alpha$, we arrive at a curious disjunction: the relations take one form when $\beta < \alpha \leq \beta + 1$ and another when $\alpha \geq \beta + 1$, with the two coinciding when $\alpha = \beta + 1$. Let

$$\Pi(u, v) = \Pi_\kappa(u, v) = \frac{up(u)}{\sigma(u)} + \kappa \int_{v-1}^{u} \frac{p(t+1)}{\sigma(t)} dt, \quad \tilde{\Pi}(u) = \Pi(u, u-1),$$

and

$$\chi(u, v) = \chi_\kappa(u, v) = \frac{uq(u)}{\sigma(u)} - \kappa \int_{v-1}^{u} \frac{q(t+1)}{\sigma(t)} dt, \quad \tilde{\chi}(u) = \chi(u, u-1),$$

for $u > 0$, $v > 1$. Then, the inner product relations at $s = \alpha$ take the form

(6.4) $\Pi(\alpha, \beta) = 2$ and $\chi(\alpha, \beta) = 0$ if $\beta_\kappa < \alpha_\kappa \leq \beta_\kappa + 1$,

or

(6.5)
$$\tilde{\Pi}(\alpha) + (\alpha - 1)p(\alpha - 1)f(\alpha - 1) = 2,$$
$$\tilde{\chi}(\alpha) - (\alpha - 1)q(\alpha - 1)f(\alpha - 1) = 0,$$
$$\text{if } \beta_\kappa + 1 \leq \alpha_\kappa,$$

where $f(\alpha - 1)$ is given by

$$(\alpha - 1)^\kappa f(\alpha - 1) = \kappa \int_{\beta}^{\alpha-1} \frac{t^{\kappa-1}}{\sigma(t-1)} dt.$$

(The coincidence when $\alpha = \beta + 1$ is now obvious.) In either case we have a simultaneous pair of equations for α and β; the chief difficulties are to prove in each case that the equations have a unique solution $\alpha = \alpha_\kappa$, $\beta = \beta_\kappa$ of the right relative size, and (even harder) to determine for each $\kappa > 1$ which of the two sets of equations is appropriate.

Numerical evidence from the pioneering investigations of [30] and [37] suggested very strongly that the set

$$\mathcal{K}_1 = \{\kappa : 1 < \kappa \leq \kappa_0 = 1.8344\ldots\}$$

corresponds to the system (6.4) and the set

$$\mathcal{K}_2 = \{\kappa : \kappa_0 \leq \kappa\}$$

to the system (6.5); but it proved very hard to establish these correspondences directly. (Obviously, κ_0 corresponds to the case when (6.4) and (6.5) coincide.)

What we did was to identify $\mathcal{K}_1$ and $\mathcal{K}_2$ in another way. We proved that

(a) $\tilde{\Pi}(s)$ is strictly decreasing in s for $s > 2$, and the equation $\tilde{\Pi}(s) = 2$ possesses a unique root, $z_{\tilde{\Pi}}(\kappa)$, to the right of 2; also,

(b) $\tilde{\chi}(s)$ is strictly increasing in u to the right of $\rho_\kappa + 1$, and the equation $\tilde{\chi}(u) = 0$ possesses a unique root, $z_{\tilde{\chi}}(\kappa)$, to the right of $\rho_\kappa + 1$.

We then conjectured that

$$\mathcal{K}_1 = \{\kappa > 1 : z_{\tilde{\chi}}(\kappa) \le z_{\tilde{\Pi}}(\kappa)\}$$

and

$$\mathcal{K}_2 = \{\kappa > 1 : z_{\tilde{\chi}}(\kappa) \ge z_{\tilde{\Pi}}(\kappa)\},$$

and we turned around our procedure: we *defined* $\mathcal{K}_1$ and $\mathcal{K}_2$ in this *latter* way, and showed that all $\kappa \ge 2$ lie in $\mathcal{K}_2$. We then succeeded in proving that the system (6.5) has solutions of the right relative size for $\kappa \in \mathcal{K}_2$, and that the system has solutions of the right relative size if $\kappa \in \mathcal{K}_1$ *and* $\kappa < 2$. Subsequently, we also proved that all $\kappa \in (1, 1.5]$ lie in $\mathcal{K}_1$, and therefore only the κ's in $(1.5, 2)$ remain to be classified. The threshold value κ_0 of κ lies in this interval and the greatest difficulty will lie in the neighborhood of κ_0. Some kind of perturbation technique is indicated, and for this we need information about the behavior of each of $z_{\tilde{\Pi}}(\kappa)$ and $z_{\tilde{\chi}}(\kappa)$. So far we know that $z_{\tilde{\Pi}}(\kappa)$ is strictly increasing, but we shall need to determine how rapidly. Corresponding results for $z_{\tilde{\chi}}(\kappa)$ have yet to be established; they depend on knowledge of $q_\kappa(s)$ as a function of κ, a topic on which we are currently making headway.

We conclude with the remark that, notwithstanding the remaining difficulties with $1.5 < \kappa < 2$, if a sieve of our type were required for a *particular* κ in this interval, it could readily be established by numerical computation.

REFERENCES

1. M. Abramowitz and I. A. Stegun, *Handbook of Mathematical Functions*, Wiley, New York, 1972.

2. N. C. Ankeny and H. Onishi, *The general sieve*, Acta Arith. **10** (1964), 31–62.

3. R. E. Bellman and K. L. Cooke, *Differential–Difference Equations*, Academic Press, New York, 1963.

4. N. G. de Bruijn, *On the number of uncancelled elements in the sieve of Eratosthenes*, Nederl. Akad. Wetensch. Proc. **53** (1950), 803–812.

5. N. G. de Bruijn, *On the number of positive integers $\le x$ and free of prime factors $> y$*, Nederl. Akad. Wetensch. Proc. **54** (1951), 50–60.

6. A. A. Buchstab, *Asymptotic estimates of a general number theoretic function*, (Russian, German summary), Mat. Sbornik **44** (1937), 1239–1246.

7. A. Y. Cheer and D. A. Goldston, *A differential delay equation arising from the sieve of Eratosthenes*, Math. Comp. **55** (1990), 129–141.

8. H. G. Diamond, H. Halberstam, and H.–E. Richert, *Combinatorial sieves of dimension exceeding one*, J. Number Theory **28** (1988), 306–346.

9. _____, *Sieve auxiliary function*, in Number Theory (R. A. Mollin, ed.), de Gruyter, Berlin, 1990, pp. 99–113.

10. _____, *Sieve auxiliary functions, II*, in A Tribute to Emil Grosswald: Number Theory and Related Analysis (M. I. Knopp, ed.), Amer. Math. Soc., Providence, RI (to appear).

11. _____, *A boundary value problem for a pair of differential delay equations related to sieve theory, I*, in Analytic Number Theory–Proc. Bateman Conf. (B. C. Berndt et al., ed.), Birkhäuser, Boston, 1990, pp. 133–157.

12. _____, *A boundary value problem for a pair of differential delay equations related to sieve theory, II*, submitted to J. Number Theory.

13. K. Dickman, *On the frequency of numbers containing primes of a certain relative magnitude*, Ark. Mat. Astr. Fys. **22** (1930), 1–14.

14. W. Feller, *An Introduction to Probability Theory and its Applications*, Vol. 1 (3rd ed.), Wiley, New York, 1968.

15. J. Friedlander and H. Iwaniec, *On Bombieri's asymptotic sieve*, Ann. Scu. Norm. Sup.–Pisa (IV) **5** (1978), 719–756.

16. D. A. Goldston and K. McCurley, *Sieving the positive integer by large primes*, J. Number Theory **28** (1988), 94–115.

17. F. Grupp, *On diffence differential equations in the theory of sieves*, J. Number Theory **24** (1986), 154–173.

18. H. Halberstam and H.–E. Richert, *Sieve Methods*, Academic Press, London, 1974.

19. G. H. Hardy and E. M. Wright, *An Introduction to the Theory of Numbers*, fifth ed., Oxford Univ. Press, London, 1979.

20. D. R. Heath–Brown, *Sieve identities and gaps between primes*, in Journées Arithmétiques (Metz 1981), vol. 94, Asterisque, Paris, 1982, pp. 61–65.

21. D. Hensley, *The convolution powers of the Dickman function*, J. London Math. Soc. (2) **33** (1986), 395–406.

22. H. Hildebrand, *On the number of positive integers $\leq x$ and free of prime factors $> y$*, J. Number Theory **22** (1986), 289–307.

23. A. Hildebrand and H. Maier, *Irregularities in the distribution of primes in short intervals*, J. Reine Angew. Math. **397** (1989), 162–193.

24. A. Hildebrand and G. Tenenbaum, *On integers free of large prime factors*, Trans. Amer. Math. Soc. **296** (1986), 265–290.

25. _____, *On a class of differential–difference equations arising in number theory*, J. d'Analyse Mathématique (to appear).

26. _____, *Integers without large prime factors*, manuscript.

27. L. K. Hua, *Estimation of an integral (Chinese)*, Chungkow Kao Hiao **2** (1951), 393–402.

28. H. Iwaniec, *The half dimensional sieve*, Acta Arith. **29** (1976), 69–95.

29. _____, *Rosser's sieve*, Acta Arith. **36** (1980), 171–202.

30. H. Iwaniec, J. van de Lune, and H. J. J. teRiele, *The limits of Buchstab's iteration sieve*, Indag. Math. (4) **42** (1980), 409–417.

31. W. B. Jurkat and H.–E. Richert, *An improvement of Selberg's small sieve method, I*, Acta Arith. **11** (1965), 217–240.

32. D. E. Knuth and L. Trabb Pardo, *Analysis of a simple factorization algorithm*, Theor. Comp. Sci. **3** (1976), 321–348.

33. H. Maier, *Primes in short intervals*, Michigan Math. J. **32** (1985), 221–225.

34. I. Niven, H. S. Zuckerman, and H. L. Montgomery, *An Introduction to the Theory of Numbers*, fifth ed., Wiley, New York, 1991.

35. K. K. Norton, *Numbers with Small Prime Factors and the Least kth Power Non Residue*, vol. 106, Memoirs of the American Math. Soc., Providence, RI, 1971.

36. R. A. Rankin, *The difference between consecutive prime numbers*, J. London Math. Soc. **13** (1938), 242–247.

37. D. A. Rawsthorne, *Improvements in the Small Sieve Estimate of Selberg by Iteration*, Ph.D. Thesis, University of Illinois, 1980.

38. A. Selberg, *Sifting problems, sifting density, and sieves*, in Number Theory, Trace Formulas and Discrete Groups (Oslo 1987), Academic Press, Boston, 1989, pp. 467–484.

39. G. Tenenbaum, *Introduction à la Théorie Analytique et Probabiliste des Nombres*, Institute Elie Cartan, Vol. 13, Nancy, France, 1990.

40. F. S. Wheeler, *Two differential–difference equations arising in number theory*, Trans. Amer. Math. Soc. **318** (1990), 491–523.

DEPARTMENT OF MATHEMATICS, UNIVERSITY OF ILLINOIS, 1409 WEST GREEN STREET, URBANA, IL 61801

E-mail: diamond@symcom.math.uiuc.edu

DEPARTMENT OF MATHEMATICS, UNIVERSITY OF ILLINOIS, 1409 WEST GREEN STREET, URBANA, IL 61801

Contemporary Mathematics
Volume **129**, 1992

A Simple Model for Price Fluctuations in A Single Commodity Market

A. M. FARAHANI AND E. A. GROVE

ABSTRACT. Let $P(t)$ denote the market price of a particular commodity. As a special case of a general model studied by Bélair and Mackey, M. C. Mackey has proposed the model

$$\frac{\dot{P}(t)}{P(t)} = \frac{a}{b + P^n(t)} - \frac{cP^m(t - \tau)}{d + P^m(t - \tau)}$$

which he calls the case of the naive consumer. We show all positive solutions are bounded from above and below. We obtain sufficient and also necessary and sufficient conditions for all positive solutions to oscillate about the unique positive steady state.

1. Introduction

In considering the dynamics of price, production, and consumption of a particular commodity, Bélair and Mackey [**1,4**] have studied the model

$$\frac{1}{P}\frac{dP}{dt} = f(P_D, P_S).$$

Under relatively mild conditions, they have determined the stability of the equilibrium solutions.

A special case of the above is the equation

$$(\Delta) \qquad \frac{\dot{P}(t)}{P(t)} = \frac{a}{b + P^n(t)} - \frac{cP^m(t - \tau)}{d + P^m(t - \tau)}$$

where

$$a, b, c, d, \tau, m \in (0, \infty) \quad \text{and} \quad n \in [1, \infty).$$

Mackey calls this the case of the naive consumer because the demand never decreases as the price increases.

1991 *Mathematics Subject Classification.* 34K15.

In this paper we obtain sufficient and also necessary and sufficient conditions for all positive solutions of Eq.(Δ) to oscillate about the positive equilibrium.

We say that the function $P(t)$ *oscillates about* $\bar{P}$ if $P(t) - \bar{P}$ has arbitrarily large zeros. If it is not the case that $P(t)$ oscillates about $\bar{P}$, then we say that $P(t)$ is *nonoscillatory about* $\bar{P}$.

2. Existence, Uniquencess, and Boundedness of Solutions of Eq.(Δ)

In this section we study some of properties of positive solutions of Eq.(Δ).

THEOREM 1. *Consider the DDE*

(1)
$$\begin{cases} \frac{\dot{P}(t)}{P(t)} = \frac{a}{b+P^n(t)} - \frac{cP^m(t-\tau)}{d+P^m(t-\tau)}, \ t \geq 0 \\ P(t) = \phi(t), \ -\tau \leq t \leq 0 \end{cases}$$

where

(2) $a, b, c, d, \tau, m \in (0, \infty), n \in [1, \infty)$ *and* $\phi \in C[[-\tau, 0], (0, \infty)]$.

Then

1) There exists a unique solution $P(t)$ of the IVP(1)-(2).

2) This solution $P(t)$ exists for all $t \geq 0$ and is positive for all $t \geq 0$.

3) There exist positive constants L and U such that

$$L \leq P(t) \leq U \ for \ all \ t \geq 0.$$

PROOF. From Eq.(1) it is clear by the method of steps that as long as P(t) exists, it satisfies the equation

$$P(t) = P(0) \exp\left(\int_0^t \left(\frac{a}{b + P^n(s)} - \frac{cP^m(s-\tau)}{d + P^m(s-\tau)} \right) ds \right)$$

and so P(t) is uniquely defined as long as it exists and is positive as long as it exists.

It will be useful to rewrite Eq.(1) as

(3) $$\frac{dP}{dt} = \frac{aP(t)}{b + P^n(t)} - P(t) \frac{cP^m(t-\tau)}{d + P^m(t-\tau)}, t \geq 0$$

We shall first show P(t) is bounded from above. For the sake of contradiction, suppose this is not the case. Then there exists $T \in (0, \infty]$ and a sequence $t_j \to T$ such that $P(t_j) \to \infty$ and $\dot{P}(t_j) \geq 0$. The contradiction will come from the consideration of the following two cases.

(a) Suppose

$$\liminf_{j \to \infty} P(t_j - \tau) > 0.$$

Then there exists $K > 0$ such that $P(t_j - \tau) \geq K$ for large j. This implies that

$$\frac{cP^m(t_j - \tau)}{d + P^m(t_j - \tau)}$$

is bounded away from zero. But since $n \geq 1$,

$$\frac{aP(t_j)}{b + P^n(t_j)}$$

is also bounded. It follows from Eq.(3) with t replaced by t_j that

$$\lim_{j \to \infty} \dot{P}(t_j) = -\infty.$$

This is impossible because $\dot{P}(t_j) \geq 0$.

(b) Suppose

$$\liminf_{j \to \infty} P(t_j - \tau) = 0.$$

By passing to a subsequence, if necessary, we may assume

(4) $$\lim_{j \to \infty} P(t_j - \tau) = 0.$$

Note that because $\varphi(t) > 0$ for $t \in [-\tau, 0]$, it follows by (4) that $T > \tau$. Integrate Eq.(3) from $t_j - \tau$ to t_j to obtain

$$
\begin{aligned}
P(t_j) - P(t_j - \tau) &= \int_{t_j - \tau}^{t_j} \frac{aP(t)}{b + P^n(t)} dt - \int_{t_j - \tau}^{t_j} \frac{cP^m(t - \tau)}{d + P^m(t - \tau)} P(t) dt \\
&\leq \int_{t_j - \tau}^{t_j} \frac{aP(t)}{b + P^n(t)} dt \\
&= \tau \frac{aP(\xi_j)}{b + P^n(\xi_j)} \quad \text{for some} \quad \xi_j \in [t_j - \tau, t_j].
\end{aligned}
$$

This is impossible because

$$\lim_{j \to \infty} \left(P(t_j) - P(t_j - \tau) \right) = \infty$$

and

$$F(p) = \frac{ap}{b + p^n}, \ p \geq 0$$

is a bounded function of p. Therefore $P(t)$ is bounded, and so in particular, $P(t)$ exists for all $t \geq 0$.

We next claim

$$\liminf_{t \to \infty} P(t) \neq 0.$$

Suppose this is not the case. Then there exists a sequence $t_j \to \infty$ such that

$$P(t_j) \to 0 \quad \text{and} \quad \dot{P}(t_j) \leq 0.$$

It follows from Eq.(1) that

$$\frac{cP^m(t_j - \tau)}{d + P^m(t_j - \tau)} \geq \frac{a}{b + P^n(t_j)} \to \frac{a}{b} \quad \text{as} \quad j \to \infty$$

and so there exists $K > 0$ such that $P(t_j - \tau) \geq K$ for all large j. By integrating Eq.(1) from $t_j - \tau$ to t_j, we obtain

$$(5) \qquad \ln \frac{P(t_j)}{P(t_j - \tau)} = \int_{t_j - \tau}^{t_j} \left(\frac{a}{b + P^n(t)} - \frac{cP^m(t - \tau)}{d + P^m(t - \tau)} \right) dt.$$

This is impossible because

$$\lim_{j \to \infty} \ln \frac{P(t_j)}{P(t_j - \tau)} = -\infty$$

while the right hand side of Eq.(5) is bounded.

3. Oscillation of Positive Solutions of Eq.(Δ)

It is easy to show that Eq.(1) has a unique positive equilibrium solution $\bar{P}$. Moreover, $\bar{P}$ satisfies the relationship

$$\frac{a}{b + \bar{P}^n} - \frac{c\bar{P}^m}{d + \bar{P}^m} = 0.$$

In this section we wish to determine conditions which insure that every solution $P(t)$ of Eq.(1) which satisfies condition (2) oscillates about the equilibrium solution $\bar{P}$; that is, $P(t) - \bar{P}$ has arbitrarily large zeroes.

The following result is adapted from a paper of Kulenovic, Ladas, and Meimaridou [3].

Consider the nonlinear DDE

$$(6) \qquad \dot{x}(t) + \sum_{i=1}^{k} c_i f_i(x(t - \tau_i)) = 0$$

and the associated "linearized equation"

$$(7) \qquad \dot{y}(t) + \sum_{i=1}^{k} c_i y(t - \tau_i) = 0$$

Suppose that for $i = 1, 2, \cdots, k$,

$$c_i \in (0, \infty), \tau_i \in [0, \infty), f_i \in C[\mathbf{R}, \mathbf{R}], \quad \text{and} \quad uf_i(u) > 0 \quad \text{for} \quad u > 0.$$

LEMMA 1. *Suppose*

$$\liminf_{u \to 0} \frac{f_i(u)}{u} \geq 1 \quad for \quad i = 1, 2, \cdots, k.$$

Suppose also that every solution of Eq.(7) oscillates about zero. Then every solution of Eq.(6) oscillates about zero.

LEMMA 2. *Suppose*

$$\lim_{u \to 0} \frac{f_i(u)}{u} = 1 \quad for \quad i = 1, 2, \cdots, k.$$

Suppose also that there exists $\delta > 0$ such that

$$either \quad f_i(u) \le u \quad for \quad 0 \le u \le \delta \quad and \quad i = 1, 2, \cdots, k.$$

$$or \quad f_i(u) \ge u \quad for \quad 0 \ge u \ge -\delta \quad and \quad i = 1, 2, \cdots, k.$$

Then every solution of Eq.(6) oscillates about zero if and only if every solution of Eq.(7) oscillates about zero.

The next result establishes sufficient conditions and necessary and sufficient conditions for every positive solution of the IVP(1) satisfying (2) to oscillate about $\bar{P}$.

THEOREM 2. *Let*

$$c_1 = \frac{an\bar{P}^n}{(b + \bar{P}^n)^2} \quad and \quad c_2 = \frac{cdm\bar{P}^m}{(d + \bar{P}^m)^2}.$$

1)

$$c_2 \tau e^{c_1 \tau} > e^{-1}$$

implies every solution of Eq.(1)-(2) oscillates about $\bar{P}$.

2) Suppose $(b - \bar{P}^n)(d - \bar{P}^m) > 0$. Then

$$c_2 \tau e^{c_1 \tau} > e^{-1}$$

if and only if every solution of Eq.(1)-(2) oscillates about $\bar{P}$.

PROOF. For the sake of contradiction, assume that $P(t)$ is a solution of Eq.(1)-(2) such that $P(t) - \bar{P}$ is eventually non–zero. Note that

$$\frac{\dot{P}(t)}{P(t)} = \frac{a}{b + P^n(t)} - \frac{a}{b + \bar{P}^n} + \frac{c\bar{P}^m}{d + \bar{P}^m} - \frac{cP^m(t - \tau)}{d + P^m(t - \tau)}$$

$$= \frac{a(\bar{P}^n - P^n(t))}{(b + P^n(t))(b + \bar{P}^n)} + \frac{cd(\bar{P}^m - P^m(t - \tau))}{(d + \bar{P}^m)(d + P^m(t - \tau))}$$

and so $P(t)$ converges monotonically to $\bar{P}$

Make the change of variables

$$P(t) = \bar{P}e^{x(t)}$$

to obtain

$$\dot{x}(t) = \frac{a}{b + \bar{P}e^{nx(t)}} - \frac{c\bar{P}^m e^{mx(t-\tau)}}{d + \bar{P}^m e^{mx(t-\tau)}}$$

which we rewrite as

(8) $$\dot{x}(t) + c_1 f_1(x(t)) + c_2 f_2(x(t - \tau)) = 0$$

where

$$f_1(u) = \left(\frac{b + \bar{P}^n}{n}\right) \frac{e^{nu} - 1}{b + \bar{P}^n e^{nu}}$$

and

$$f_2(u) = \left(\frac{d + \bar{P}^m}{m}\right)\frac{e^{mu} - 1}{d + \bar{P}^m e^{mu}}.$$

Clearly, every solution of Eq.(1) oscillates about $\bar{P}$ if and only if every solution of Eq.(8) oscillates about zero.

i) Observe that Eq.(8) satisfies all the hypotheses of Lemma 1. Hence every solution of Eq.(8) oscillates provided every solution of the linear equation

$$(9) \qquad\qquad \dot{y}(t) + c_1 y(t) + c_2 y(t - \tau) = 0$$

oscillates. However, it is known that every solution of Eq.(9) oscillates about zero if and only if

$$(10) \qquad\qquad \lambda + c_1 + c_2 e^{-\lambda\tau} = 0$$

has no real roots. See Györi and Ladas [2] and the references cited therein.

Set

$$G(\lambda) = \lambda + c_1 + c_2 e^{-\lambda\tau}$$

and observe that $G(\lambda) = 0$ has no real roots if and only if

$$\min_{\lambda \in \mathbf{R}} G(\lambda) = \frac{1}{\tau}\ln(c_2\tau e^{c_1\tau}e) > 0$$

which is the case if and only if

$$c_2\tau e^{c_1\tau} > e^{-1}$$

ii) Consider

$$f_1(u) - u = \frac{be^{nu} - b + \bar{P}^n e^{nu} - \bar{P}^n - nub - nu\bar{P}^n e^{nu}}{n(b + \bar{P}^n e^{nu})}.$$

Set

$$H(u) = be^{nu} - b + \bar{P}^n e^{nu} - \bar{P}^n - nub - nu\bar{P}^n e^{nu}$$

and observe that $f_1(u) - u$ and $H(u)$ have the same sign. Furthermore,

$$H(0) = 0 = \dot{H}(0), \ddot{H}(0) = n^2(b - \bar{P}^n).$$

If $b > \bar{P}^n$, then $H(u)$ is positive in some deleted neighborhood of zero, and so there exists $\delta_1 > 0$ such that

$$f_1(u) > u \quad\text{for}\quad u \in [-\delta_1, 0).$$

If $b < \bar{P}^n$, then there exists $\delta_2 > 0$ such that

$$f_2(u) < u \quad\text{for}\quad u \in (0, \delta_1].$$

A similar argument holds for $f_2 - u$. The result follows by Lemma 2.

References

1. J. Bélair and M. C. Mackey, *Consumer memory and price fluctuations in commodity markets: an integrodifferential model*, J. Dynamics Diff. Equations **3** (1989), 299–325.

2. I. Györi and G. Ladas, *Oscillation Theory of Delay Difference Equations with Applications*, Oxford University Press, 1991.

3. M. R. S. Kulenovic, G. Ladas and A. Meimaridou, *Oscillations of nonlinear delay equations*, Quart. Appl. Math. **XLV** (1987), 155–164..

4. M. C. Mackey, *Commodity price fluctuations: price dependent delays and nonlinearities as explanatory factors*, Jour. of Econ. Theory **48** (1989), 497–508.

DEPARTMENT OF MATHEMATICS, THE UNIVERSITY OF RHODE ISLAND, KINGSTON, RI 02881-0816

DEPARTMENT OF MATHEMATICS, THE UNIVERSITY OF RHODE ISLAND, KINGSTON, RI 02881-0816

Contemporary Mathematics
Volume **129**, 1992

Some Results on the Asymptotic Behavior of the Solutions of a Second Order Nonlinear Neutral Delay Equation

JOHN R. GRAEF, MYRON K.
GRAMMATIKOPOULOS, AND PAUL W. SPIKES

ABSTRACT. The authors consider the second order nonlinear neutral delay
equation

(E)
$$[y(t) + P(t)y(t - \tau)]'' - F(t, y(h(t))) = 0$$

where $uF(t, u) \geq 0$ for $u \neq 0$, and obtain results on the asymptotic behavior
of the nonoscillatory solutions of (E). Examples illustrating their results are
included as is a discussion comparing the results here to those obtained by
other authors in the case where $uF(t, u) \leq 0$ for $u \neq 0$.

1. Introduction

In [5] the present authors studied the asympototic properties of the solutions
of the second order nonlinear neutral delay differential equation

(E')
$$[y(t) + P(t)y(t - \tau)]'' + Q(t)f(y(t - \sigma)) = 0$$

where $P, Q : [t_0, \infty) \to R$ are continuous with neither P nor Q identically zero
on any half line $[t, \infty)$, $Q(t) \geq 0$, τ and σ are nonnegative constants, $f : R \to R$
is continuous, and $uf(u) > 0$ for $u \neq 0$. In Section 2 of this paper we obtain
some analogous results for the equation

(E)
$$[y(t) + P(t)y(t - \tau)]'' - F(t, y(h(t))) = 0$$

with P and τ as above, $h : [t_0, \infty) \to R$ is continuous with $h(t) \leq t$ and $h(t) \to \infty$
as $t \to \infty$, and $F : [t_0, \infty) \times R \to R$ is continuous, $uF(t, u) \geq 0$ for $u \neq 0$
and $t \geq t_0$, and $F(t, u) \not\equiv 0$ on $[t_1, \infty) \times R \setminus \{0\}$ for every $t_1 \geq t_0$. Section 3
compares the asymptotic properties of the solutions of (E) obtained here with

---

1991 *Mathematics Subject Classification*. Primary 34K15, 34K40, 34C11, 34C15.

those of the solutions of (E') obtained in [5]. As will be noted in Section 3, there are significant differences in the behavior of the solutions of (E) and (E'). These differences will be demonstrated by a combination of theorems and examples.

Recently, other authors have studied the asymptotic and oscillatory properties of first and higher order neutral delay differential equations, i.e., equations in which the highest order derivative of the unknown function appears both with and without delays. Some results of this type for neutral delay equations and how they differ from the corresponding results for delay equations, as well as discussions of existence and uniqueness of solutions and some applications for neutral equations, can be found in [1–21]. Most previous results for second order equations are for the special cases of (E) when $F(t, u) = Q(t) f(u)$ or $F(t, u) = Q(t)u$, and a substantial number of them require $P(t)$ and $Q(t)$ to be constant functions. Consequently, the results here extend or generalize many of those previously obtained. Since our purpose is to obtain asymptotic properties of the solutions of (E), every solution $y(t)$ mentioned here will be understood to be nontrivial and continuable to the right, i.e., $y(t)$ is defined on $[t_y, \infty)$ for some $t_y \geq t_0$ and $\sup \{|y(t)| : t \geq t_1\} > 0$ for every $t_1 \geq t_y$. Furthermore, we will, say that such a solution is oscillatory if its set of zeros is unbounded from above, and say that it is nonoscillatory otherwise.

2. Main Results

In order to carry out the proofs of our results, it will be convenient to define

$$z(t) = y(t) + P(t)y(t - \tau)$$

so that (E) becomes

$$z'' = F(t, y(h(t))).$$

Also, for some results we will need the condition that if $u(t) > 0$ $(u(t) < 0)$ is a continuous function such that $\liminf_{t \to \infty} |u(t)| > 0$, then

$$(1) \qquad \int^{\infty} F(s, u(s))ds = \infty \ (-\infty).$$

We begin with a lemma that will be utilized in the proof of our first theorem.

LEMMA 1. *Suppose that (1) holds and that there exists a constant $P_1 < 0$ such that*

$$(2) \qquad P_1 \leq P(t) \leq 0.$$

(a) *If $y(t)$ is an eventually positive solution of (E), then $z'(t)$ is increasing and either*

$$(3) \qquad \lim_{t \to \infty} z(t) = \lim_{t \to \infty} z'(t) = +\infty,$$

or

$$(4) \qquad \lim_{t \to \infty} z(t) = \lim_{t \to \infty} z'(t) = 0, \ z'(t) < 0 \ and \ z(t) > 0.$$

(b) *If $y(t)$ is an eventually negative solution of (E), then $z'(t)$ is decreasing and either*

(5)
$$\lim_{t \to \infty} z(t) = \lim_{t \to \infty} z'(t) = -\infty,$$

or

(6)
$$\lim_{t \to \infty} z(t) = \lim_{t \to \infty} z'(t) = 0, \quad z'(t) > 0, \quad and \quad z(t) < 0.$$

PROOF. Let $y(t)$ be an eventually positive solution of (E). Then there exists $t_1 \geq t_0$ such that $y(t - \tau)$ and $y(h(t))$ are both positive on $[t_1, \infty)$. Hence, $z''(t) = F(t, y(h(t))) \geq 0$ for $t \geq t_1$ and $z''(t) \not\equiv 0$ on any half line $[t_2, \infty), t_2 \geq t_1$. Thus $z'(t)$ is increasing and eventually has fixed sign. Integrating (E) we have

$$z'(t) = \int_{t_1}^{t} F(s, y(h(s))) ds + z'(t_1).$$

There are two cases to consider. First, if

$$\int_{t_1}^{t} F(s, y(h(s))) ds \to \infty \text{ as } t \to \infty,$$

then $z'(t) \to \infty$ as $t \to \infty$ and clearly (3) holds. On the other hand, if

$$\int_{t_1}^{t} F(s, y(h(s))) ds \to M < \infty \text{ as } t \to \infty,$$

then $z'(t) \to L$ as $t \to \infty$. Furthermore,

$$\int_{t_1}^{\infty} F(s, y(h(s))) ds < \infty$$

implies,

(7)
$$\liminf_{t \to \infty} y(t) = 0$$

since (1) holds. Now if $L > 0$, then again $z(t) \to \infty$ as $t \to \infty$ contradicting (7) since $y(t) \geq z(t)$. If $L < 0$, then $z(t) \to -\infty$ as $t \to \infty$ and we have

$$P_1 y(t - \tau) \leq p(t) y(t - \tau) < z(t).$$

Thus $y(t) \to \infty$ as $t \to \infty$ which, as noted above, is impossible. Therefore $L = 0$, and, since $z'(t)$ is eventually of fixed sign, $z'(t) < 0$ on $[t, \infty)$ and hence $z(t)$ is decreasing. Now suppose there exists t_3 such that $z(t_3) \leq 0$; then there exists t_4 such that $z(t) < z(t_4) < 0$ for $t \geq t_4 > t_3$. So $y(t) < z(t_4) - P_1 y(t - \tau)$ for $t \geq t_4$. In view of (7), there is an increasing sequence $\{s_n\}$ such that $s_1 \geq t_4$ and $y(s_n - \tau) \to 0$ as $n \to \infty$. Clearly, there exists N such that $y(s_N) < 0$ contradicting $y(t) > 0$. Therefore $z(t) > 0$ on $[t_1, \infty)$ and hence $z(t) \to M_1 \geq 0$. If $M_1 > 0$, then $y(t) \geq z(t) \geq M_1$ contradicting (7). Thus $z(t) \to 0$ as $t \to \infty$ and (a) is proved.

The proof of (b) is similar and will be omitted.

REMARK. Notice that Lemma 1 is the best possible under its hypothesis. This is demonstrated by the equation

$$(F) \quad \left[y(t) - \left(\frac{e^{a\tau}}{2} \right) y(t - \tau) \right]''$$

$$- \frac{a^2 e^{a(\sigma - 1)}}{2} y(t - \sigma) exp \left\{ \frac{1}{t - \sigma} \ln \left[\frac{|y(t - \sigma)| + e^{a(t - \sigma)}}{2} \right] \right\} = 0$$

where $a \neq 0$ is a constant. Here condition (1) is satisfied and $y_1(t) = e^{at}$ and $y_2(t) = -e^{at}$ are both solutions of (F). Moreover, $z_1(t) = y_1(t) - (e^{a\tau}/2) y(t-\tau) = \left(\frac{1}{2} \right) e^{at}$ satisfies (3) if $a > 0$ and satisfies (4) if $a < 0$. Similarly, $z_2(t) = - \left(\frac{1}{2} \right) e^{at}$ satisfies (5) if $a > 0$ and (6) if $a < 0$.

REMARK. Results analogous to Lemma 1 were obtained by the authors [5, **Lemma 1**] for (E'). Lemma 1 generalizes Lemma 3 in [9].

THEOREM 2. *Suppose (1) holds.*

(a) *If (2) holds with $P_1 > -1$, i.e.,*

$$(8) \qquad\qquad -1 < P_1 \leq P(t) \leq 0,$$

then every nonoscillatory solution $y(t)$ of (E) satisfies either $|y(t)| \to \infty$ or $y(t) \to 0$ as $t \to \infty$.

(b) *If*

$$(9) \qquad\qquad P_2 \leq P(t) \leq -1,$$

then every nonoscillatory solution $y(t)$ of (E) satisfies $|y(t)| \to \infty$ as $t \to \infty$.

PROOF. If $y(t)$ is an eventually positive solution of (E) such that $y(t) \nrightarrow \infty$ as $t \to \infty$, then (3) cannot hold since $z(t) \leq y(t)$. Thus, by Lemma 1, (4) holds, i.e.,

$$\lim_{t \to \infty} z(t) = \lim_{t \to \infty} z'(t) = 0, \ z'(t) < 0, \text{ and } z(t) > 0.$$

Choose $t_1 \geq t_0$ so that $y(h(t)) > 0$ and $y(t - \tau) > 0$ for $t \geq t_1$. Since $y(t) \leq z(t) - P_1 y(t - \tau)$, it follows, by iterating, that

$$y(t + n\tau) \leq z(t + n\tau) - P_1 z(t + (n-1)\tau) + (-P_1)^2 z(t + (n-2)\tau)$$

$$+ \cdots + (-P_1)^{n-1} z(t - \tau) + (-P_1)^n y(t)$$

for each positive integer n. Let

$$m = \begin{cases} \dfrac{n}{2}, & \text{if } n \text{ is even} \\[2mm] \dfrac{(n+1)}{2}, & \text{if } n \text{ is odd.} \end{cases}$$

Then

$$y(t + n\tau) \le z(t + n\tau) - P_1 z(t + (n-1)\tau) + \cdots$$
$$+ (-P_1)^m z(t + m\tau) + (-P_1)^{m+1} z(t + (m-1)\tau) + \cdots$$
$$+ (-P_1)^{n-1} z(t + \tau) + (-P_1)^n y(t).$$

Now (4) implies that $z(t)$ is positive and decreasing, which, together with (8), implies that

$$S_{1n} = z(t + n\tau) - P_1 z(t + (n-1)\tau) + \cdots + (-P_1)^m z(t + m\tau)$$
$$\le z(t + n\tau) \sum_{k=0}^{m} (-P_1)^k$$

and

$$S_{2n} = (-P_1)^{m+1} z(t + (m-1)\tau) + \cdots + (-P_1)^{n-1} z(t + \tau)$$
$$\le z(t + \tau) \sum_{k=m+1}^{n-1} (-P_1)^k$$
$$\le L_1 \sum_{k=m+1}^{n-1} (-P_1)^k$$

for some constant $L_1 > 0$. Since $0 < -P_1 < 1$, the series $\sum_{k=0}^{\infty} (-P_1)^k$ converges. Thus, by the definition of m, it follows that for fixed t both S_{1n} and S_{2n} tend to zero as $n \to \infty$. But

$$y(t + n\tau) \le S_{1n} + S_{2n} + (-P)^n y(t).$$

so, for each fixed t, $y(t + n\tau) \to 0$ as $n \to \infty$. Therefore $y(t) \to 0$ as $t \to \infty$. This proves (a) for $y(t)$ eventually positive.

For the proof of (b) when $y(t) > 0$ we again assume that $y(t)$ does not tend to ∞ as $t \to \infty$. Hence, as in the proof of (a), (4) holds. Integrating (E) over $[t_1, t]$ and then letting $t \to \infty$ we see that

$$\int_{t_1}^{\infty} F(s, y(h(s))) ds = -z'(t_1) < \infty.$$

It then follows from (1) that (7) holds. This leads to a contradiction, since, by (4) and (9),

$$0 < z(t) = y(t) + P(t) y(t - \tau) \le y(t) - y(t - \tau).$$

The proofs of (a) and (b) are similar when $y(t)$ is eventually negative and will be omitted.

REMARK. With appropriate choices of a and τ, equation (F) satisfies all the hypotheses of Theorem 2 and has nonoscillatory solutions tending to $+\infty$, $-\infty$, and 0. Specifically we have the following:

(A_1) If $0 < a\tau < \ln 2$, then (8) holds and the solutions $y_1(t) = e^{at}$ and $y_2(t) = -e^{at}$ tend to $+\infty$ and $-\infty$ respectively as $t \to \infty$. Condition (8) is also satisfied if $a < 0$, and in this case $y_1(t)$ and $y_2(t)$ both tend to zero as $t \to \infty$.

(A_2) When $a\tau \geq \ln 2$, equation (F) satisfies (9) and again the solutions $y_1(t)$ and $y_2(t)$ tend to $+\infty$ and $-\infty$ respectively as $t \to \infty$.

REMARK. Theorem 2 generalizes Theorem 4 in [**9**] and Theorem 2(b) includes Theorem 1 in [**11**] as a special case.

Theorem 2(b) can be restated as an oscillation result as follows.

COROLLARY 3. *If (1) and (9) are satisfied, then every bounded solution of (E) is oscillatory.*

Our final two theorems describe the behavior of the unbounded solutions of (E) without imposing condition (1) on $F(t, u)$.

THEOREM 4. *If (8) holds, then every unbounded solution $y(t)$ of (E) is either oscillatory or satisfies $|y(t)| \to \infty$ as $t \to \infty$.*

PROOF. Suppose $y(t)$ is an unbounded solution of (E) that is eventually positive, say $y(h(t)) > 0$ and $y(t-\tau) > 0$ for $t \geq t_1 \geq t_0$. Thus $z'(t)$ is increasing and $z(t)$ is montonic. Since $F(t, y(h(t))) \not\equiv 0$ on $[t_1, \infty)$, both $z(t)$ and $z'(t)$ eventually have fixed sign. If $z(t) < 0$ for $t \geq t_2 \geq t_1$, then (8) implies that

$$y(t) = z(t) - P(t)y(t-\tau) < y(t-\tau)$$

contradicting the fact that $y(t)$ is unbounded. Hence, there exists $t_3 \geq t_1$ such that $z(t) > 0$ for $t \geq t_3$. Now suppose that $z'(t) < 0$ on $[t_3, \infty)$. Then $z(t) \leq z(t_3)$ and we have

$$y(t) \leq z(t_3) - P_1 y(t-\tau).$$

But $y(t)$ unbounded ensures the existence of an increasing sequence $\{T_n\}$ such that $T_1 \geq t_3$, $T_n \to \infty$ and $y(T_n) \to \infty$ as $n \to \infty$, and $y(T_n) \geq \max_{t_3 \leq t \leq T_n} y(t)$. Therefore, for each $n \geq 1$,

$$y(T_n) \leq z(t_3) - P_1 y(T_n - \tau) \leq z(t_3) - P_1 y(T_n),$$

and hence

$$(1 + P_1)y(T_n) \leq z(t_3)$$

which, in view of (8), is a contradiction. A similar proof handles the case when $y(t) < 0$.

REMARK. As noted above, the hypotheses of Theorem 4 do not require condition (1) as do the hypotheses of Theorem 2(a). In particular, the equation

(G) $\qquad \left[y(t) + \dfrac{bt}{t-1} y(t-1) \right]'' - \dfrac{2(b+1)}{(t-\sigma)^2} e^{-|a|t(t-\sigma)^2} y(t-\sigma) e^{t|y(t-\sigma)|} = 0$

where $a \neq 0$, $-1 < b < 0$, and $t \geq t_0 = 1/(b+1)$ satisfies the hypotheses of Theorem 4, and has the nonoscillatory solution $y_3(t) = at^2$. However, Theorem 2(a) cannot be applied to (G) since (1) is not satisfied for every $u(t)$ with $\liminf_{t \to \infty} |u(t)| > 0$.

All of our previous results have required $P(t) \leq 0$. Our final result is with $P(t)$ nonnegative.

THEOREM 5. *If there exists a constant P_3 such that*

(10) $\qquad\qquad\qquad 0 \leq P(t) \leq P_3 < 1,$

then any unbounded solution $y(t)$ of (E) is either oscillatory or satisfies $|y(t)| \to \infty$ as $t \to \infty$.

PROOF. Suppose $y(t)$ is a nonoscillatory solution of (E) that satisfies $y(t - \tau) > 0$ and $y(h(t)) > 0$ for $t \geq t_1 \geq t_0$. Then, as in the proof of Theorem 4, $z'(t)$ is increasing and $z(t)$ is monotonic. Since $y(t) \leq z(t)$ and $y(t)$ is unbounded, then clearly $z'(t)$ is eventually positive and $z(t) \to \infty$ as $t \to \infty$. By (10),

$$(1 - P_3)z(t) \leq y(t),$$

and therefore $y(t) \to \infty$ as $t \to \infty$. The proof is similar for the case $y(t) < 0$.

The equation

$(H_1) \qquad \left[y(t) + \dfrac{1}{2} y(t-\tau) \right]'' - \dfrac{a^2 e^{5a\sigma}(1 + 2e^{a\tau})}{2e^{a\tau} e^{4at}} y^5(t-\sigma) = 0$

satisfies the hypotheses of Theorem 5 for any constant $a \neq 0$ and any positive constatnts τ and σ. Observe that $y_4(t) = e^{at}$ is a nonoscillatory solution of (H_1). If $a > 0$, then $y_4(t)$ satisfies the conclusion of Theorem 5. Also notice that when $a < 0, y_4(t)$ is a bounded nonoscillatory solution of (H_1), so (E) may have both bounded and unbounded nonoscillatory solutions under the hypotheses of Theorem 5. Notice that if $a > 0$, (1) does not hold. In view of this observation, it seems reasonable to ask whether adding condition (1) to the hypotheses of Theorem 5 would resuslt in all nonoscillatory solutions being bounded, or even converging to zero as $t \to \infty$? Another question to raise would be if (10) holds and condition (1) fails for every $u(s)$, could we conclude that any nonoscillatory solution $y(t)$ of (E) satisfies $|y(t)| \to \infty$ as $t \to \infty$? The following equation shows that the answer to both these questions is no. Consider

$(H_2) \; [y(t) + Py(t-\tau)]'' - \dfrac{a^2 (1 + Pe^{a\tau}) \left(e^{2a(t-\sigma)} + 1 \right) e^{2a\left(t - \frac{\sigma}{2}\right)/3} y^{\frac{1}{3}}(t-\sigma)}{y^2(t-\sigma) + 1} = 0$

with a, τ, and σ as in (H_1) and $0 < P < 1$. The hypotheses of Theorem 5 are satisfied by (H_2) and $y_4(t)$ is also a solution of (H_2). When $a > 0$, (1) is satisfied and $y_4(t) \to \infty$ as $t \to \infty$. On the other hand, $y_4(t) \to 0$ as $t \to \infty$ when $a < 0$ and, moreover, condition (1) fails for every $u(t)$.

3. Comparisons

The present authors [5] studied the asymptotic properties of the solutions of (E') with Q and f satisfying $Q(t) \geq 0$, $uf(u) > 0$ for $u \neq 0$, as well as requiring that

(1′) $f(u)$ is bounded away from zero if u is bounded away from zero,

and

(2′)
$$\int_{t_0}^{\infty} Q(s)\,ds = \infty.$$

Notice that (1′) and (2′) together imply (1) when $F(t, u) = Q(t)f(u)$. Hence, the hypotheses in [5] and those here are consistent. It is interesting to compare the results obtained here to those in [5] with similar conditions on $P(t)$. From [5] we know that:

   (i) if (8) holds, then any nonoscillatory solution $y(t)$ of (E') satisfies $y(t) \to 0$ as $t \to \infty$;
   (ii) if $-1 \leq P(t) \leq 0$, then every unbounded solution of (E') is oscillatory;
   (iii) if $P_4 \leq P(t) \leq P_5 < -1$, then any bounded solution $y(t)$ of (E') is either oscillatory or satisfies $y(t) \to 0$ as $t \to \infty$;
   (iv) if $P(t)$ is not eventually negative, then any solution $y(t)$ of (E') is either oscillatory or satisfies $\liminf_{t \to \infty} |y(t)| = 0$.

Clearly there are significant differences in the behavior of the solutions of (E) and (E') that are brought into sharp focus by (i)–(iv) and the examples and theorems in Section 2. In particluar, as illustrated by (F), (E) can have unbounded nonoscillatory solutions with (1) and (8) holding, whereas (i)–(ii) show that this is not the case for (E'). If (9) holds, we see from (iii) that (E') may have both oscillatory and nonoscillatory solutions that are bounded, while Theorem 2(b) shows that (E) cannot have a bounded nonoscillatory solution.

Finally, to see that (iv) does not hold for (E), we need only observe that the equation

$$[y(t) + (\sin t - 1)y(t - 3)]'' - e^{\frac{2}{3}}[1 + e^{-3}(2\cos t - 1)]e^{\frac{2t}{3}}y^{\frac{1}{3}}(t - 2) = 0$$

has the nonoscillatory solution $y(t) = e^t$. Notice that this equation satisfies (1′) and (2′).

REFERENCES

1. R. Bellman and K. L. Cooke, *Differential-difference Equations*, Academic Press, New York, 1963.
2. R. D. Driver, *Existence and continuous dependence of solutions of a neutral function - differential equation*, Arch. Rational Mech. Anal. **19** (1965), 149–166.
3. R. D. Driver, *A mixed neutral system*, Nonlinear Anal. **8** (1984), 155–158.
4. J. K. Hale, *Theory of Functional Differential Equations*, Springer-Verlag, 1977.
5. J. R. Graef, M. K. Grammatikopoulos, and P. W. Spikes, *Asymptotic properties of solutions of nonlinear neutral delay differential equations of the second order*, Rad. Mat. **4** (1988), 133–149.
6. J. R. Graef, M. K. Grammatikopoulos, and P. W. Spikes, *Behavior of the nonoscillatory solutions of first order neutral delay differential equations*, in Differential Equations, Proceedings of the EQUADIFF Conference, Dekker, 1989, pp. 265–272.
7. J. R. Graef, M. K. Grammatikopoulos, and P. W. Spikes, *On the behavior of solutions of a first order nonlinear neutral delay differential equation*, Applicable Anal. **40** (1991), 111–121.
8. J. R. Graef, M. K. Grammatikopoulos, and P. W. Spikes, *Asymptotic and oscillatory behavior of solutions of first order nonlinear neutral delay differential equations*, J. Math. Anal. Appl. **155** (1991), 562–571.
9. J. R. Graef, M. K. Grammatikopoulos, and P. W. Spikes, *On the asymptotic behavior of solutions of a second order nonlinear neutral delay differential equation*, J. Math. Anal. Appl. **156** (1991), 23–39.
10. M. K. Grammatikopoulos, E. A. Grove, and G. Ladas, *Oscillation and asymptotic behavior of neutral differential equations with deviating arguments*, Applicable Anal. **22** (1986), 1–19.
11. M. K. Grammatikopoulos, E. A. Grove, and G. Ladas, *Oscillation and asymptotic behavior of second order neutral differential equations with deviating arguments*, Canadian Math. Soc. Conference Proceedings **8** (1987), 153–161.
12. M. K. Grammatikopoulos, G. Ladas, and A. Meimaridou, *Oscillation of second order neutral delay differential equations*, Rad. Mat. **1** (1985), 267–274.
13. M. K. Grammatikopoulos, G. Ladas, and A. Meimaridou, *Oscillation and asymptotic behavior of second order neutral differential equations*, Ann. Mat. Pura Appl. (4) **148** (1987), 29–40.
14. M. K. Grammatikopoulos, G. Ladas, and A. Meimaridou, *Oscillation and asymptotic behavior of higher order neutral equations with variable coefficients*, Chinese Ann. Math. Ser. B **9** (1988), 322–328.
15. M. K. Grammatikopoulos, G. Ladas, and Y. G. Sficas, *Oscillation and asymptotic behavior of neutral equations with variable coefficients*, Rad. Mat. **2** (1986), 279–303.
16. M. K. Grammatikopoulos, Y. G. Sficas, and I. P. Stavroulakis, *Necessary and sufficient conditions for oscillations of neutral equations with several coefficients*, J. Differential Equations **76** (1988), 294–311.
17. G. Ladas and Y. G. Sficas, *Oscillations of higher-order neutral equations*, J. Austral. Math. Soc. Ser. B **27** (1986), 502–511.
18. G. Ladas and Y. G. Sficas, *Oscillations of neutral delay differential equations*, Canad. Math. Bull. **29** (1986), 438–445.
19. Y. G. Sficas and I. P. Stavroulakis, *Necessary and sufficient conditions for oscillations of neutral differential equations*, J. Math. Anal. Appl. **123** (1987), 494–507.
20. Y. T. Xu, *Asymptotic behavior of nonoscillatory solutions of higher-order neutral equations*, Ann. Diff. Eqs. **5** (1989), 199–209.
21. A. I. Zahariev and D. D. Bainov, *On some oscillation criteria for a class of neutral type functional differential equations*, J. Austral. Math. Soc. Ser. B **28** (1986), 229–239.

DEPARTMENT OF MATHEMATICS AND STATISTICS, MISSISSIPPI STATE UNIVERSITY, MISSISSIPPI STATE, MS 39762

E-mail: graef@math.msstate.edu

DEPARTMENT OF MATHEMATICS, UNIVERSITY OF IOANNINA, IOANNINA, GREECE.

DEPARTMENT OF MATHEMATICS AND STATISTICS, MISSISSIPPI STATE UNIVERSITY, MIS-
SISSIPPI STATE, MS 39762

Contemporary Mathematics
Volume **129**, 1992

Linearized Oscillations For Nonautonomous Delay Difference Equations

G. LADAS AND C. QIAN

ABSTRACT. We obtain a linearized oscillation result for quite general nonlinear and nonautonomous delay difference equations.

1. Introduction

Our aim in this paper is to establish a linearized oscillation result for the following quite general nonlinear and nonautonomous delay difference equation

$$(1) \qquad x_{n+1} - x_n + f(n, x_{n-k_1}, \cdots, x_{n-k_m}) = 0, n = 0, 1, \cdots$$

where

$$(2) \qquad \begin{cases} k_1, \cdots, k_m \in \mathbf{N} = \{0, 1, \cdots\}, k = \max\{k_1, \cdots, k_m\} > 0, \\ f : \mathbf{N} \times \mathbf{R} \to \mathbf{R}, f(n, u_1, \cdots, u_m) \text{ is continuous for every} \\ \text{fixed } n, \quad f(n, u_1, \cdots, u_m) \geq 0 \text{ for } u_1, \cdots, u_m \geq 0 \quad \text{and} \\ f(n, u_1, \cdots, u_m) \leq 0 \quad \text{for} \quad u_1, u_2, \cdots, u_m \leq 0. \end{cases}$$

We show that, under appropriate hypotheses, the oscillatory behavior of Eq.(1) is characterized by the oscillatory behavior of an associated linear difference equation with variable coefficients of the form

$$(3) \qquad y_{n+1} - y_n + \sum_{i=1}^{m} p_i(n) y_{n-k_i} = 0, n = 0, 1, \cdots.$$

Sufficient conditions for the oscillation of all solutions of Eq.(3) have been established in [1] and [5-8].

Linearized oscillation results for autonomous difference equations have been established in [2] and [3].

1991 *Mathematics Subject Classification.* 39A12.

By a *solution* of Eq.(1), we mean a sequence $\{x_n\}$ which is defined for $n \geq -k$ and which satisfies (1) for $n \geq 0$. Let

$$a_{-k}, a_{-k+1}, \cdots, a_0$$

be given numbers. Then it is easily seen that Eq.(1) has a unique solution $\{x_n\}$ which satisfies the initial conditions

$$x_i = a_i \quad \text{for} \quad i = -k, \cdots, 0.$$

A sequence $\{x_n\}$ is said to *oscillate about the sequence* $\{x_n^*\}$ if the terms $x_n - x_n^*$ of the sequence $\{x_n - x_n^*\}$ are neither all eventually positive nor all eventually negative. When $x_n^* \equiv 0$, we say that $\{x_n\}$ *oscillates about zero* or that it simply *oscillates*.

2. Linearized Oscillations for Eq.(1)

The following lemma, which is interesting in its own right, will be needed in the proof of the main result.

LEMMA 1. *Assume that there exist $N^* \geq 0$ and $\delta > 0$ such that for $n \geq N^*$ and for $u_1, u_2, \cdots, u_m \in [0, \delta]$,*

$$(4) \quad \begin{cases} f(n, u_1, \cdots, u_m) \text{ is positive for } (u_1, \cdots, u_m) \neq 0 \text{ and} \\ \text{increasing in } u_1, \cdots, u_m \text{ in the sense that if } u_i' \leq u_i'' \\ \text{for } i = 1, \cdots, m, \text{ then } f(n, u_1', \cdots, u_m') \leq f(n, u_1'', \cdots, u_m''). \end{cases}$$

Suppose also that for $n \geq N^*$ the inequality

$$(5) \qquad x_{n+1} - x_n + f(n, x_{n-k_1}, \cdots, x_{n-k_m}) \leq 0, n = 0, 1, \cdots$$

has a positive solution $\{x_n^*\}$ with $x_n^* \leq \delta$. Then Eq.(1) has an eventually positive solution $\{x_n\}$ with $x_n \leq x_n^*$ for n sufficiently large.

PROOF. Let $k = \max\{k_1, \cdots, k_m\}$ and set $N = N^* + k$. Then

$$(6) \qquad \{x_n^*\} \quad \text{is strictly decreasing for} \quad n \geq N - k$$

and so

$$\lim_{n \to \infty} x_n^* = l \in [0, \infty), \quad \text{exists}.$$

By summing up both sides of (5) from n to ∞, we see that

$$(7) \qquad l + \sum_{j=n}^{\infty} f(j, x_{j-k_1}^*, \cdots, x_{j-k_m}^*) \leq x_n^*, \; n \geq N.$$

Now, define the set of nonnegative sequences

$$(8) \qquad \Lambda = \{\{y_n\} : 0 \leq y_n \leq x_n^*, n \geq N\}$$

and for each $\{y_n\} \in \Lambda$, define an associated sequence $\{Y_n\}_{n=N-k}^{\infty}$ by

$$Y_n = \begin{cases} y_n, n \geq N, \\ y_N + x_n^* - x_N^*, N - k \leq n < N. \end{cases}$$

Form (8) and (6), we see that

(9)
$$0 \leq Y_n \leq x_n^* \quad \text{for} \quad n \geq N - k$$

and

(10)
$$Y_n > 0 \quad \text{for} \quad N - k \leq n < N.$$

Next, we define the mapping T on Λ as follows:

$$Ty_n = l + \sum_{j=n}^{\infty} f(j, Y_{j-k_1}, \cdots, Y_{j-k_m}), \ n \geq N.$$

In view of (4), we see that if $\{y_n'\}, \{y_n''\} \in \Lambda$ with $y_n' \leq y_n''$ for $n \geq N$ then $Ty_n' \leq Ty_n''$ for $n \geq N$. Note that by (7), $Tx_n^* \leq x_n^*, n \geq N$. Hence, for any $\{y_n\} \in \Lambda, Ty_n \leq Tx_n^* \leq x_n^*$ for $n \geq N$ and so we see that $T : \Lambda \to \Lambda$. Now, consider the following sequences

$$\{x_n^{(0)}\} = \{x_n^*\} \quad \text{and} \quad \{x_n^{(i)}\} = \{Tx_n^{(i-1)}\}, i = 1, 2, \cdots.$$

Then one can see by induction that for any $n \geq N$,

(11)
$$0 \leq x_n^{(i+1)} \leq x_n^{(i)} \leq x_n^*, i = 1, 2, \cdots.$$

Thus

(12)
$$x_n = \lim_{i \to \infty} x_n^{(i)}, \ n \geq N$$

exists and $\{x_n\} \in \Lambda$. Also,

(13)
$$x_n = l + \sum_{j=n}^{\infty} f(j, x_{j-k_1}, \cdots, x_{j-k_m}), \ n \geq N.$$

and so it follows that $\{x_n\}$ satisfies Eq.(1) for $n \geq N$. Now we claim that

(14)
$$x_n > 0 \quad \text{for} \quad n \geq N.$$

First, assume that $l > 0$. Then from (13) it is clear that (14) holds. Next, assume that $l = 0$. In view of (10), if (14) were false there would exist some $N' \geq N$ such that

$$x_{N'} = 0 \quad \text{and} \quad x_n > 0 \quad \text{for} \quad N - k \leq n < N'.$$

But, from (13) and (4),

$$x_N' = \sum_{j=N'}^{\infty} f(j, x_{j-k_1}, \cdots, x_{j-k_m}) > 0.$$

This contradiction implies that (14) holds. Clearly, $\{x_n\}$ can be extended as a solution of Eq.(1). Finally, from (11) and (12), we see that $x_n \leq x_n^*$ for $n \geq N$. The proof is complete.

By a similar argument, we can establish the following dual result of Lemma 1.

LEMMA 1'. *Assume that there exist $N^* \geq 0$ and $\delta > 0$ such that for $n \geq N^*$ and for $u_1, u_2, \cdots, u_m \in [-\delta, 0]$, $f(n, u_1, \cdots, u_m)$ is negative for $(u_1, \cdots, u_m) \neq 0$ and increasing. Suppose also that for $n \geq N^*$ the inequality*

$$x_{n+1} - x_n + f(n, x_{n-k_1}, \cdots, x_{n-k_m}) \geq 0, n = 0, 1, \cdots$$

has a negative solution $\{x_n^*\}$ with $x_n^* \geq -\delta$. Then Eq.(1) has an eventually negative solution with $x_n \geq x_n^*$ for n sufficiently large.

In the proof of the main result we also need the following result from [10].

LEMMA 2. *Consider the linear delay difference equations*

$$(15) \qquad A_{n+1} - A_n + \sum_{i=1}^{m} p_i(n)A_{n-k_i} = 0$$

and

$$(16) \qquad B_{n+1} - B_n + \sum_{i=1}^{m} q_i(n)B_{n-k_i} = 0$$

where for each $i = 1, 2, \cdots, n$

$(17) \qquad \{p_i(n)\}$ *and $\{q_i(n)\}$ are nonnegartive sequences and $k_i \in \mathbf{N}$.*

Let

$$Z_i = \{n \in \mathbf{N} : p_i(n) = 0\} \text{ and } Z_i' = \{n \in \mathbf{N} : q_i(n) = 0\}.$$

Assume that for each $i = 1, 2, \cdots, m$, the sets Z_i and Z_i' are equal and that

$$(18) \qquad \lim_{n \to \infty, n \in N - Z_i} \frac{p_i(n)}{q_i(n)} = 1.$$

Then every solution of Eq.(15) oscillates if and only if every solution of Eq.(16) oscillates.

The following result gives sufficient conditions for the oscillation of all solutions of Eq.(1) in terms of the oscillations of all solutions of Eq.(3).

THEOREM 1. *Assume that in every interval $[a, b]$ with $ab > 0$,*

$$(19) \qquad \sum_{n=0}^{\infty} c_n = \infty \text{ where } c_n = \min_{u_1, \cdots, u_m \in [a,b]} |f(n, u_1, \cdots, u_m)|,$$

*and that there exists a positive constant δ and nonnegative sequences $\{p_1(n)\}$,
$\cdots, \{p_m(n)\}$ such that one of the following hypotheses (20) or (21) holds:*

(20)
$$\begin{cases} f(n, u_1, \cdots, u_m) \geq \sum_{i=1}^m p_i(n)u_i > 0 \quad for \ \ 0 < u_1, \cdots, u_m \leq \delta, \\ f(n, u_1, \cdots, u_m) \leq \sum_{i=1}^m p_i(n)u_i < 0 \quad for \ \ 0 > u_1, \cdots, u_m \geq -\delta. \end{cases}$$

(21)
$$\lim_{(u_1, \cdots, u_m) \to 0, u_i u_j > 0, i,j,=1,\cdots,m} \frac{f(n, u_1, \cdots, u_m)}{p_1(n)u_1 + \cdots + p_m(n)u_m} \equiv 1.$$

*Suppose that every solution of Eq.(3) oscillates. Then every solution of Eq.(1)
also oscillates.*

PROOF. Otherwise, Eq.(1) has a nonoscillatory solution $\{x_n\}$. We assume
that $\{x_n\}$ is eventually positive. The proof when $\{x_n\}$ is eventually negative is
similar and will be omitted.

Choose $N \geq 0$ such that $x_{n-k} > 0$ for $n \geq N$ where $k = \max\{k_i : 1 \leq i \leq m\}$.
From (1) we see that $\{x_n\}$ is decreasing for $n \geq N$ and so $\lim_{n \to \infty} x_n = l \in [0, \infty)$
exists. We claim that l is zero. Otherwise, $l > 0$. Then by summing up both
sides of Eq.(1) from N to ∞, we obtain

$$l - x_N + \sum_{j=N}^{\infty} f(j, x_{j-k_1}, \cdots, x_{j-k_m}) = 0$$

which, clearly, contradicts (19). Hence, $\lim_{n \to \infty} x_n = 0$.

Now, we first assume that (20) holds. Then it follows from (1) and (20) that
for $n \geq N$

$$x_{n+1} - x_n + \sum_{i=1}^n p_i(n)x_{n-k_i} \leq 0.$$

By Lemma 1 (with $f(n, u_1, \cdots, u_m) = p_1(n)u_1 + \cdots + p_m(n)u_m$) we see that
Eq.(3) has an eventually positive solution. This is a contradiction and so the
proof is complete when (20) holds. Next, assume that (21) holds. Set

$$P_i(n) = p_i(n)\frac{f(n, x_{n-k_1}, \cdots, x_{n-k_m})}{p_1(n)x_{n-k_1} + \cdots + p_m(n)x_{n-k_m}}.$$

Then in view of the hypotheses, for each $i = 1, 2, \cdots, m, P_i(n) \geq 0$, the sequences
$\{p_i(n)\}$ and $\{P_i(n)\}$ have the same set of zeros Z_i and

$$\lim_{n \to \infty, n \in N - Z_i} \frac{p_i(n)}{P_i(n)} = 1.$$

Observe that $\{x_n\}$ is a nonoscillatory solution of the equation

$$z_{n+1} - z_n + \sum_{i=1}^m P_i(n)z_{n-k_i} = 0.$$

Then, it follows, by Lemma 2, that Eq.(3) has a nonoscillatory solution. This
contradicts the hypothesis and completes the proof of the theorem.

REMARK 1. The hypothesis (21) can be replaced by the following stronger but easier to check condition:

There is a positive number $\delta > 0$ such that the partial derivatives

$$(21') \qquad \frac{\partial f(n, u_1, \cdots, u_m)}{\partial u_i}, i = 1, 2, \cdots, m,$$

are continuous for $n \geq 0$ and $u_1, \cdots, u_m \in [-\delta, \delta]$, and

$$\frac{\partial f(n, 0, \cdots, 0)}{\partial u_i} = p_i(n), i = 1, 2, \cdots, m.$$

The following theorem establishes sufficient conditions for the existence of a nonoscillatory solution of Eq.(1).

THEOREM 2. *Assume that there exists a positive number δ and nonnegative sequences $\{p_1(n)\}, \cdots, \{p_m(n)\}$ such that either*

$$(22) \qquad \begin{cases} 0 < f(n, u_1, \cdots, u_m) \leq \sum_{i=1}^{m} p_i(n)u_i \quad and \ f \ is \\ increasing \ in \ u_1, \cdots, u_m \quad for \ 0 < u_1, \cdots, u_m \leq \delta \end{cases}$$

or

$$(23) \qquad \begin{cases} 0 > f(n, u_1, \cdots, u_m) \geq \sum_{i=1}^{m} p_i(n)u_i \quad and \ f \ is \\ increasing \ in \ u_1, \cdots, u_m \quad for \ 0 > u_1, \cdots, u_m \geq -\delta. \end{cases}$$

Assume that Eq.(3) has a nonoscillatory solution. Then Eq.(1) also has a nonoscillatory solution.

PROOF. Suppose that (22) holds. The proof when (23) holds is similar and will be omitted. As the negative of a solution of Eq.(3) is also a solution, we will assume that Eq.(3) has an eventually positive solution $\{y'_n\}$. Choose $N \geq 0$ such that $y'_{n-k} > 0$ for $n \geq N$, Then from Eq.(3), we see that $\{y'_n\}$ is decreasing for $n \geq N$ and so $\{y'_n\}$ is bounded. Therefore, for M sufficiently large,

$$y_n = \frac{y'_n}{M} \leq \delta \quad \text{for} \quad n \geq N.$$

Clearly, $\{y_n\}$ is also a positive solution of Eq.(3) for $n \geq N$. From (3) and (22), it follows that

$$y_{n+1} - y_n + f(n, y_{n-k_1}, \cdots, y_{n-k_m}) \leq 0, \ n \geq N.$$

Then by Lemma 1, Eq.(1) has an eventually positive solution and the proof is complete.

By combining Theorems 1 and 2, we obtain the following linearized oscillation result for Eq.(1).

THEOREM 3. *Assume that (19) holds and that there exists a positive number δ and nonegative sequences $\{p_1(n)\}, \cdots, \{p_m(n)\}$ such that (21) holds and such that either (22) or (23) is satisfied. Then every solution of Eq.(1) oscillates if and only if every solution of Eq.(3) oscillates.*

The following result about difference equations with seperable delays is an immediate consequence of Theorem 3.

COROLLARY 1. *Consider the delay difference equation*

$$(24) \qquad x_{n+1} - x_n + \sum_{i=1}^{m} p_i(n) f_i(x_{n-k_i}) = 0, n = 0, 1, \cdots$$

where for $i = 1, 2, \cdots, m, \{p_i(n)\}$ are nonnegative sequences, $f_i \in C[\mathbf{R}, \mathbf{R}]$, $u f_i(u) > 0$ for $u \neq 0$, and

$$\lim_{u \to 0} \frac{f_i(u)}{u} = 1.$$

Assume that

$$\sum_{i=1}^{m} \sum_{n=0}^{\infty} p_i(n) = \infty \quad and \quad \sum_{i=1}^{\infty} p_i(n) k_i > 0 \quad for \quad n = 0, 1, \cdots.$$

Furthermore, suppose that there exists a positive number δ such that for all $i = 1, 2, \cdots, m$ either

$$f_i(u) \leq u \ and \ f_i \ is \ increasing \ for \ 0 < u < \delta,$$

or

$$f_i(u) \geq u \ and \ f_i \ is \ increasing \ for \ 0 > u > -\delta.$$

Then every solution of Eq.(24) oscillates if and only if every solution of Eq.(3) oscillates.

3. Applications of Theorem 3

Consider the nonautonomous, discrete delay logistic equation

$$(25) \qquad x_{n+1} = x_n [A(n) - \sum_{i=0}^{m} B_i(n) x_{n-i}]$$

where

$$\{A(n)\}, \{B_1(n)\}, \cdots, \{B_m(n)\}$$

are positive sequences. The following result gives a characterization of the oscillation of all positive solutions of Eq.(25) about a "fixed" positive solution $\{x_n^*\}$ of the same equation.

THEOREM 4. *Let $\{x_n^*\}$ be a positive solution of Eq.(25) and assume that*

(26)
$$\sum_{i=0}^{m}\sum_{n=0}^{\infty}\frac{x_n^*}{x_{n+1}^*}B_i(n)x_{n-i}^* = \infty.$$

Then every positive solution of Eq.(25) oscillates about $\{x_n^*\}$ if and only if every solution of the linear equation

(27)
$$y_{n+1} - y_n + \frac{x_n^*}{x_{n+1}^*}\sum_{i=0}^{m}B_i(n)x_{n-i}^*y_{n-i} = 0$$

oscillates.

PROOF. The change of variable $x_n = x_n^*e^{z_n}$ reduces Eq.(25) to the equation

(28)
$$z_{n+1} - z_n - \ln\frac{x_n^*}{x_{n+1}^*}[A(n) - \sum_{i=0}^{m}B_i(n)x_{n-i}^*e^{z_{n-i}}] = 0.$$

Clearly, every positive solution of Eq.(25) oscillates about $\{x_n^*\}$ if and only if every solution of Eq.(28) oscillates (about zero). Set

$$f(n, u_0, \cdots, u_m) = -\ln\frac{x_n^*}{x_{n+1}^*}[A(n) - \sum_{i=0}^{m}B_i(n)x_{n-i}^*e^{u_i}].$$

and

$$g(n, u_0, \cdots, u_m) = f(n, u_0, \cdots, u_m) - \frac{x_n^*}{x_{n+1}^*}\sum_{i=0}^{m}B_i(n)x_{n-i}^*u_i.$$

By noting that $\{x_n^*\}$ satisfies Eq.(25), it follows that

$$\frac{\partial g}{\partial u_j} = \frac{B_j(n)x_{n-j}^*e^{u_j}}{A(n) - \sum_{i=0}^{m}B_i(n)x_{n-i}^*e^{u_i}} - \frac{x_n^*}{x_{n+1}^*}B_j(n)x_{n-j}^*$$

$$= \frac{B_j(n)x_{n-j}^*e^{u_j}}{A(n) - \sum_{i=0}^{m}B_i(n)x_{n-i}^*e^{u_i}} - \frac{B_j(n)x_{n-j}^*}{A(n) - \sum_{i=0}^{m}B_i(n)x_{n-i}^*}.$$

From this it is easy to see that

(29)
$$\frac{\partial g}{\partial u_j} < 0 \quad \text{for} \quad u_0, \cdots, u_m < 0$$

and because $g(n, 0, \cdots 0) \equiv 0$, we see that

$$f(n, u_0, \cdots, u_m) \geq \frac{x_n^*}{x_{n+1}^*}\sum_{i=0}^{m}B_i(n)x_{n-i}^*u_i \quad \text{for} \quad u_0, \ldots, u_m < 0.$$

In addition, for $j = 0, 1, \cdots, m$ and $u_0, \cdots, u_m < 0$,

$$\frac{\partial f(n, u_0, \cdots, u_m)}{\partial u_j} = \frac{B_j(n)x_{n-j}^*e^{u_j}}{A(n) - \sum_{i=0}^{m}B_i(n)x_{n-i}^*e^{u_i}} > 0$$

and

$$\frac{\partial f(n, 0, \cdots, 0)}{\partial u_j} = \frac{B_j(n)x_{n-j}^*}{A(n) - \sum_{i=0}^{m}B_i(n)x_{n-i}^*} = \frac{x_n^*}{x_{n+1}^*}B_j(n)x_{n-j}^*$$

and so $f(n, u_0, \cdots, u_m)$ is increasing, $f(n, u_0, \cdots, u_m) < 0$ for $u_0, \cdots, u_m < 0$ and

$$(30) \qquad \lim_{(u_0, \cdots, u_m) \to 0, u_i u_j > 0, i, j, = 0, \cdots, m} \frac{f(n, u_0, \cdots, u_m)}{\frac{x_n^*}{x_{n+1}^*} \sum_{i=0}^{m} B_i(n) x_{n-i}^* u_i} \equiv 1.$$

Hence, by Theorem 3, every solution of Eq.(28) oscillates if and only if every solution of Eq.(27) oscillates. The proof is complete.

By applying Theorem 4 to the difference equation

$$(31) \qquad x_{n+1} = A x_n \left[1 - \sum_{i=0}^{m} B_i x_{n-i} \right]$$

where

$$(32) \qquad m \in N, A \in (1, \infty) \quad \text{and} \quad B_i \in (0, \infty) \quad \text{for} \quad i = 0, 1, \cdots, m$$

and by taking the fixed solution of Eq.(31) to be the positive equilibrium

$$(33) \qquad x^* = \frac{A - 1}{A \sum_{i=0}^{m} B_i}$$

we obtain the following immediate consequence of Theorem 4.

COROLLARY 2. *Assume that (32) holds. Then every positive solution of Eq.(31) oscillates about the positive equilibrium x^* of Eq.(31) if and only if every solution of the linear equation*

$$(34) \qquad y_{n+1} - y_n + A x^* \sum_{i=1}^{m} B_i(n) y_{n-i} = 0$$

oscillates.

REMARK 2. The above corollary, with some additional hypotheses, was established in [3].

REMARK 3. Eq.(25) with $m = 0, 1$, and 2 was studied in [9] and has application in neural networks.

In a recent paper [4], the following nonlinear delay difference equation

$$(35) \qquad x_{n+1} = \frac{\alpha x_n}{1 + \beta x_{n-1}}$$

has been studied and the following result has been obtained.

LEMMA 3. *Assume that*

$$(36) \qquad \alpha \in (1, \infty) \quad \text{and} \quad \beta \in (0, \infty).$$

Then every positive solution of Eq.(35) converges to the positive equilibrium $x^* = \frac{\alpha - 1}{\beta}$.

By using this fact and our linearized oscillation Theorem 3, we obtain the following interesting oscillation result for Eq.(35).

THEOREM 5. *Assume that (36) holds. Then any two positive solutions of Eq.(35) oscillate about each other if and only if $\alpha > 4/3$.*

PROOF. Let $\{x_n^*\}$ be a positive solution of Eq.(35). It suffices to show that every positive solution of Eq.(35) oscillates about $\{x_n^*\}$ if and only if $\alpha > 4/3$.

The change of variable $x_n = x_n^* e^{z_n}$ reduces Eq.(35) to the equation

$$(37) \qquad z_{n+1} - z_n + \ln \frac{x_{n+1}^*}{\alpha x_n^*}[1 + \beta x_{n-1}^* e^{z_{n-1}}] = 0.$$

Clearly, every positive solution of Eq.(35) oscillates about $\{x_n^*\}$ if and only if every solution of Eq.(37) oscillates. Set

$$(38) \qquad f(n, u) = \ln \frac{x_{n+1}^*}{\alpha x_n^*}[1 + \beta x_{n-1}^* e^u]$$

and

$$(39) \qquad g(n, u) = f(n, u) - \frac{\beta x_{n-1}^* x_{n+1}^*}{\alpha x_n^*} u.$$

By noting that $\{x_n^*\}$ satisfies Eq.(35), we see that

$$(40) \qquad \frac{dg}{du} = \frac{\beta x_{n-1}^* e^u}{1 + \beta x_{n-1}^* e^u} - \frac{\beta x_{n-1}^* x_{n+1}^*}{\alpha x_n^*} = \frac{\beta x_{n-1}^* e^u}{1 + \beta x_{n-1}^* e^u} - \frac{\beta x_{n-1}^*}{1 + \beta x_{n-1}^*}.$$

Since $x/(x+1)$ is an increasing function for $x > 0$, it follows that

$$\frac{dg}{du} < 0 \quad \text{for} \quad u < 0.$$

As $g(n, 0) \equiv 0$, we see that for $u < 0$,

$$f(n, u) \geq \frac{\beta x_{n-1}^* x_{n+1}^*}{\alpha x_n^*} u.$$

In addition, observe that

$$\frac{df}{du} = \frac{\beta x_{n-1}^* e^u}{1 + \beta x_{n-1}^* e^u} > 0$$

and

$$\frac{df(n, 0)}{du} = \frac{\beta x_{n-1}^*}{1 + \beta x_{n-1}^*} = \frac{\beta x_{n-1}^* x_{n+1}^*}{\alpha x_n^*}.$$

Hence, f(n,u) is increasing in u, $f(n, u) < 0$ for $u < 0$, and

$$\lim_{u \to 0} \frac{f(n, u)}{\frac{\beta x_{n-1}^* x_{n+1}^*}{\alpha x_n^*} u} \equiv 1.$$

Then by Theorem 3, every solution of Eq.(37) oscillates if and only if every solution of the linear equation

$$(41) \qquad w_{n+1} - w_n + \frac{\beta x_{n-1}^* x_{n+1}^*}{\alpha x_n^*} w_{n-1} = 0$$

oscillates. But, by the result in [4] which was mentioned above,

$$x_n^* \to \frac{\alpha - 1}{\beta} \quad \text{as} \quad n \to \infty$$

and so

$$\frac{\beta x_{n-1}^* x_{n+1}^*}{\alpha x_n^*} \to \frac{\alpha - 1}{\alpha} \quad \text{as} \quad n \to \infty.$$

Hence, by Lemma 2, every solution of Eq.(41) oscillates if and only if every solution of the equation

$$(42) \qquad\qquad y_{n+1} - y_n + \frac{\alpha - 1}{\alpha} y_{n-1} = 0$$

oscillates. However, it is well–known that every solution of Eq.(41) oscillates if and only if $(\alpha - 1)/\alpha > 1/4$, that is,

$$\alpha > 4/3.$$

The proof is complete.

REFERENCES

1. L. H. Erbe and B. G. Zhang, *Oscillation of discrete analogues of delay equations*, Differential Integral Equations **2** (1989), 300–309.
2. I. Györi and G. Ladas, *Linearized oscillations for equations with piecewise constant arguments*, Differential Integral Equations **2** (1989), 123–131.
3. V. Lj. Kocic and G. Ladas, *Linearized oscillations for difference equations*, Hiroshima Math. J. (to appear).
4. S. A. Kuruklis and G. Ladas, *Oscillation and global attractivity in a discrete delay logistic model*, Quart. Appl. Math. (to appear).
5. G. Ladas, *Explicit conditions for the oscillation of difference equations*, J. Math. Anal. Appl. **153** (1990), 276–287.
6. G. Ladas, *Recent developments in the oscillation of delay difference equations*, in Differential Equations (Colorado Springs, CO, 1989), Lecture Notes in Pure and Applied Math., Vol. 127, Dekker, 1991, pp. 321–332.
7. G. Ladas, Ch. G. Philos and Y. G. Sficas, *Sharp conditions for the oscillation of delay difference equations*, J. Applied Math. Simu. **2** (1989), 101–112.
8. G. Ladas, Ch. G. Philos and Y. G. Sficas, *Necessary and sufficient conditions for the oscillation of difference equations*, Libertas Math. **9** (1989), 121–125.
9. U. Morimoto, *Bifurcation diagram of recurrence equation $X(t + 1) = AX(t)(1 - X(t) - X(t - 1) - X(t - 2))$*, J. Phys. Soc. Japan **53** (1983), 2460–2463.
10. J. Yan and C. Qian, *Oscillation and comparison results for delay difference equations*, J. Math. Anal. Appl. (to appear).

DEPARTMENT OF MATHEMATICS, THE UNIVERSITY OF RHODE ISLAND, KINGSTON, RI 02881-0816

DEPARTMENT OF MATHEMATICS, THE UNIVERSITY OF RHODE ISLAND, KINGSTON, RI 02881-0816

Contemporary Mathematics
Volume **129**, 1992

Small Solutions and Completeness for Linear Functional Differential Equations

S. M. VERDUYN LUNEL

ABSTRACT. In this paper we present a class of unbounded operators that in particular contains the generator for functional differential equations. It is possible for this class of operators to state necessary and sufficient conditions in order to have a complete system of eigenfunctions and generalized eigenfunctions. As an illustration of the abstract results, we consider both autonomous and periodic functional differential equations. In particular, if the delays are integer multiples of the period, we present necessary and sufficient conditions in order to have a complete system of Floquet solutions.

0. Introduction

For linear functional differential equations precise information is known about the spectral data of the infinitesimal generator, such as, necessary and sufficient conditions for completeness of the system of eigenfunctions and generalized eigenfunctions, an explicit representation for the eigenfunctions and generalized eigenfunctions etc. (see [9]). Two ingredients played a crucial role in this analysis: a representation of the resolvent as a quotient of entire functions of exponential type and the existence of a characteristic matrix such that the spectrum of the generator is precisely given by the roots of the determinant of this characteristic matrix.

Based on the results for functional differential equations, we have been working to define classes of unbounded operators for which similar results can be proven. In [11] we presented a class of infinitesimal generators such that the resolvent of an element from this class can be represented as the quotient of entire functions of finite exponential type. For elements of this class we proved a complete characterization of the closure of the generalized eigenspace. In [7] we gave conditions which define an abstract class of unbounded operators such that

---

1991 *Mathematics Subject Classification*. Primary 34K05, 45E10.

to each element of this class one can associate a characteristic matrix. For operators of this class the spectral data of the generator and the characteristic matrix are equivalent. In particular, the eigenfunctions and generalized eigenfunctions of the generator can be obtained from the Jordan chains of the characteristic matrix. The results can be directly applied to different types of problems, like neutral functional differential equations, models from age-dependent population dynamics and certain hyperbolic partial differential equations (see the examples in [7] and [11]).

In several applications one would like to study the closure of the generalized eigenspace for unbounded operators that do not necessarily generate a semigroup. For example, given a compact operator T on a complex Banach space X, the question of completeness of the system of eigenfunctions and generalized eigenfunctions arises naturally. If T is one-to-one, the inverse $A = T^{-1}$ is an unbounded operator and it suffices to analyse the spectral data of A. In this paper we again consider the class of unbounded operators presented in [7] and show that there are natural additional assumptions such that the resolvent of an element of this class can be represented as the quotient of entire functions of exponential type. This will clarify the relation between the operators studied in [7] and [11] and we will again prove a characterization of the closure of the generalized eigenspace for this type of operator.

The motivating example will be retarded functional differential equations. We will consider autonomous equations of the form

$$(0.1) \qquad\qquad \dot{x}(t) = Lx_t, \qquad x_0 = \varphi \quad t \geq 0,$$

where φ belongs to X, the state space of continuous functions on the interval $[-h, 0]$ endowed with the supremum norm. The state x_t is given by $x_t(\theta) = x(t + \theta)$, $-h \leq \theta \leq 0$ and $L : X \to \mathbb{C}^n$ is a continuous mapping. From the Riesz representation theorem we can write

$$L\varphi = \int_0^h d\zeta(\theta)\varphi(-\theta),$$

where ζ is an $n \times n$-matrix valued function of bounded variation. In Section 5 we shall apply the results to the period map of a system of periodic functional differential equations given by

$$(0.2) \qquad\qquad \dot{x}(t) = \sum_{j=0}^{m} B_j(t)x(t - j\omega), \qquad x_s = \varphi, \quad t \geq s,$$

where B_j are continuous matrix valued functions such that $B_j(t + \omega) = B_j(t)$ and $\varphi \in C[-m\omega, 0]$. In the literature this equation has been studied before. In [4] Hahn proved that the system of Floquet solutions is complete if for all values of t the eigenvalues of $B_m(t)$ are simple and positive. Using the abstract theory we find necessary and sufficient conditions which guarantee that the system of

Floquet solutions is complete. Furthermore, we study the existence of small solutions for this type of equation.

The solution map of (0.1), given by $T(t)\varphi = x_t(\,\cdot\,;\varphi)$, is a strongly continuous semigroup with infinitesimal generator

$$D(A) = \{\varphi \in X : \dot{\varphi} \in X,\ \dot{\varphi}(0) = L\varphi\}, \quad A\varphi = \dot{\varphi}.$$

Note that the action of the generator is independent of the equation. Hence perturbing equation (0.1) means perturbing the domain of A. To develop a good perturbation theory for (0.1) one would like the domain of the generator to be independent of the equation. A natural approach is to try to extend the operator A to an operator $\widehat{A}$ defined on a larger space Z such that $D(\widehat{A}) \subset X$ and A is the part of $\widehat{A}$ in X, i.e.,

$$D(A) = \{\varphi \in X : \varphi \in D(\widehat{A}),\ \widehat{A}\varphi \in X\}, \quad A\varphi = \widehat{A}\varphi.$$

A simple choice would be $Z = \mathbb{R}^n \times C[-h, 0]$ provided with the product topology and with the embedding

$$i : X \to Z, \qquad \varphi \mapsto (\varphi(0), \varphi).$$

For the operator $\widehat{A}(Z \to Z)$ we take

$$(0.3) \qquad D(\widehat{A}) = \{(c, \varphi) \in Z : \dot{\varphi} \in X,\ c = \varphi(0)\}, \qquad \widehat{A}(c, \varphi) = (L\varphi, \dot{\varphi}).$$

Then the domain of $\widehat{A}$ is independent of the equation and A is precisely the part of $\widehat{A}$ in X. For the operator $\widehat{A}$ there exists a perturbation theory. The mapping

$$B : Z \to Z, \quad (c, \varphi) \mapsto (L\varphi, 0)$$

is a bounded perturbation of the unperturbed generator

$$D(\widehat{A}_0) = \{(c, \varphi) \in Z : \dot{\varphi} \in X,\ c = \varphi(0)\}, \qquad \widehat{A}_0(c, \varphi) = (0, \dot{\varphi})$$

corresponding to the equation $\dot{x}(t) = 0$ considered as a delay equation. Let $\widehat{T}_0(t)$ denote the strongly continuous semigroup generated by $\widehat{A}_0$ on Z. The variation of constants formula for the semigroup generated by $\widehat{A}$ is then given by

$$(0.4) \qquad \widehat{T}(t)(c, \varphi) = \widehat{T}_0(t)(c, \varphi) + \int_0^t \widehat{T}_0(t - s)B\widehat{T}(s)(c, \varphi)\, ds.$$

The integral on the right hand side belongs to $D(\widehat{A}_0)$ and we can restrict (0.4) to initial values $(\varphi(0), \varphi)$ belonging to X to obtain a variation of constants formula for $T(t)$ on X.

A systematic approach using perturbed dual semigroups is given by Clément, Diekmann et al [2]. Let $T_0(t)$ denote an unperturbed semigroup. The dual semigroup $T_0^*(t)$ is weak$*$ continuous. Now take the largest invariant subspace $X^\odot$ on which $T_0^*(t)$ is strongly continuous ($X^* \cong X^\odot$ if and only if X is reflexive). Let $T_0^\odot$ denote the restriction of $T_0^*(t)$ to $X^\odot$. Repeating the process we find a weak$*$ continuous semigroup $T_0^{\odot*}(t)$ on $X^{\odot*}$. Again take the largest invariant subspace $X^{\odot\odot}$ on which $T_0^{\odot*}(t)$ is strongly continuous. From the construction $X^{\odot\odot} \subset X$ and X is called sun-reflexive with respect to $T_0(t)$ if $X \cong X^{\odot\odot}$. If X is sun-reflexive, then any bounded perturbation from X into $X^{\odot*}$ of $A_0^{\odot*}$ generates a strongly continuous semigroup on X given by

$$(0.5) \qquad T(t) = T_0(t) + \int_0^t T_0^{\odot*}(t - s)BT(s)\,ds.$$

Here the integral must be interpreted as a weak* integral. Even though the map B takes X into $X^{\odot*}$, the convolution integral is in X. If $T_0(t)$ denotes the unperturbed semigroup associated with the equation $\dot{x}(t) = 0$ on X. Then $X^{\odot*} = \mathbb{R}^n \times L^\infty[-h, 0]$ and X is sun-reflexive with respect to $T_0(t)$. So if

$$B : X \to X^{\odot*}, \quad \varphi \mapsto (L\varphi, 0),$$

then the variation of constants formula (0.5) yields solutions of (0.1) and we can write (0.1) as an abstract evolutionary system

$$(0.6) \qquad \frac{du}{dt} = A_0^{\odot*}u + Bu,$$

where $u(t) = (x(t), x_t)$.

We will use the abstract evolutionary system (0.6) to study the completeness of the system of eigenfunctions and generalized eigenfunctions for the generator of (0.1) directly. Some of our results have been obtained earlier by rewriting the delay equation (0.1) as a Volterra convolution equation (see [8], [10]). Here we will only use the resolvent of the generator. This has the advantage that the approach becomes direct and that it becomes clear how to apply the techniques to different types of problems. As a first step towards more general problems, we study the period map for periodic delay equations, where the delays are integer multiples of the period, and prove a spectral structure similar to the one for autonomous equations.

The organization of this paper is as follows. In Section 1 we recall a general scheme to construct characteristic matrices. In Section 2 we take the class of unbounded operators from Section 1 and prove the characterization of the closure of the generalized eigenspace for elements of this class. Section 3 contains an illustration of the abstract results. It is shown that the infinitesimal generator of a delay equation satisfies the hypotheses. In Section 4 the results from the previous sections will be combined and necessary and sufficient conditions for completeness of the system of eigenfunctions and generalized eigenfunctions for the infinitesimal generator of a delay equation are given. Furthermore, the relation between completeness and the existence of small solutions is discussed. Finally, in Section 5, the abstract results will be applied to periodic delay equations and we prove necessary and sufficient conditions for completeness of the system of Floquet solutions.

1. Characteristic Matrices and Unbounded Operators

In this section we recall a general scheme to construct characteristic matrices for a rather general class of unbounded operators as was first described in [7], where we made use of three auxiliary operators D, L and M.

The operator $D(X \to X)$ is a closed linear operator acting on a complex Banach space X and D is assumed to satisfy the following two conditions:

(H1) $\mathcal{N} := Ker(D)$ is finite dimensional and $\mathcal{N} \neq \{0\}$;

(H2) The operator D has a restriction $D_0(X \to X)$ such that

    (i) $D(D) = \mathcal{N} \oplus D(D_0)$,

    (ii) $\rho(D_0) = \mathbb{C}$.

Apart from D, we need two bounded linear operators

$$(1.1) \qquad L : X_D \to \mathbb{C}^n, \qquad M : X \to \mathbb{C}^n.$$

Here $n = \dim \mathcal{N}$ and X_D is the domain of D endowed with the graph norm. One may think about D as a maximal operator and about L and M as generalized boundary value operators.

With D, L and M we associate two operators $A(X \to X)$ and $\widehat{A}(Z \to Z)$ with

$$(1.2) \qquad Z = \mathbb{C}^n \times X = \left\{ \begin{pmatrix} c \\ \varphi \end{pmatrix} : c \in \mathbb{C}^n, \ \varphi \in X \right\}.$$

Their definitions are as follows

$$(1.3) \qquad D(A) = \{ \varphi \in D(D) : MD\varphi = L\varphi \}, \quad A\varphi = D\varphi$$

and

$$(1.4) \qquad D(\widehat{A}) = \left\{ \begin{pmatrix} c \\ \varphi \end{pmatrix} \in Z : \varphi \in D(D), \ c = M\varphi \right\}, \quad \widehat{A} \begin{pmatrix} c \\ \varphi \end{pmatrix} = \begin{pmatrix} L\varphi \\ D\varphi \end{pmatrix}.$$

The operators A and $\widehat{A}$ are well defined closed linear operators and are closely related. In fact, A is similar to the part of $\widehat{A}$ in the graph of M. We shall refer to A and $\widehat{A}$, respectively, as the *first* and *second operator associated with* D, L and M.

Next, we define a candidate for the characteristic matrix function Δ. Let $j : \mathbb{C}^n \to \mathcal{N}$ be some isomorphism, and set

$$(1.5) \qquad \Delta(z) = -(zM - L)D_0(z - D_0)^{-1}j, \quad z \in \mathbb{C}.$$

Here D_0 is the operator appearing in Hypothesis (H2).

THEOREM 1.1. *Suppose that $\widehat{A}(Z \to Z)$ is the second operator associated with D, L and M. Then the matrix function Δ defined in (1.5) is a characteristic matrix for $\widehat{A}$ and the equivalence is given by*

$$(1.6) \qquad F(z)(z - \widehat{A})E(z) = \begin{pmatrix} \Delta(z) & 0 \\ 0 & I_X \end{pmatrix}, \quad z \in \mathbb{C},$$

where $E : \mathbb{C} \to \mathcal{L}(Z, Z_{\widehat{A}})$ *and* $F : \mathbb{C} \to \mathcal{L}(Z)$ *are holomorphic operator functions whose values are bijective mappings. Furthermore, these operators and their inverses have the following representations*

$$E(z) \begin{pmatrix} c \\ \varphi \end{pmatrix} = \begin{pmatrix} -MD_0(z - D_0)^{-1}jc + M(z - D_0)^{-1}\varphi \\ -D_0(z - D_0)^{-1}jc + (z - D_0)^{-1}\varphi \end{pmatrix},$$

$$E(z)^{-1} \begin{pmatrix} M\psi \\ \psi \end{pmatrix} = \begin{pmatrix} j^{-1}(\psi - D_0^{-1}D\psi) \\ (z - D)\psi \end{pmatrix}$$

and

$$F(z) \begin{pmatrix} c \\ \varphi \end{pmatrix} = \begin{pmatrix} c - zM(z - D_0)^{-1}\varphi + L(z - D_0)^{-1}\varphi \\ \varphi \end{pmatrix},$$

$$F(z)^{-1} \begin{pmatrix} M\varphi \\ \varphi \end{pmatrix} = \begin{pmatrix} c + zM(z - D_0)^{-1}\varphi - L(z - D_0)^{-1}\varphi \\ \varphi \end{pmatrix}.$$

To prove Theorem 1.1, the condition (H2)(ii) can be relaxed to $\Omega := \rho(D_0) \neq \emptyset$ and the equivalence relation (1.6) would hold for $z \in \Omega$ (see [**7**]). In our application we need the equivalence relation (1.6) on the whole complex plane. Consequently, the operator $\widehat{A}$ only has a point spectrum. For $\lambda \in \sigma(\widehat{A})$ the Riesz spectral projection

$$P_\lambda = \frac{1}{2\pi i} \int_{\Gamma_\lambda} (z - \widehat{A})^{-1} \, dz,$$

where $\Gamma_\lambda = \{z : |z - \lambda| = \eta\}$ encloses only λ of the discrete set $\sigma(\widehat{A})$, induces a direct sum decomposition

$$Z = \mathcal{M}_\lambda \oplus \mathcal{R}_\lambda.$$

Here $\mathcal{M}_\lambda = Im(P_\lambda)$ and $\mathcal{R}_\lambda = Ker(P_\lambda)$ are closed subspaces. The subspace $\mathcal{M}_\lambda$ is contained in $D(A)$ and $A\mathcal{M}_\lambda \subset \mathcal{M}_\lambda$. The dimension of $\mathcal{M}_\lambda$ is called the *algebraic multiplicity* of λ. In particular, the point spectrum of A equals the point spectrum of $\widehat{A}$ and the space $\mathcal{M}_\lambda$ is the same for both A and $\widehat{A}$. The subspace $\mathcal{M}$ generated by the subspaces $\mathcal{M}_\lambda$, $\lambda \in \sigma(\widehat{A})$ is called the *generalized eigenspace*.

COROLLARY 1.2. *The spectrum of the operator* $\widehat{A} : Z \to Z$ *defined by* (1.4) *consists of eigenvalues of finite type only,*

(1.7) $\sigma(\widehat{A}) = \{\lambda : \det \Delta(\lambda) = 0\},$

where Δ *is given by* (1.5). *For* $\lambda \in \sigma(\widehat{A})$, *the algebraic multiplicity of the eigenvalue* λ *equals the order of* λ *as a zero of* $\det \Delta$.

REMARK 1.3. The equivalence relation (1.6) yields, in fact, a complete equivalence for the spectral data of $\widehat{A}$ and Δ. In particular, the partial multiplicities of an eigenvalue λ are equal to the zero-multiplicities of λ as a characteristic value of Δ, and the largest partial multiplicity (the ascent) of λ equals the order of λ as a pole of Δ^{-1}. Furthermore, a canonical basis of eigenvectors and generalized eigenvectors for $\widehat{A}$ at λ may be obtained from the canonical system of Jordan chains for Δ at λ. See [**7**] for details.

2. Completeness for Unbounded Operators

Recall the order and exponential type of an entire function. Let X be a complex Banach space. An entire function $F : \mathbb{C} \to X$ is of order ρ if and only if

$$\limsup_{r \to \infty} \frac{\log \log M(r)}{\log r} = \rho,$$

where

$$M(r) = \max_{0 \leq \theta \leq 2\pi} \left\{ \|F(re^{i\theta})\| \right\}.$$

An entire function of order at most 1 is of exponential type if and only if

$$\limsup_{r \to \infty} \frac{\log M(r)}{r} = E(F),$$

where $0 \leq E(F) < \infty$. In that case, $E(F)$ is called the exponential type of F.

In a recent paper [11], we have studied the closure of the generalized eigenspace for a class of (unbounded) infinitesimal generators. The main hypothesis was a condition on the resolvent of the operator. To be precise, if $A(X \to X)$ denotes a (semigroup) generator, we assumed that the resolvent of A satisfies

$$R(z, A)\varphi = \frac{1}{q(z)} P(z)\varphi,$$

where $P(z) : X \to X$ is a holomorphic operator function and $q : \mathbb{C} \to \mathbb{C}$ is an entire function of finite exponential type with at most polynomial growth on the imaginary axis. This condition was motivated by our studies of delay equations. (See Corollary 3.3.). Given this assumption, the closure of the generalized eigenspace of A is given by

$$(2.1) \qquad \overline{\mathcal{M}} = \left\{ \varphi \in X : E(z \mapsto P(z)\varphi) \leq E(q) \right\}.$$

In this section we consider the class of operators from Section 1 and show that this class gives rise to a condition on the resolvent as above in a natural way.

Suppose that $\widehat{A} : Z \to Z$ is the second operator associated with D, L and M as we defined in Section 1. To characterize the closure of the generalized eigenspace of $\widehat{A}$ we need two additional hypotheses.

(H3) For all $\varphi \in X$, the function $z \mapsto (z - D_0)^{-1}\varphi$ is an entire function of finite exponential type that is bounded on the imaginary axis.

(H4) For every $\epsilon > 0$, there exists a $C > 0$ such that $\det \Delta$ satisfies the inequality

$$| \det \Delta(z) | \geq C e^{-E(\det \Delta(z))\Re z} \qquad \text{as} \quad | z | \to \infty,$$

outside circles of radius ϵ centered around the zeros of $\det \Delta$.

The third hypothesis implies that the resolvent of $\widehat{A}$ can be represented as a quotient of two entire functions of finite exponential type

$$(2.2) \qquad R(z, \widehat{A})\varphi = \frac{1}{\det \Delta(z)} P(z)\varphi,$$

where $P(z) : Z \to Z$ is a holomorphic operator function. So, if $\widehat{A}$ is the generator of a strongly continuous semigroup, then (2.1) yields a characterization for the closure of the generalized eigenspace of $\widehat{A}$. In several applications one would like to study the closure of the generalized eigenspace for unbounded operators that do not necessarily generate a semigroup (see Section 5). It turns out that the generator assumption can be easily removed and that (2.1) holds for $\widehat{A}$ as well. As an illustration, we shall prove this result in a simple case and for this reason we assume the (redundant) hypothesis (H4). A similar approach as followed in [**11**], using the estimates for Blaschke products, can be used to prove the theorem without assuming (H4).

Define

$$\mathcal{E} = \{\varphi \in Z : E(z \mapsto P(z)\varphi) \leq E(\det \Delta)\}.$$

We will show that the fourth hypothesis implies that for $\varphi \in \mathcal{E}$ the function $z \mapsto R(z, \widehat{A})\varphi$ is $O(|z|^n)$ as $|z| \to \infty$. This enables us to prove the following theorem.

THEOREM 2.1. *Let* $\widehat{A}(Z \to Z)$ *be the second operator associated with* D, L *and* M. *If* (H1)-(H4) *are satisfied and* $\dim \mathcal{N} = n$, *then:*

(i) *If* $\varphi \in \mathcal{E}$ *and* $\psi \in R(\mu, \widehat{A})^n \varphi$, $\mu \in \rho(\widehat{A})$, *then*

$$\psi = \lim_{r \to \infty} \sum_{\lambda \in \sigma(\widehat{A}), |\lambda| \leq r} P_\lambda \psi;$$

(ii) $\overline{\mathcal{M}} = \mathcal{E};$

(iii) *Define* $\mathcal{S} = \cap_{\lambda \in \sigma(\widehat{A})} Ker(P_\lambda) = \{\varphi : z \mapsto R(z, \widehat{A})\varphi$ *is entire* $\}$. *Then*

$$Z = \overline{\mathcal{M} \oplus \mathcal{S}}.$$

PROOF. (i). The cofactors of Δ are polynomials in z with coefficients that are entire functions of finite exponential type. Using the equivalence relation (1.6) we represent the resolvent of $\widehat{A}$ by

$$R(z, \widehat{A}) = E(z) \begin{pmatrix} \Delta(z)^{-1} & 0 \\ 0 & I \end{pmatrix} F(z), \quad z \in \rho(\widehat{A}).$$

Hence, it follows from (H4) and the definition of $\mathcal{E}$ that for $\varphi \in \mathcal{E}$

(2.3) $$\|R(z, A)\varphi\| \leq C_1 |z|^n \quad \text{as} \quad |z| \to \infty.$$

From the resolvent equation

(2.4) $$R(\lambda_1, A) - R(\lambda_2, A) = (\lambda_2 - \lambda_1)R(\lambda_1, A)R(\lambda_2, A), \quad \lambda_1, \lambda_2 \in \rho(\widehat{A}),$$

it follows that for $\lambda \in \rho(\widehat{A})$

$$\varphi \in \mathcal{E} \quad \text{if and only if} \quad R(\lambda, \widehat{A})\varphi \in \mathcal{E}$$

(see Lemma 2.2 in [11]). Therefore, repeated applications of the resolvent equation (2.4) yields

$$(2.5) \qquad R(z, \widehat{A})\psi = (\lambda - z)^{-(n+1)} \Big(R(z, \widehat{A})\varphi - \sum_{j=0}^{n} (\lambda - z)^j R(\lambda, A)^{j+1}\varphi \Big).$$

This implies that $\|R(z, \widehat{A})\psi\| \leq C_2 |z|^{-1}$ and, by [3, 7.3.1],

$$\lim_{R \to \infty} \Big\| \int_{|z|=R} R(z, \widehat{A})\psi \, dz \Big\| = 0.$$

This completes the proof of (i).

(ii). Theorem 2.3 of [11] yields that $\mathcal{E}$ is a closed subset of X. Since for $\varphi \in \mathcal{M}$, the function $z \mapsto R(z, \widehat{A})\varphi$ is rational, we conclude that $\overline{\mathcal{M}} \subset \mathcal{E}$. On the other hand, part (i) gives

$$R(\lambda, \widehat{A})^n \mathcal{E} \subset \overline{\mathcal{M}} \cap D(\widehat{A}^n).$$

Thus

$$\mathcal{E} \subset (\lambda - \widehat{A})^n \big(\overline{\mathcal{M}} \cap D(\widehat{A}^n) \big) \subset \overline{\mathcal{M}}.$$

(iii). Suppose $\varphi \in \overline{\mathcal{M}} \cap \mathcal{S}$ and $\varphi \neq 0$. Using the invariance of $\overline{\mathcal{M}}$ and $\mathcal{S}$

$$R(\lambda, \widehat{A})^n \varphi \in \overline{\mathcal{M}} \cap \mathcal{S}.$$

Therefore, it follows from part (ii) that

$$R(\lambda, \widehat{A})^n \varphi = \sum_{\mu \in \sigma(\hat{A})} P_\mu R(\lambda, \widehat{A})^n \varphi = R(\lambda, \widehat{A})^n \sum_{\mu \in \sigma(\hat{A})} P_\mu \varphi,$$

since P_μ and $R(\lambda, \widehat{A})$ commute. However $\varphi \in \mathcal{S}$ and hence $P_\mu \varphi = 0$ for every $\lambda \in \sigma(\hat{A})$. This proves $\overline{\mathcal{M}} \cap \mathcal{S} = \{0\}$. To prove the density of the direct sum, we use duality. From the Neumann series for the resolvent

$$\mathcal{S} = \bigcap_{\lambda \in \sigma(\hat{A})} Ker(P_\lambda).$$

So the density of $\overline{\mathcal{M}} \oplus \mathcal{S}$ is equivalent to $\overline{\mathcal{M}}^* \cap \mathcal{S}^* = \{0\}$, where $\mathcal{M}^*$ is the linear subspace generated by $Im(P_\lambda^*)$, $\lambda \in \sigma(\widehat{A}^*)$, and

$$\mathcal{S}^* = \bigcap_{\lambda \in \sigma(\hat{A}^*)} Ker(P_\lambda^*).$$

Since the characteristic matrix for $\widehat{A}^*$ is $\Delta(z)^T$ and the hypotheses are invariant under duality, a similar argument yields $\overline{\mathcal{M}}^* \cap \mathcal{S}^* = \{0\}$, from which the theorem follows. $\square$

3. The Characteristic Matrix for FDE

Recall from the introduction that the retarded functional differential equation

$$(3.1) \qquad \dot{x}(t) = Lx_t, \qquad x_0 = \varphi, \quad t \geq 0$$

can be realized as an abstract Cauchy problem on the space $X^{\odot *} = \mathbb{R}^n \times L^\infty[-h, 0]$,

$$(3.2) \qquad \frac{du}{dt}(t) = A^{\odot *}u(t) = A_0^{\odot *}u(t) + Bu(t)$$

with $u(0) = u_0 \in D(A_0^{\odot *})$ and where

$$D(A_0^{\odot *}) = \{(c, \varphi) : \dot{\varphi} \in L^\infty[-h, 0], \ c = \varphi(0)\}, \quad A_0^{\odot *}(c, \varphi) = (0, \dot{\varphi})$$

and

$$D(B) = \{(c, \varphi) : \varphi \in C[-h, 0], \ c = \varphi(0)\}, \quad B(c, \varphi) = (L\varphi, 0).$$

Furthermore, the semigroup $\{T(t)\}$ associated with (3.1) defined by translation along the solution is generated by the part of $A_0^{\odot *} + B$ in X.

To apply the results of Section 1 to $A_0^{\odot *} + B$, we start with a lemma.

LEMMA 3.1. *Let* $D : D(D) \to L^\infty[-h, 0]$ *be the second operator defined by*

$$D(D) = \{\varphi \in L^\infty[-h, 0] : \dot{\varphi} \in L^\infty[-h, 0]\}, \quad D\varphi = \dot{\varphi}.$$

Then D *satisfies the hypothesis* (H1), (H2). *Furthermore, the operator* $A_0^{\odot *} + B$ *is the second operator associated with* D, L *and* M, *where*

$$(3.3) \qquad M\varphi = \varphi(0), \quad L\varphi = \int_0^h d\zeta(\theta)\varphi(-\theta) \quad and \quad Z = X^{\odot *}.$$

PROOF. Put $Y = L^\infty[-h, 0]$. Clearly, the kernel of D consists of the constant functions. It follows that $\mathcal{N} = Ker(D)$ has dimension n. For D_0 we take the operator $D_0 : D(D_0) \to Y$ defined by

$$D(D_0) = \{\varphi \in D(D) : \varphi(0) = 0\}, \quad D_0\varphi = D\varphi.$$

We have $D(D) = \mathcal{N} \oplus D(D_0)$ and for each $z \in \mathbb{C}$ the operator $z - D_0$ is invertible. The resolvent of D_0 is given by

$$(3.4) \qquad ((z - D_0)^{-1}\varphi)(\theta) = -\int_0^\theta e^{(\theta - \sigma)z}\varphi(\sigma)d\sigma, \quad -h \leq \theta \leq 0.$$

Thus D satisfies (H1),(H2) with $\Omega = \mathbb{C}$. The second operator $\widehat{A}$ associated with D, L and M is given by

$$(3.5) \qquad D(\widehat{A}) = \{\begin{pmatrix} c \\ \varphi \end{pmatrix} \in Z : \varphi \in D(D), \ c = \varphi(0)\}, \quad \widehat{A}\begin{pmatrix} c \\ \varphi \end{pmatrix} = \begin{pmatrix} L\varphi \\ D\varphi \end{pmatrix},$$

where L, M and Z are defined by (3.3). So, $A_0^{\odot *} + B$ is the second operator associated with D, L and M. $\square$

THEOREM 3.2. *The matrix function* $\Delta : \mathbb{C} \to \mathcal{L}(\mathbb{C}^n)$ *defined by*

(3.6) $$\Delta(z) = zI - \int_0^h e^{-z\theta} d\zeta(\theta)$$

is a characteristic matrix for $A_0^{\odot *} + B$. *The equivalence relation is given by*

$$\begin{pmatrix} \Delta(z) & 0 \\ 0 & I \end{pmatrix} = F(z)(z - A_0^{\odot *} - B)E(z), \quad z \in \mathbb{C},$$

where $E : \mathbb{C} \to \mathcal{L}(X^{\odot *}, X^{\odot *}_{A_0^{\odot *}+B})$ *is given by*

$$E(z) \begin{pmatrix} c \\ \varphi \end{pmatrix} = \begin{pmatrix} c \\ \psi \end{pmatrix}, \quad \psi(\theta) = e^{\theta z} c - \int_0^\theta e^{(\theta-\sigma)z} \varphi(\sigma) d\sigma,$$

$$E(z)^{-1} \begin{pmatrix} \psi(0) \\ \psi \end{pmatrix} = \begin{pmatrix} \psi(0) \\ (z - D)\psi \end{pmatrix}$$

and $F : \mathbb{C} \to \mathcal{L}(X^{\odot *})$ *is given by*

$$F(z) \begin{pmatrix} c \\ \varphi \end{pmatrix} = \begin{pmatrix} c + L(z - D_0)^{-1}\varphi \\ \varphi \end{pmatrix},$$

$$F(z)^{-1} \begin{pmatrix} c \\ \varphi \end{pmatrix} = \begin{pmatrix} c - L(z - D_0)^{-1}\varphi \\ \varphi \end{pmatrix}.$$

PROOF. Let D and D_0 be as in the proof of Lemma 3.1. Define $j : \mathbb{C}^n \to \mathcal{N}$, $\mathcal{N} = Ker(D)$, by

$$(jc)(\theta) = c, \quad -h \le \theta \le 0.$$

From Theorem 1.1 we know that

$$\Delta(z) = (zM - L)(j - z(z - D_0)^{-1}j), \quad z \in \mathbb{C}$$

is a characteristic matrix for A. To verify the concrete representation (3.6) we use the resolvent formula (3.4) for D_0 and calculate

$$\begin{aligned}
(jc - z(z - D_0)^{-1}jc)(\theta) &= c + z \int_0^\theta e^{(\theta-\sigma)z} c\, d\sigma \\
&= c - e^{(\theta-\sigma)z} c \big|_0^\theta \\
&= e^{\theta z} c, \quad -h \le \theta \le 0.
\end{aligned}$$

The concrete representations for E and F are verified in a similar way. $\square$

COROLLARY 3.3. *The resolvent* $R(z, A_0^{\odot *} + B)$ *has the following representation*

(3.7) $$R(z, A_0^{\odot *} + B) \begin{pmatrix} c \\ \varphi \end{pmatrix} = \begin{pmatrix} \Delta(z)^{-1} c \\ e^z \, \Delta(z)^{-1} c \end{pmatrix} + \begin{pmatrix} \Delta(z)^{-1} \zeta(zI - D_0)^{-1}\varphi \\ (zI - D_0)^{-1}\varphi \end{pmatrix}$$
$$+ \begin{pmatrix} \Delta(z)^{-1} \zeta(zI - D_0)^{-1}\varphi \\ e^z \, \Delta(z)^{-1} \int_0^h d\zeta(\theta)((zI - D_0)^{-1}\varphi)(\theta)\, d\theta \end{pmatrix}.$$

We like to remark that in addition to the equivalence relation between $A_0^{\odot*}+B$ and Δ, one can also establish a more elaborate equivalence relation between A and Δ directly (see [7]). Since the spectral data of A and $A_0^{\odot*}+B$ are identical, we do not need this additional equivalence relation in this paper. Note that the equivalence relation yields a new proof for the "folk theorem" that the algebraic multiplicity of λ as an eigenvalue of A equals the multiplicity of λ as a root of $\det\Delta$. (See Levinger). Further, it yields a simple construction of the elementary solutions, i.e., $x(t) = p(t)e^{\lambda t}$, where p is a polynomial, of (3.1). (See Hale [5].) The precise details are given in [7].

Of course one can also directly compute a representation for the resolvent of the generator of (3.1) on X. From the definition

$$D(A) = \{\varphi \in X : \dot{\varphi} \in X, \ \dot{\varphi}(0) = \int_0^h d\zeta(\theta)\varphi(-\theta)\}, \quad A\varphi = \dot{\varphi},$$

an easy calculation shows that

$$\big(R(z,A)\varphi\big)(\theta) = e^{\lambda\theta}\big\{\Delta(z)^{-1}K(z)(\varphi) + \int_\theta^0 e^{-zs}\varphi(s)\,ds\big\}, \qquad (3.8)$$

$$\big(K(z)\varphi\big)(\theta) = \varphi(0) + \int_0^h d\zeta(\theta)\int_0^\theta e^{-zs}\varphi(s-\theta)\,ds.$$

Here Δ is given by (3.6).

4. Small Solutions or Completeness

A solution x of (3.1) is called small if

$$\lim_{t\to\infty} x(t)e^{kt} = 0 \qquad \text{for every} \quad k \in \mathbb{R}.$$

The zero solution is a trivial small solution; small solutions that are not identically zero are called nontrivial.

THEOREM 4.1. *The system*

$$(4.1) \qquad \dot{x}(t) = \int_0^h d\zeta(\tau)x(t-\tau), \quad x_0 = \varphi, \ \varphi \in X = C[-h,0]$$

has no small solutions if and only if $E(\det\Delta) = nh$. *Here*

$$\Delta(z) = zI - \int_0^h e^{-z\tau}d\zeta(\tau).$$

We divide the proof into several lemma's. First we analyse the small solutions of (4.1). The Laplace transform of a small solution converges everywhere, hence it is an entire function. If $x = x(\cdot\,;\varphi)$ is a small solution, then clearly

$$\lim_{t\to\infty} \|x_t(\theta)e^{k\theta}\| = 0 \qquad \text{for every} \quad k \in \mathbb{R}.$$

So

$$\int_0^\infty e^{-zt}T(t)\varphi\,dt$$

is an entire function as well. Let $S \subset X$ denote the subspace of initial conditions $\varphi \in S$ such that

$$\int_0^\infty e^{-zt} T(t)\varphi \, dt \quad \text{is entire.}$$

LEMMA 4.2. *If* $A : D(A) \to X$ *is the generator of a strongly continuous semigroup* $\{T(t)\}$, *then*

(4.2) $$S = \{\varphi \in X : z \mapsto R(z, A)\varphi \quad \text{is an entire function}\}.$$

PROOF. Let ω denote the growth bound of $T(t)$. For $\Re z > \omega$, the Laplace transform of the semigroup equals the resolvent of the generator

(4.3) $$R(z, A)\varphi = \int_0^\infty e^{-zt} T(t)\varphi \, dt \qquad \text{for} \quad \Re z > \omega, \ \varphi \in X.$$

For $\varphi \in S$, the Laplace transform of $T(t)\varphi$ is an entire function and hence (4.3) yields that the function $z \mapsto R(z, A)\varphi$ has an analytic continuation to the whole complex plane. On the other hand, if $z \mapsto R(z, A)\varphi$ is entire, then (4.3) shows that the Laplace transform of $T(t)\varphi$ has an analytic continuation to the whole complex plane. Hence $\varphi \in S$. $\square$

In particular, if the infinitesimal generator has an empty spectrum then for every φ, the solution $t \mapsto T(t)\varphi$ is a small solution.

LEMMA 4.3. *Let* $A(X \to X)$ *be the generator of a strongly continuous semigroup* $\{T(t)\}$. *Suppose*

$$R(z, A)\varphi = \frac{1}{q(z)} P(z)\varphi,$$

where $P(z) : X \to X$ *is a holomorphic operator function and* $q : \mathbb{C} \to \mathbb{C}$ *is an entire function of finite exponential type with at most polynomial growth on the imaginary axis. Then*

$$S = Ker\big(T(\eta)\big),$$

where

(4.4) $$\eta = \max_{\varphi \in Z} \{E(z \mapsto P(z)\varphi) - E(q)\}.$$

PROOF. It follows from (4.3) that $Ker\big(T(\eta)\big) \subset S$. On the other hand, for $0 \neq \varphi \in S$, the Laplace transform of $T(t)\varphi$ is given by

(4.5) $$\int_0^\infty e^{-zt} T(t)\varphi \, dt = \frac{1}{q(z)} P(z)\varphi$$

and is an entire function. However, the right hand side of (4.5) shows that this Laplace transform has finite exponential type. Therefore, there exists a $\sigma \leq \eta$, such that

$$\int_0^\infty e^{-zt} T(t)\varphi \, dt = \int_0^\sigma e^{-zt} T(t)\varphi \, dt$$

and $\varphi \in Ker\big(T(\eta)\big)$. $\square$

LEMMA 4.4. *Let $A(X \to X)$ be the generator of a strongly continuous semi-group $\{T(t)\}$. Suppose*

$$R(z, A)\varphi = \frac{1}{q(z)}P(z)\varphi,$$

where $P(z) : X \to X$ is a holomorphic operator function and $q : \mathbb{C} \to \mathbb{C}$ is an entire function of finite exponential type with at most polynomial growth on the imaginary axis. Then $T(t)$ is one-to-one if and only if $E(z \mapsto P(z)\varphi) = E(q)$ for every $\varphi \in X$.

PROOF. Since A is the generator of $T(t)$,

$$R(z, A)\big(e^{zt} - T(t)\big)\varphi = \int_0^t e^{z(t-s)}T(s)\varphi \, ds, \qquad t > 0, \ z \in \mathbb{C}.$$

So using the representation for the resolvent we find

$$(4.2) \qquad P(z)\big(1 - e^{-zt}T(t)\big)\varphi = q(z)\int_0^t e^{-zs}T(s)\varphi \, ds, \quad t \geq 0.$$

If we assume that $T(t)$ is not one-to-one, then for $\varphi \in Ker\big(T(t_0)\big)$ and $t \geq t_0$, the equation (4.2) could be written as

$$P(z)\varphi = q(z)\int_0^{t_0} e^{-zs}T(s)\varphi \, ds$$

and $E(z \mapsto P(z)\varphi) > E(q)$, a contradiction. On the other hand, suppose that $T(t)$ is one-to-one. Applying the exponential type calculus (Corollary 4.11 of [9]) to the left and right hand side of (4.2) yields

$$t + E(z \mapsto P(z)\varphi) = t + E(q).$$

Hence $E(z \mapsto P(z)\varphi) = E(q)$. $\square$

Next we analyse the exponential type of the characteristic matrix given by (2.6). Since the coefficients of Δ have exponential type of at most h, the exponential type of $\det \Delta$ is at most nh. Let $\text{adj} \, \Delta$ denote the matrix of cofactors of Δ and

$$E(\text{adj} \, \Delta) = \max_{i,j=1,\dots,n} E(\text{adj} \, \Delta_{ij}).$$

The cofactors of Δ are of exponential type at most $(n-1)h$. So $E(\text{adj} \, \Delta) \leq (n-1)h$.

LEMMA 4.5. $E(adj \, \Delta) + h > E(\det \Delta)$ *if and only if* $E(\det \Delta) < nh$.

PROOF. From the definition $\det \text{adj} \, \Delta = (\det \Delta)^{n-1}$. So, the exponential type calculus (Corollary 4.11 in [9]) implies

$$(4.3) \qquad E(\det \text{adj} \, \Delta) = (n-1)E(\det \Delta).$$

Suppose that $E(\text{adj} \, \Delta) + h \leq E(\det \Delta)$. Then

$$E(\det \text{adj} \, \Delta) \leq nE(\det \Delta) - nh,$$

and, together with (4.3), this implies $E(\det \Delta) \geq nh$ which is a contradiction. $\square$

PROOF OF THEOREM 4.1. Recall from (3.8) that the resolvent for the generator of the semigroup $T(t)\varphi = x_t(\,\cdot\,;\varphi)$ associated with (4.1) is given by

$$(4.6) \qquad R(z, A)\varphi = \frac{1}{\det \Delta(z)} P(z)\varphi,$$

where

$$(P(z)\varphi)(\theta) = e^{z\theta} \Big\{ \operatorname{adj} \Delta(z)[\varphi(0) + \int_0^h d\zeta(\tau) \int_0^\tau e^{-z\sigma}\varphi(\sigma - \tau)\,d\sigma]$$
$$+ \det \Delta(z) \int_\theta^0 e^{-z\sigma}\varphi(\sigma)\,d\sigma \Big\}.$$

To compute the exponential type of $z \mapsto P(z)$ write

$$P(z)\varphi = \operatorname{adj} \Delta(z)C(z)\varphi,$$

where

$$(C(z)\varphi)(\theta) = \varphi(0) + \int_0^h d\zeta(\tau) \int_0^\tau e^{-z(\sigma-\theta)}\varphi(\sigma - \tau)\,d\sigma$$
$$+ \Delta(z) \int_0^{-\theta} e^{-z\sigma}\varphi(\sigma + \theta)\,d\sigma$$
$$= \varphi(0) + \int_0^h d\zeta(\tau) \int_0^\tau e^{-z(\sigma-\theta)}\varphi(\sigma - \tau)\,d\sigma$$
$$+ z \int_0^{-\theta} e^{-z\sigma}\varphi(\sigma + \theta)\,d\sigma$$
$$- \int_0^h d\zeta(\tau)e^{-z\tau} \int_0^{-\theta} e^{-z\sigma}\varphi(\sigma + \theta)\,d\sigma.$$

Since the exponential type of $z \mapsto z \int_0^{-\theta} e^{-z\sigma}\varphi(\sigma + \theta)\,d\sigma$ cannot be cancelled by the other terms, the exponential type of $z \mapsto C(z)\varphi$ is at least θ. To compute the exponential type of $z \mapsto C(z)\varphi$, it suffices to compute the exponential type of

$$(4.7) \qquad \begin{aligned} z \mapsto & \int_0^h d\zeta(\tau) \int_0^\tau e^{-z(\sigma-\theta)}\varphi(\sigma - \tau)\,d\sigma \\ & - \int_0^h d\zeta(\tau)e^{-z\tau} \int_0^{-\theta} e^{-z\sigma}\varphi(\sigma + \theta)\,d\sigma. \end{aligned}$$

Changing the order of integration and substituting $s = \sigma - \theta$, we find

$$\int_0^h d\zeta(\tau) \int_0^\tau e^{-z(\sigma-\theta)}\varphi(\sigma - \tau)\,d\sigma = \int_0^h e^{-z(\sigma-\theta)} \int_\sigma^h d\zeta(\tau)\varphi(\sigma - \tau)\,d\sigma$$
$$= \int_{-\theta}^{h-\theta} e^{-zs} \int_{s+\theta}^h d\zeta(\tau)\varphi(s + \theta - \tau)\,ds.$$

For the second integral in (4.7), we take $s = \tau + \sigma$ and τ as new integration variables.

$$
\int_0^h d\zeta(\tau) e^{-z\tau} \int_0^{-\theta} e^{-z\sigma} \varphi(\sigma + \theta)\, d\sigma = \int_0^{h-\theta} e^{-zs} \int_{-\frac{1}{2}s}^{-\frac{1}{2}s+h} d\zeta(\tau) \varphi(s + \theta - \tau)\, ds
$$
$$
= \int_0^{h-\theta} e^{-zs} \int_{s+\theta}^h d\zeta(\tau) \varphi(s + \theta - \tau)\, ds,
$$

where the last identity follows from the fact that ζ is constant for $\tau \geq h$ and $\varphi(\sigma) = 0$ for $\sigma \geq 0$. Using the Paley-Wiener theorem it follows that the exponential type of (4.7) equals

$$
E\left(\int_0^{-\theta} e^{-zs} \int_{s+\theta}^h d\zeta(\tau) \varphi(s + \theta - \tau)\, ds\right) = \theta
$$

and $E(z \mapsto P(z)\varphi) \leq nh$. Furthermore, from Lemma 4.5 we conclude that $E(z \mapsto P(z)\varphi) > E(\det \Delta)$ if and only if $E(\det \Delta) < nh$. Thus the theorem follows from Lemma 4.4. $\square$

REMARK 4.6. In [8] a different proof was given by constructing a small solution if $E(\det \Delta) < nh$. It is not difficult to find kernels such that (4.4) has small solutions (see [8] and [10]).

Using the representation (3.7) it is easy to see that the same proof applies to the operator $\widehat{A}$ acting on $\mathbb{C}^n \times L^\infty[-h, 0]$. Since the condition $E(\det \Delta) = nh$ is invariant under duality, i.e., the characteristic matrix of $\widehat{A}^*$ is Δ^T and $\det \Delta = \det \Delta^T$, we have the following corollary.

COROLLARY 4.7. *The system of eigenfunctions and generalized eigenfunctions of A is complete if and only if $E(\det \Delta) = nh$.*

PROOF. From the proof of Theorem 4.1, it follows that $E(z \mapsto P(z)\varphi) > E(\det \Delta)$ if and only if $E(\det \Delta) < nh$. So the corollary follows from the characterization (2.1) for the closure of the generalized eigenspace of A. $\square$

COROLLARY 4.8. *The semigroups $T(t), T^*(t), T^\odot(t)$ and $T^{\odot *}(t)$ associated with*

$$
\dot{x}(t) = \int_0^h d\zeta(\tau) x(t - \tau)
$$

are one-to-one if and only if $E(\det \Delta) = nh$.

COROLLARY 4.9. *The system of eigenfunctions and generalized eigenfunctions of $A, A^*, A^\odot$ and $A^{\odot *}$ is complete if and only if $E(\det \Delta) = nh$.*

REMARK 4.10. From Lemma 4.5 we find that $\overline{\mathcal{M}} \cap \mathcal{S} = \{0\}$ and it is not difficult to prove the "almost decomposition"

$$
X = \overline{\mathcal{M} \oplus \mathcal{S}}.
$$

(See [11].)

We end this section with an application to oscillations. For linear scalar delay equations, it is well known that all solutions oscillate if and only if $\det \Delta$ has no real roots. A proof using Laplace transforms proceeds as follows. Suppose that a solution $x = x(\,\cdot\,;\varphi)$ does not oscillate; then we can assume that it becomes eventually positive (negative), i.e. $x(t) \geq 0$ for $t \geq T$. From Doetsch [3, 3.4.1] it follows that the Laplace transform of an eventually positive function converges in a right half plane $\Re z > \gamma$ and has a pole at the boundary point $z = \gamma$. The representation of the resolvent yields

$$\int_0^\infty e^{-zt} x(t)\, dt = \int_0^\infty e^{-zt} \big(T(t)\varphi\big)(0)\, dt = \frac{(P(z)\varphi)(0)}{\det \Delta(z)}.$$

So the poles of the Laplace transform are roots of $\det \Delta$. Hence, if $\det \Delta$ has no real roots, the Laplace transform must converge in every half plane and consequently is an entire function.

LEMMA 4.11. *Suppose* $\det \Delta$ *has no real roots. Then eventually positive solutions are small solutions.*

The connection between small solutions and eventually positive solutions is not unnatural. The condition "$\det \Delta$ has no real roots" is a condition on the spectrum and in order to control the behaviour of all solutions one needs completeness of the system of eigenfunctions and generalized eigenfunctions.

From the above results, we know that there are no small solutions if and only if $E(\det \Delta) = nh$.

COROLLARY 4.12. *The statement "All solutions oscillate if and only if* $\det \Delta$ *has no real roots" is true if and only if* $E(\det \Delta) = nh$.

It is easy to construct systems such that $E(\det \Delta) < nh$ (see [10]). Since we saw before that for autonomous systems small solutions are identically zero after finite time, we find that if $\det \Delta$ has no real roots, then all positive solutions are identically zero after finite time and one can say that eventually all solutions either oscillate or become identically zero if and only if $\det \Delta$ has no real roots.

5. Periodic Delay Equations

In this section we extend our results about the geometric structure of the solution map of linear autonomous delay equations to the period map of the following system of linear periodic delay equations

$$(5.1) \qquad \dot{x}(t) = \sum_{j=0}^m B_j(t) x(t - j\omega), \qquad x_s = \varphi, \quad t \geq s,$$

where B_j are continuous matrix valued functions such that $B_j(t + \omega) = B_j(t)$. The state at time s is given by $x_s(\theta) = x(s + \theta)$ for $-m\omega \leq \theta \leq 0$. Further φ belongs to $\mathcal{C} = C\left([-m\omega, 0]; \mathbb{R}\right)$, the space of continuous functions on the interval $[-m\omega, 0]$ provided with the supremum norm.

For every $s \in \mathbb{R}$ and $\varphi \in \mathcal{C}$, there is a unique solution $x = x(\,\cdot\,; \varphi; s)$ of (5.1) defined on $[s, \infty)$. The solution map $T(t,s) : \mathcal{C} \to \mathcal{C}$ associated with (5.1) is given by

$$T(t,s)\varphi = x_t(\,\cdot\,; \varphi; s), \quad t \geq s.$$

The family of operators $T(t,s)$ defines a evolutionary system and periodicity of (5.1) implies

$$T(t + \omega, s) = T(t,s)T(s + \omega, s).$$

Throughout this section we assume that the solution map $T(t,s)$ is one-to-one, i.e., $\det B_m$ has only isolated zeros. Let $U : \mathcal{C} \to \mathcal{C}$ denote the period map

$$U\varphi = T(\omega, 0).$$

From the general theory (e.g., Hale [5]) it follows that U^m is compact and hence the spectrum of U is at most countable. It is a compact set in the complex plane with the only possible accumulation point being zero. If $\mu \neq 0$ is in $\sigma(U)$, then μ is in the point spectrum of U and is called a *characteristic multiplier* of U. Let $\mathcal{M}_\mu$ denote the (generalized) eigenspace corresponding to μ. For $\mu \in \sigma_p(U)$ define

$$P_\mu \varphi = \frac{1}{2\pi i} \int_\Gamma R(z, U) dz$$

to be the spectral projection onto $\mathcal{M}_\mu$. This projection induces the direct sum into closed U-invariant subspaces and since U^m is compact

(i) $X = \mathcal{M}_\mu(s) \oplus Q_\mu$;
(ii) $\dim \mathcal{M}_\mu(s) < \infty$;
(iii) $\sigma(U | \mathcal{M}_\mu(s)) = \{\mu\}$.

Moreover, the characteristic multipliers are independent of s and $T(t,s)|_{\mathcal{M}_\mu(s)}$ is a diffeomorphism onto $\mathcal{M}_\mu(t)$. Solutions of (5.1) with initial value in $\mathcal{M}_\mu(s)$ are of the Floquet type, namely of the form $e^{\lambda t} p(t)$, where $\mu = e^\lambda$ and $p(t) = p(t+\omega)$.

Define $\Omega_s^t(z)$ to be the fundamental matrix solution of the periodic ODE

$$(5.3) \qquad\qquad \dot{y}(t) = \sum_{j=0}^{m} z^j B_j(t) y(t)$$

with $\Omega_s^s(z) = I$. The period map of (5.1) is given by
(5.4)

$$(U\varphi)(\theta) = \Omega_{-\omega}^\theta(0)\varphi(0) + \sum_{j=1}^{m} \int_{-\omega}^\theta \Omega_\tau^\theta(0) B_j(\tau)\varphi(\tau + 1 - j)\,d\tau, \quad -\omega \leq \theta \leq 0.$$

THEOREM 5.1. *Suppose that* $\det B_m$ *has only isolated zeros. The spectrum of the operator* $U : C[-m\omega, 0] \to C[-m\omega, 0]$ *defined by* (5.4) *consists of eigenvalues of finite type only,*

$$(5.5) \qquad\qquad \sigma(U) = \{\mu : \det \Delta(\mu) = 0\},$$

where $\Delta(z) = z - \Omega_0^\omega(z^{-1})$. For $\mu \in \sigma(U)$, the algebraic multiplicity of the eigenvalue μ equals the order of μ as a zero of $\det \Delta$.

For a proof of the theorem, we are going to apply the abstract results from Section 1. Before we can do so, we have to make some preparations. Since U^m is compact and one-to-one, the inverse of U is a well-defined unbounded closed operator on $\mathcal{C}$. In order to analyse this operator, we define $\widehat{\mathcal{C}} = C[-\omega, 0]^m$ and we make the following state space identification

$$\mathcal{C} \cong \left\{ (\varphi_1, \dots, \varphi_m)^T \in \widehat{\mathcal{C}} \mid \varphi_i(-\omega) = \varphi_{i+1}(0), \ i = 1, 2, \dots, m-1 \right\}$$

$$\varphi \mapsto \varphi_1(\cdot)|_{\chi[-\omega, 0]} + \varphi_2(-\omega + \cdot)|_{\chi[-\omega, 0]} + \cdots + \varphi_m(-(m-1)\omega + \cdot)|_{\chi[-\omega, 0]}.$$

Here $\chi[-\omega, 0]$ denotes the characteristic function on $[-\omega, 0]$.

The period map becomes $U : \mathcal{C} \to \mathcal{C}$

$$U \begin{pmatrix} \varphi_1 \\ \vdots \\ \varphi_m \end{pmatrix} = \begin{pmatrix} \Omega_{-\omega}^\theta(0)\varphi_1(0) + \int_{-\omega}^\theta \Omega_\tau^\theta \sum_{j=1}^m B_j(\tau)\varphi_j(\tau)\,d\tau \\ \varphi_1 \\ \vdots \\ \varphi_{m-1}. \end{pmatrix}$$

We shall show that $A = U^{-1}$ is the first operator associated with D, L and M. Define $D(\widehat{\mathcal{C}} \to \widehat{\mathcal{C}})$ to be the operator, $D(D) = \{\psi \in \widehat{\mathcal{C}} : D\psi \in \mathcal{C}\}$ with

(5.6) $$D \begin{pmatrix} \psi_1 \\ \vdots \\ \psi_m \end{pmatrix} = \begin{pmatrix} \psi_2 \\ \vdots \\ \psi_m \\ B_m^{-1}(\dot\psi_1 - B_0\psi_1 - B_1\psi_2 - \cdots - B_{m-1}\psi_m) \end{pmatrix}.$$

The kernel of D is given by

$$Ker(D) = \left\{ \psi \in \widehat{\mathcal{C}} : \psi_1(\theta) = \Omega_{-\omega}^\theta(0)\psi_1(-\omega), \ \psi_i = 0, \ i = 2, \dots, m \right\}$$

and (H1) is satisfied. Next we define the restriction $D_0(\widehat{\mathcal{C}} \to \widehat{\mathcal{C}})$

(5.7) $$D(D_0) = \left\{ \psi \in \widehat{\mathcal{C}} : \psi \in D(D), \ \psi_1(-\omega) = 0 \right\}, \qquad D_0\psi = D\psi.$$

To find the resolvent of D_0, put $(z - D_0)^{-1}\varphi = \psi$. This implies

$$\varphi_1 = z\psi_1 - \psi_2$$

(5.8) $$\vdots$$

$$\varphi_{m-1} = z\psi_{m-1} - \psi_m$$

$$\varphi_m = z\psi_m - B_m^{-1}(\dot\psi_1 - B_0\psi_1 - B_1\psi_2 - \cdots - B_{m-1}\psi_m).$$

Hence

(5.9) $$\psi_l = z^{l-1}\psi_1 - \sum_{j=0}^{l-2} z^j \varphi_{l-1-j}, \quad l = 1, 2, \dots, m,$$

and from (5.8) and (5.9) we find

$$\dot{\psi}_1 = \left(\sum_{j=0}^{m} z^j B_j \right) \psi_1 - \sum_{l=1}^{m} B_l \sum_{j=0}^{l-1} z^j \varphi_{l-j}.$$

The initial condition is $\psi_1(-\omega) = 0$ and we can solve for ψ_1. The corresponding formulae for $\psi_2, \ldots, \psi_m$ follow from the substitution of ψ_1 into (5.9), and $\psi = (z - D_0)^{-1}$ is given by

$$(5.10) \quad \psi_l(\theta) = -z^{l-1} \int_{-\omega}^{\theta} \Omega_\tau^\theta(z) \kappa(\tau) \, d\tau - \sum_{j=0}^{l-2} z^j \varphi_{l-1-j}(\theta), \quad l = 1, 2, \ldots, m,$$

and κ is given by

$$\kappa(\tau) = \sum_{l=1}^{m} B_l(\tau) \sum_{j=0}^{l-1} z^j \varphi_{l-j}(\tau).$$

Consequently, $\Omega = \rho(D_0) = \mathbb{C}$ and (H2) is satisfied. Finally, put $Z = \mathbb{C}^n \times \widehat{\mathcal{C}}$

$$L : \widehat{\mathcal{C}} \to \mathbb{C}^n, \qquad \varphi \mapsto \varphi_1(-\omega) \qquad\qquad (5.11)$$

$$M : \widehat{\mathcal{C}} \to \mathbb{C}^n, \qquad \varphi \mapsto \varphi_1(0). \qquad\qquad (5.12)$$

Then the operator $A = U^{-1}$ can be represented by

$$(5.13) \qquad D(A) = \left\{ \psi \in \widehat{\mathcal{C}} : \psi \in D(D), \quad MD\psi = L\psi \right\}, \qquad A\psi = D\psi.$$

Hence the operator A defined by (5.13) is the first operator associated with D, L and M. The second operator $\widehat{A}(Z \to Z)$ associated with D, L and M is given by

$$(5.14) \quad D(\widehat{A}) = \left\{ \begin{pmatrix} c \\ \psi \end{pmatrix} \in Z : \psi \in D(D), \ c = M\psi \right\}, \quad \widehat{A} \begin{pmatrix} c \\ \psi \end{pmatrix} = \begin{pmatrix} L\psi \\ D\psi \end{pmatrix}.$$

THEOREM 5.2. *The matrix function $\Delta : \mathbb{C} \to \mathcal{L}(\mathbb{C}^n)$ defined by*

$$(5.15) \qquad\qquad\qquad \Delta(z) = z\Omega_0^\omega(z) - I$$

is a characteristic matrix for $\widehat{A}$. The equivalence relation is given by

$$(5.16) \qquad\qquad \begin{pmatrix} \Delta(z) & 0 \\ 0 & I \end{pmatrix} = F(z)(z - \widehat{A})E(z),$$

where $E : \mathbb{C} \to \mathcal{L}(Z, Z_{\widehat{A}})$ is given by

$$E(z) \begin{pmatrix} c \\ \psi \end{pmatrix} = \begin{pmatrix} \Omega_{-\omega}^0(z)c - \int_{-\omega}^0 \Omega_\tau^0(z)\kappa(\tau) \, d\tau \\ \chi \end{pmatrix},$$

$$\chi_l(\theta) = z^{l-1}\Omega_{-\omega}^\theta(z)c - z^{l-1} \int_{-\omega}^{\theta} \Omega_\tau^\theta(z)\kappa(\tau) \, d\tau - \sum_{j=0}^{l-2} z^j \psi_{l-1-j}(\theta),$$

and $F : \mathbb{C} \to \mathcal{L}(Z)$ is given by

$$F(z)\begin{pmatrix} c \\ \psi \end{pmatrix} = \begin{pmatrix} c + z \int_{-\omega}^{0} \Omega_{\tau}^{0}(z)\kappa(\tau)\,d\tau \\ \psi \end{pmatrix}.$$

Here κ equals

$$\kappa(\tau) = \sum_{l=1}^{m} B_{l}(\tau) \sum_{j=0}^{l-1} z^{j}\,\psi_{l-j}(\tau).$$

PROOF. From Theorem 1.1 it follows that the characteristic matrix for $\widehat{A}$ is given by

$$\Delta(z) = (zM - L)(j - z(z - D_{0})^{-1}j),$$

where $j : \mathbb{C}^{n} \to Ker(D)$

$$c \mapsto \begin{pmatrix} \Omega_{-\omega}^{\theta}(0)c \\ 0 \\ \vdots \\ 0 \end{pmatrix}.$$

To compute $(z - D_{0})^{-1}j$ note that

$$\dot{\psi}_{1} = \Big(\sum_{j=0}^{m} B_{j}z^{j}\Big)\psi_{1} - \sum_{l=1}^{m} z^{l-1} B_{l}\Omega_{-\omega}^{\theta}(0).$$

A particular solution is given by $z^{-1}\Omega_{-\omega}^{\theta}(0)$ and from (5.7)

$$\psi_{1} = \frac{1}{z}\big(\Omega_{-\omega}^{\theta}(0) - \Omega_{-\omega}^{\theta}(z)\big).$$

So from (5.10)

$$\psi_{l} = -z^{l-2}\Omega_{-\omega}^{\theta}(z), \qquad l = 2, 3, \ldots, m.$$

Hence

(5.17)
$$j - z(z - D_{0})^{-1}j = \begin{pmatrix} \Omega_{-\omega}^{\theta}(z) \\ z(\Omega_{-\omega}^{\theta}(z)) \\ \vdots \\ z^{m-1}(\Omega_{-\omega}^{\theta}(z)) \end{pmatrix}$$

and

$$\Delta(z) = z\Omega_{-\omega}^{0}(z) - I.$$

The concrete representations for E and F are verified in a similar way. $\square$

PROOF OF THEOREM 5.1. Since U^{-1} is A, an eigenvalue μ of U corresponds to an eigenvalue μ^{-1} of A and the spectral data of U and A are the same. But this implies that the spectral data of U and $\widehat{A}$ are the same. Hence the theorem follows from Theorem 5.2 and Corollary 1.2. $\square$

In the special case that the matrices B_{j}, $j = 0, \ldots, m$, are diagonal, we find a characteristic equation for the exponents:

$$\det\big(e^{\omega\lambda} - e^{\omega}\sum_{j=0}^{m}\overline{B}_{j}e^{-j\omega\lambda}\big) = 0,$$

where

$$\overline{B}_j = \frac{1}{\omega} \int_0^\omega B_j(s)\, ds, \quad j = 0, \dots, m$$

denotes the average of B_j. So the characteristic matrix for the exponents becomes

$$\lambda = \sum_{j=0}^m \overline{B}_j e^{-j\omega\lambda}$$

which is just the characteristic matrix for the autonomous equation

$$\dot{y}(t) = \sum_{j=0}^m \overline{B}_j x(t - j\omega).$$

(See Hale [5].)

Theorem 5.2 leads to a situation where we can apply the results from Section 3. The complete results will be published elsewhere. Here we restrict our attention to the scalar case. Consider a scalar linear periodic delay equation

$$(5.18) \qquad \dot{x}(t) = \sum_{j=0}^m b_j(t) x(t - j\omega), \qquad t \geq s, \quad x_s = \varphi,$$

where b_j, $j = 0, \dots, m$, are continuous periodic functions with period ω. As before we will assume that the solution map $T(t,s) : \mathcal{C} \to \mathcal{C}$ for (5.18) is one-to-one, i.e., the zeros of b_m are isolated.

THEOREM 5.3. *Suppose that the zeros of b_m are isolated. Then the system of Floquet solutions is complete if and only if b_m has no sign change.*

As in the autonomous case there exists a relation between the completeness of the system of Floquet solutions and the existence of small solutions.

THEOREM 5.4. *Suppose that the zeros of b_m are isolated. Then (5.18) has small solutions if and only if b_m has a sign change.*

Put

$$\mathcal{S} = \{\varphi \in \mathcal{C} : T(t,s)\varphi \quad \text{is a small solution of (5.18)}\}.$$

COROLLARY 5.5. *The system of Floquet solutions is complete if and only if there are no small solutions. Furthermore,*

$$\mathcal{C} = \overline{\mathcal{M} \oplus \mathcal{S}}.$$

The equation

$$\dot{x}(t) = \left(\frac{1}{2} + \sin(2\pi t)\right) x(t - 1)$$

has small solutions, although there are infinitely many independent Floquet solutions

$$x(t) = e^{\lambda t}, \qquad \text{with} \quad \lambda = \frac{1}{2} e^{-\lambda}.$$

In the remaining part of this section we shall prove the above results. To verify the hypotheses (H3) and (H4) recall that in the scalar case

$$\Omega^\theta_{-\omega}(z) = e^{\sum_{j=0}^m z^j \int_{-\omega}^\theta b_j(s)\,ds}$$

and (H3) follows from (5.10). Furthermore

(5.19) $$\Delta(z) = \det \Delta(z) = z e^{-\omega \sum_{j=0}^m \bar{b}_j z^j} - I, \quad \bar{b}_j = \frac{1}{\omega}\int_{-\omega}^0 b_j(s)\,ds,$$

and (H4) is satisfied as well. From (5.16) we find

$$\begin{pmatrix} d \\ \psi \end{pmatrix} = R(z,\widehat{A}) \begin{pmatrix} c \\ \varphi \end{pmatrix}$$

with

$$d = \Omega^0_{-\omega}(z)\Delta(z)^{-1}\Big[c + z\int_{-\omega}^0 \Omega^0_\tau(z)\kappa(\tau)\,d\tau\Big] - \int_{-\omega}^0 \Omega^0_\tau(z)\kappa(\tau)\,d\tau$$

$$\psi_l(\theta) = z^{l-1}\Omega^\theta_{-\omega}\Delta(z)^{-1}\Big[c + z\int_{-\omega}^0 \Omega^0_\tau(z)\kappa(\tau)\,d\tau\Big] - z^{l-1}\int_{-\omega}^\theta \Omega^\theta_\tau(z)\kappa(\tau)\,d\tau$$

$$- \sum_{j=0}^{l-2} z^j \varphi_{l-1-j}(\theta), \qquad l = 1,2,\ldots,m, \quad -\omega \le \theta \le 0.$$

Here,

(5.20) $$\kappa(\tau) = \sum_{l=1}^m B_l(\tau)\sum_{j=0}^{l-1} z^j \varphi_{l-j}(\tau).$$

So, the representation (2.2) holds with P given by

$$P(z)\begin{pmatrix} c \\ \varphi \end{pmatrix} = \begin{pmatrix} p_1(z,c,\varphi,0) \\ p(z,c,\varphi,\theta) \end{pmatrix},$$

where

(5.21)
$$\begin{aligned} p_l(z,c,\varphi,\theta) =\ & z^{l-1}\big(\Omega^\theta_{-\omega}(z)\big[c + z\int_{-\omega}^0 \Omega^0_\tau(z)\kappa(\tau)\,d\tau\big] \\ & - \det\Delta(z)\int_{-\omega}^\theta \Omega^\theta_\tau(z)\kappa(\tau)\,d\tau \\ & - \det\Delta(z)\sum_{j=0}^{l-2} z^j \varphi_{l-1-j}(\theta), \qquad l = 1,2,\ldots,m. \end{aligned}$$

Here, κ is given by (5.20). Therefore, using Theorem 2.1 (ii), we conclude that the generalized eigenspace of $\widehat{A}$ is given by

(5.22) $$\mathcal{M} = \left\{ \begin{pmatrix} c \\ \varphi \end{pmatrix} \in Z : E(z \mapsto p_1(z,c,\varphi,\theta)) \le |\overline{\omega}b_m| \right\}.$$

Hence, it suffices to compute the exponential type of $z \mapsto p_1(z, c, \varphi, \theta)$. From (5.21)

$$
\begin{aligned}
p_1(z, c, \varphi, \theta) = {}& \Omega^\theta_{-\omega}(z)c + z\Omega^\theta_{-\omega}(z) \int_{-\omega}^0 \Omega^0_\tau(z)\kappa(\tau)\,d\tau \\
& - z\Omega^0_{-\omega}(z) \int_{-\omega}^\theta \Omega^\theta_\tau(z)\kappa(\tau)\,d\tau \\
& + \int_{-\omega}^\theta \Omega^\theta_\tau(z)\kappa(\tau)\,d\tau - \det \Delta(z) \sum_{j=0}^{l-2} z^j \varphi_{l-1-j}(\theta).
\end{aligned}
$$

Since

$$
\begin{aligned}
z\Omega^\theta_{-\omega}(z) \int_{-\omega}^0 \Omega^0_\tau(z)\kappa(\tau)\,d\tau - z\Omega^0_{-\omega}(z) \int_{-\omega}^\theta \Omega^\theta_\tau(z)\kappa(\tau)\,d\tau = {} \\
= z\Omega^\theta_{-\omega}(z)\Big[\int_{-\omega}^0 \Omega^0_\tau(z)\kappa(\tau)\,d\tau - z\Omega^0_\theta(z) \int_{-\omega}^\theta \Omega^\theta_\tau(z)\kappa(\tau)\,d\tau\Big] \\
= z\Omega^\theta_{-\omega}(z) \int_\theta^0 \Omega^0_\tau(z)\kappa(\tau)\,d\tau,
\end{aligned}
$$

we find

$$
\begin{aligned}
p_1(z, c, \varphi, \theta) = {}& \Omega^\theta_{-\omega}(z)c + z\Omega^\theta_{-\omega}(z) \int_\theta^0 \Omega^0_\tau(z)\kappa(\tau)\,d\tau \\
& + \int_{-\omega}^\theta \Omega^\theta_\tau(z)\kappa(\tau)\,d\tau - \det \Delta(z) \sum_{j=0}^{l-2} z^j \varphi_{l-1-j}(\theta).
\end{aligned}
$$
(5.23)

First suppose that b_m does not change sign, say $b_m(s) \geq 0$. The entire function

$$
z \mapsto \int_{-\omega}^\theta e^{\sum_{j=0}^m z^j \int_\tau^\theta b_j(s)ds}\kappa(\tau)\,d\tau
$$

is of order m, has maximal type in the left half plane, and is bounded in the right half plane. Using the Paley-Wiener theorem we find

$$
\begin{aligned}
E\Big(z \mapsto \int_{-\omega}^\theta e^{\sum_{j=0}^m z^j \int_\tau^\theta b_j(s)ds}\kappa(\tau)\,d\tau\Big) &= \max_{-\omega \leq \tau \leq \theta} \int_\tau^\theta b_m(s)ds \\
&\leq \omega \bar{b}_m.
\end{aligned}
$$

Furthermore,

$$
\begin{aligned}
E\Big(z \mapsto z\Omega^\theta_{-\omega}(z) \int_\theta^0 \Omega^0_\tau(z)\kappa(\tau)\,d\tau\Big) &= \int_{-\omega}^\theta b_m(s)\,ds + \max_{\theta \leq \tau \leq 0} \int_\tau^0 b_m(s)\,ds \\
&= \omega \bar{b}_m.
\end{aligned}
$$

Similarly, for $b_m(s) \leq 0$, the functions have exponential growth in the right half plane and are bounded in the left half plane, and the same exponential estimates (with b_m replaced by $-b_m$) hold. In both cases it follows from (5.22) that $\overline{\mathcal{M}} = Z$. On the other hand, if b_m does change sign, the maximum of the function

$$\tau \mapsto \int_{-\omega}^{\theta} b_m(s)\, ds + \int_{\tau}^{0} b_m(s)\, ds$$

on $[\theta, 0]$ is larger than $\omega \overline{b}_m$. Since $z \mapsto z\Omega_{-\omega}^{\theta}(z) \int_{\theta}^{0} \Omega_{\tau}^{0}(z)\kappa(\tau)\, d\tau$ can not be cancelled by any other term in (5.23) the type of $z \mapsto p_1(z, c, \varphi, \theta)$ is larger than $\omega \overline{b}_m$. Hence (5.22) implies that $Z \neq \overline{\mathcal{M}}$. Since the eigenfunctions and generalized eigenfunctions of $\widehat{A}$ correspond to the Floquet solutions of (5.1) we obtain Theorem 5.3. To prove the remaining results it suffices to prove that the subspace

$$\widehat{\mathcal{S}} = \{\begin{pmatrix} c \\ \varphi \end{pmatrix} : z \mapsto R(z, \widehat{A}) \begin{pmatrix} c \\ \varphi \end{pmatrix} \quad \text{is entire } \}$$

corresponds to the space of initial conditions $\mathcal{S}$ that yield small solutions of (5.1). It is clear from the exponential estimates (see Hale [5, 8.1.1]) that $x(t) = T(t, s)\varphi$ is a small solution of (5.1) if and only if

$$P_\mu \varphi = 0 \qquad \text{for all} \quad \mu \in \sigma(U)$$

or, equivalently, $z \mapsto z(I - zU)^{-1}\varphi$ is an entire function. But $U = A^{-1}$ and hence $z \mapsto -zA(z - A)^{-1}$ must be entire. Hence $x(t) = T(t, s)\varphi$ is a small solution of (5.1) if and only if

$$P_\lambda \varphi = 0 \qquad \text{for all} \quad \lambda \in \sigma(A).$$

Since the spectral projections for A and $\widehat{A}$ are the same, the mapping

$$\varphi \mapsto \begin{pmatrix} M\varphi \\ \varphi \end{pmatrix}$$

defines an isomorphism between $\mathcal{S}$ and $\widehat{\mathcal{S}}$ and this completes the proof of Theorem 5.4 and Corollary 5.5.

REFERENCES

1. Y. Cao, *On the scalar linear differential delay equations with periodic coefficients*, Preprint, University of Georgia, (1990).
2. Ph. Clément, O. Diekmann, M. Gyllenberg, H. J. A. M. Heijmans, and H. R. Thieme, *Perturbation theory for dual semigroups. I. The sun-reflexive case*, Math. Ann. **277** (1988), 709–725.
3. G. Doetsch, *Handbuch der Laplace-Transformation Band I*, Birkhäuser, Basel, 1971.
4. W. Hahn, *On difference differential equations with periodic coefficients*, J. Math. Anal. Appl. **3** (1961), 70–101.
5. J. K. Hale, *Theory of Functional Differential Equations*, Springer-Verlag, Berlin, 1977.
6. D. Henry, *Small solutions of linear autonomous functional differential equations*, J. Differential Equations **8** (1970), 494–501.
7. M. A. Kaashoek and S. M. Verduyn Lunel, *Characteristic matrices and spectral properties of evolutionary systems*, IMA preprint series **707** (1990), (to appear in Trans. Amer. Math. Soc.).
8. S. M. Verduyn Lunel, *A sharp version of Henry's theorem on small solutions*, J. Differential Equations **62** (1986), 266–274.
9. S. M. Verduyn Lunel, *Exponential type calculus for linear delay equations*, Centre for Mathematics and Computer Science, Tract No. 57, Amsterdam, 1989.

10. S. M. Verduyn Lunel, *Series expansions and small solutions for Volterra equations of convolution type*, J. Differential Equations **85** (1990), 17–53.
11. S. M. Verduyn Lunel, *The closure of the generalized eigenspace of a class of infinitesimal generators*, Proc. Roy. Soc. Edinburgh Sect. A **117A** (1991), 171–192.
12. A. M. Zverkin, *The completeness of a system of Floquet type solutions for equations with retardations*, Differential Equations **4** (1968), 249–251.

FACULTEIT WISKUNDE EN INFORMATICA, VRIJE UNIVERSITEIT, DE BOELELAAN 1081A, 1081 HV AMSTERDAM, THE NETHERLANDS

SCHOOL OF MATHEMATICS, GEORGIA INSTITUTE OF TECHNOLOGY, ATLANTA, GA 30332, U.S.A.

Contemporary Mathematics
Volume **129**, 1992

Periodic Solutions of Differential Delay Equations with Threshold-Type Delays

H. L. SMITH AND Y. KUANG

ABSTRACT. Periodic solutions are shown to exist for the differential delay equation $x'(t) = -\nu x(t) - e^{-\eta\tau} f(x(t-\tau))$ where τ is determined implicitly by the solution $x(t)$ through the threshold relation $\int_{t-\tau}^{t} k(x(t), x(s))\, ds = m$. These periodic solutions have two simple zeros in the period interval. If k is independent of the second variable then $\tau = \sigma(x(t))$ while if k is independent of the first variable then the delay is of threshold-type. The latter arise in many models in the biological sciences. The proof consists of modifying the classical proof for the case of constant delays.

0. Introduction

In this paper the existence of periodic solutions for the state-dependent delay equation with threshold-type delay of the form

$$x'(t) = -\nu x(t) - e^{-\eta\tau} f(x(t-\tau))$$

(0.1)

$$\int_{t-\tau}^{t} k(x(t), x(s))\, ds = m$$

is established. In (0.1), ν and η are nonnegative constants, m is a positive constant and $k(x, y)$ is a positive-valued locally Lipschitz function on $\mathbf{R}^2$. The delay τ is determined implicitly by the integral relation in (0.1). The function f is assumed to be locally Lipschitz and satisfy

$$x f(x) > 0, \qquad x \neq 0.$$

Thus (0.1) is of negative feedback type.

In the special case that k is independent of y, that is, $k(x, y) = (\sigma(x))^{-1} m$ where $\sigma : \mathbf{R} \to (0, \infty)$ is locally Lipschitz, then the delay τ is given by

(0.2)
$$\tau = \sigma(x(t)).$$

1991 *Mathematics Subject Classification*. Primary 34K15.
Research supported by NSF Grant DMS #8922654 .

If $k(x, y) = k(y)$, then the delay τ is determined implicitly by

$$(0.3) \qquad \int_{t-\tau}^{t} k(x(s))\, ds = m.$$

Delays of the form (0.3) have appeared frequently in applications (see below) and are called threshold delays. By allowing k to depend on $x(t)$ in addition to $x(s)$, we include most of the state-dependent delays which have been proposed in the literature. In fact, our main result is proved for a more general delay functional $\tau = \tau(x_t)$ which includes the threshold delay in (0.1) as a special case.

The term $e^{-\eta\tau}$ in (0.1) (which is absent if $\eta = 0$) requires a word of motivation. This term is not merely a constant since $\tau = \tau(x_t)$ depends on past history. It is present in many equations arising from mathematical models in the biological sciences (see, e.g., [7], [28], [26]) due to an exponential death or decay rate which discounts a delayed population size. This term has a stabilizing effect on the trivial solution of (0.1) and a large η precludes the existence of periodic solutions of (0.1) as we shall see. From a purely mathematical point of view, the term could be replaced by any positive locally Lipschitz function of $\tau(x_t)$. Moreover, the term $e^{-\eta\tau} f(x(t-\tau))$ could be replaced by $f(e^{-\eta\tau} x(t-\tau))$ without affecting our main result.

In stating our main results, it will be convenient to scale variables and eliminate unnecessary parameters. Obviously, the positive constant m can be incorporated into the function k so we may as well assume that $m = 1$. Scaling time by $\bar{t} = \delta t$ leaves the form of (0.1) unchanged but with $\bar{\nu} = \nu\delta$, $\bar{\eta} = \eta\delta$, $\bar{f} = \delta f$, $\bar{k} = \delta k$. Choosing $\delta = k(0,0)^{-1}$ has the effect of making $\bar{k}(0,0) = 1$. Henceforth, we assume that

$$(0.4) \qquad m = 1, \qquad k(0,0) = 1.$$

Observe that $x \equiv 0$ is a solution of (0.1) and, by virtue of (0.4), the delay τ corresponding to the trivial solution is unity.

It turns out to be possible to relate small solutions of (0.1) to the solutions of the linear equation with constant delay $\tau = \tau(0) = 1$

$$(0.5) \qquad y'(t) = -\nu y(t) - e^{-\eta} f'(0) y(t-1),$$

associated to which is the characteristic equation

$$(0.6) \qquad \nu + \lambda + \alpha e^{-\lambda} = 0$$

where $\alpha = e^{-\eta} f'(0)$. If $\alpha > \alpha_\nu$, where α_ν is the smallest positive solution of

$$\nu + \alpha \cos\sqrt{\alpha^2 - \nu^2} = 0,$$

then (0.6) has a root $\lambda = \mu + i\gamma$ satisfying $\mu > 0$ and $\frac{\pi}{2} < \gamma < \pi$ (see, e.g., [15], [18]).

The following is a special case of our main result.

THEOREM. *Suppose that*

(i) $e^{-\eta} f'(0) > \alpha_{\nu}$.

(ii) *There exist* $M, N > 0$ *such that* $F_{\nu}(x) \equiv -\nu^{-1} f(x)$ *maps* $[-M, N]$ *into itself.*

(iii) $\frac{\partial k}{\partial x}(x, y)$ *exists for* $x \neq 0$, $x, y \in [-M, N]$, *is continuous and*

$$(0.7) \qquad \nu x \frac{\partial k}{\partial x}(x, y) < k(x, y) k(x, x), \qquad x, y \in [-M, N], \ x \neq 0.$$

Then (0.1) *has a nonconstant periodic solution* $x(t) = x(t + \omega)$ *satisfying:*

(a) $x(t)$ *has exactly two zeros on* $[0, \omega]$, *namely* z_1 *and* z_2, $0 < z_1 < z_2 < \omega$ *with* $x'(z_1) < 0$ *and* $x'(z_2) > 0$.

(b) $-M < x(t) < N$, $0 \leq t \leq \omega$.

(c) $e^{\nu t} x(t)$ *has exactly two extrema in* $(0, \omega]$, *namely* t_1 *and* $t_2 = \omega$, $0 < z_1 < t_1 < z_2 < t_2$ *with* $e^{\nu t} x(t)$ *increasing on* (t_1, t_2) *and decreasing on* $(0, t_1)$.

Several remarks regarding the Theorem will be helpful. In (ii), it is assumed that $\nu > 0$. If $\nu = 0$, a different assumption is required (see (B) of Section 1). The assumption (0.7) is crucial to our proof of the theorem for it allows us to establish that $t - \tau \in (z_1, z_2)$ if and only if $t_1 < t < t_2$. Obviously, (0.7) holds if $k(x, y) = k(y)$ is independent of x so (0.7) is not a restriction for the threshold delay (0.3). It may be a restriction, however, for delays of the form (0.2) for which (0.7) requires that $\sigma(x)$ be continuously differentiable for $x \in [-M, N] \backslash 0$ and satisfy

$$(0.8) \qquad \nu x \sigma'(x) > -1, \qquad x \neq 0, \ -M \leq x \leq N.$$

For example, (0.8) holds for $\sigma(x) = 1 + |x|$ as $x\sigma'(x) > 0$, but (0.8) may fail to hold, depending on the values of M and N, for $\sigma(x) = \frac{1}{1+|x|}$ if $\nu \geq 4$.

We can estimate the period ω, but only clumsily in the full generality of the threshold delay of (0.1). On the other hand, if the delay is given by (0.2), then $z_2 - z_1 > 1$ and $\omega = t_2 > 2$. In this case, the periodic solution of the Theorem could be called a "slowly oscillating" solution of (0.1).

All the assumptions of the Theorem except (i) can be weakened at the expense of a more complicated and lengthy statement (see Section 1).

Threshold delays of the form (0.3) have been used in epidemiological modeling [34], in modeling of the immune response system [34], [11] and in modeling of respiration [16]. The basic idea is that the delay is due to the time required to accumulate an appropriate dosage of infection or antigen concentration. In the context of epidemiology, $x(t)$ may represent the proportion of a population which is infective at time t and (0.3) may reflect that an individual who is first exposed to the disease at time $t - \tau$ becomes infectious at time t if, during the interval from $t - \tau$ to t, a threshold level of exposure is accumulated where the per unit time exposure depends on the infective fraction $x(s)$ via $k(x(s))$. See [34] for a description of such models and [31] for the existence of periodic solutions for such models.

Well-posedness results for systems of differential delay equations with threshold delays given by a slightly more general expression than (0.1) (k is allowed to depend explicitly on t) have been proved in [12], [13]. These results are not immediately applicable in our setting, however, since the authors consider initial data for $x(t)$ to be simply the assignment of a value for $x(0)$ and they set $\tau \equiv t$ until such time $t_0 > 0$ when $\int_0^{t_0} k(x(t), x(s))\, ds = 1$. For $t > t_0$ the delay is determined by (0.1) and hence the equation is viewed as an ordinary differential equation for $0 < t < t_0$ and a delay equation for $t > t_0$. On the other hand, we view (0.1) as providing a delay immediately since we specify $x(t)$ for $t \le 0$. However, for large t, assuming the threshold time t_0 exists, the equations are identical and the influence of differing initial data should wane. Hence, the existence of stable periodic solutions will be of interest.

Delay differential equations containing delays of the threshold type (0.3) arise from attempts to simplify structured population models, which usually take the form of hyperbolic partial differential equations. As an example, Nisbet and Gurney [28] consider insect populations which have several life stages (instars). They construct a mathematical model consisting of an equation for the mass density function for the population. Then, under the assumption that the population in any life stage is homogeneous, they are able to reduce the model to a system of delay differential equations for the size of the population in each life stage. These equations have threshold delays (see [28], Appendix 2) due to the assumption that the insect must spend an amount of time in the larval stage sufficient to accumulate a threshold amount of food, where food density is also a dynamical variable. The system of delay equations considered in [28] is considerably more complicated than (0.1).

Metz and Diekmann ([26], p. 236–237) discuss a modification of a structured model for the control of the bone marrow stem cell population which supplies the circulating red blood cell population, due to Kirk, Orr and Forrest [21]. The maturing stem cell population is structured by a maturation variable and the rate of maturation is assumed to depend only on the total mature red blood cell population. Since a threshold level of maturation is required in order for an immature cell to enter the mature population, a threshold delay differential equation arises naturally. See ([26], Exercises 4.30 and 4.31) for the equations of the reduced delay equation model. While there is an error in the equations of Exercise 4.31 $(V(P(t))n(t,1) = (V(P(t))/V(P(t-\tau)))e^{-\delta\tau}\gamma P(t-\tau))$, it does not affect our argument that delay equations containing threshold type delays arise naturally from structured population models are therefore worthy of study.

Bélair and Mackey [3] consider a state-dependent delay equation modeling commodity prices. Bélair [4] discusses a number of models involving state-dependent delay equations. Cao and Freedman [7] consider a model of a stage structured population where the length of the immature stage (a delay in the model) is a function of total population size. In these models the delay takes the form (0.2).

Admittedly, (0.1) is extremely simple compared to the equations which arise from the population models described above. We view our efforts here as only a beginning towards understanding the more complicated equations which arise in the applications. In this regard, we hasten to point out that the first steps in this program were taken by Alt [2] who earlier recognized the importance of considering delay equations containing threshold delays. Alt [2] has established the existence of periodic solutions of the equation

$$x'(t) = -f\left(x(t), \int_{-r}^{0} g(x(t - \tau + \theta))\, d\eta(\theta)\right)$$

where τ is determined by (0.3) and f and g satisfy suitable conditions, making his equation more general, although his delay is a special case of ours.

J. Mallet-Paret and R. Nussbaum have announced very interesting work related to ours at the International Conference on Differential Equations and Applications to Biology and Population Dynamics (Claremont, California, January 1990). They consider a singularly perturbed equation of the form

$$\epsilon x'(t) = -f(x(t), x(t - r))$$

where $r = r(x(t)) \geq 0$ and f and r are given functions. They are able to describe the asymptotic shape of slowly oscillating solutions as ϵ tends to zero under appropriate conditions on f and r. In a private communication, J. Mallet-Paret indicates that they obtain, as part of their work, an existence result for periodic solutions of this equation [27].

The proof of our theorem follows more or less standard lines (see in particular [15] and also [1], [6], [8], [10], [15], [18], [19], [29], [30], [32], [33], [35]). We set up a Poincaré map from a compact convex subset K of C_B into itself where C_B is the space of bounded and continuous functions on $(-\infty, 0]$. The set K contains the zero function which we show to be an ejective fixed point of the mapping. A theorem of Browder [5] asserts the existence of at least one nonejective fixed point, proving our theorem.

As a simple illustration of the theorem, consider the equation

(0.9)
$$x'(t) = -\nu x(t) + g(x(t - \tau))$$
$$\int_{t-\tau}^{t} k(x(t), x(s))\, ds = m$$

where g is a continuously differentiable nonnegative function on $x \geq 0$ and $g'(x) < 0$ for $x > a$ for some $a \geq 0$. Assume $G_\nu(x) = \nu^{-1} g(x)$ has a period two orbit $G_\nu(x_1) = x_2$, $G_\nu(x_2) = x_1$, where $a < x_1 < x_2$. Then G_ν has a unique fixed point x_0 satisfying $x_1 < x_0 < x_2$, which is an equilibrium for (0.9). The change of variables $y = x - x_0$ transforms (0.9) to (0.1) where

$$f(y) = g(x_0) - g(x_0 + y).$$

For simplicity, assume that $m = k(x_0, x_0)$ so that on dividing through by m, the transformed equations are as in (0.1) with (0.4) holding for $\bar{k}(u, v) \equiv k(x_0 + u, x_0 + v)/m$. It is easily checked that (ii) of the theorem holds ($M = x_1 - x_0, N = x_2 - x_0$) for the transformed system and (i) holds provided $g'(x_0) < -\alpha_\nu$. Provided $\bar{k}$ has property (iii), the theorem implies the existence of a periodic solution $x(t)$ of (0.9) which oscillates about $x = x_0$ and satisfies $x_1 < x(t) < x_2$.

The main result of this paper represents an improvement of earlier results of the authors [23] who considered (0.1) in the case that $\nu = \eta = 0$. In an earlier paper [22] the authors considered (0.1) in case $\nu = \eta = 0$ and (0.2) holds. This paper contains some numerical simulations and some results on the oscillatory properties of solutions.

Our main result (Theorem 1.9) is described in the next section. In that section, the delay is assumed to be specified by a functional $\tau : C_B \to (0, \infty)$ where C_B is the Banach space of bounded continuous functions on $(-\infty, 0]$ with the uniform norm $|| \cdot ||$. Four hypotheses are required to be satisfied by the functional τ in order that our result applies. In a brief final section, the threshold delay appearing in (0.1) is shown to have the required properties.

1. Main Results

Consider the delay differential equation

$$(1.1) \qquad x'(t) = -\nu x(t) - e^{-\eta \tau(x_t)} f(x(t - \tau(x_t)))$$

where $x_t(s) = x(t + s)$, $-\infty \le s \le 0$, and ν, η are nonnegative constants. Alternatively, the term $e^{-\eta \tau}$ could multiply $x(t - \tau)$ inside f.

The following are assumed to hold throughout this section.

(F) $f : \mathbf{R} \to \mathbf{R}$ is locally Lipschitz continuous,

$$x f(x) > 0, \qquad x \neq 0,$$

and $f'(0)$ exists.

The functional $\tau : C_B \to (0, \infty)$ satisfies

(D1) $\tau(0) = 1$

(D2) If $\varphi, \psi \in C_B$ satisfy $\varphi|_{[-\tau(\varphi), 0]} = \psi|_{[-\tau(\varphi), 0]}$ then $\tau(\varphi) = \tau(\psi)$.

(D3) Given $L > 0$ there exists $V > 0$ such that

$$|\tau(\phi) - \tau(\psi)| \le V \max_{-Q \le s \le 0} |\phi(s) - \psi(s)|$$

if $||\phi||, ||\psi|| \le L$ where $Q = \max\{\tau(\varphi), \tau(\psi)\}$.

Note that (D3) implies that τ is bounded on bounded subsets of C_B. Indeed, using (D1) and (D3), the estimate $\tau(\varphi) \le 1 + V(L)||\varphi||$ holds for $||\phi|| \le L$. (D1) is simply a normalization which can be assumed without loss of generality.

LEMMA 1.1. *For each bounded, Lipschitz function* $\phi : (-\infty, 0] \to \mathbf{R}$, *there exists a unique noncontinuable solution* $x(t, \phi)$ *of* (1.1) *defined for* $t \in [0, \omega)$, $0 < \omega \leq \infty$, *satisfying* $x(t) = \phi(t)$ *for* $t \leq 0$. *If* $\omega < \infty$, *then there exists* $t_n \nearrow \omega$ *such that* $|x(t_n)| \to \infty$ *as* $n \to \infty$.

PROOF. These are well-known results. See [17], [14], [9]. Uniqueness of solutions follows from ([17], Prop. 1.3) and the fact that f and τ are locally Lipschitz. Since the right-hand side of (1.1) is bounded on bounded subsets of the space of bounded, Lipschitz functions on $(-\infty, 0]$, the behavior of $x(t)$ as $t \to \omega$ follows from ([17], Prop. 1.2). In the case of (1.1), this is easy to see since, if $|x(t)|$ is bounded, then its derivative is bounded and thus $\lim_{t \to \omega-} x(t)$ exists. The solution can then be extended to a larger interval.

Define
$$F^+(x) = \max\{f(s) : 0 \leq s \leq x\}$$
$$F^-(x) = \max\{-f(s) : -x \leq s \leq 0\}$$

for $x \geq 0$. Given $M, N > 0$ let

$$R = \sup\{\tau(\phi) : \phi \in C_B, \ -M \leq \phi(s) \leq N\}$$
$$R_M = \sup\{\tau(\phi) : \phi \in C_B, \ \phi(0) = -M \leq \phi(s) \leq N\}$$
$$R_N = \sup\{\tau(\phi) : \phi \in C_B, \ -M \leq \phi(s) \leq N = \phi(0)\}.$$

R, R_M and R_N depend on both M and N but we suppress this dependence for notational convenience. Obviously, $R_M, R_N \leq R < \infty$, the latter by (D3), and $R \geq 1$ by (D1).

We now make two hypotheses which determine M and N such that solutions corresponding to suitably restricted initial data exist for all $t \geq 0$, take values in the interval $[-M, N]$ and oscillate about $x = 0$.

(B) There exist $M, N > 0$ such that

(1.2a)
$$F^+(N) \left[\frac{1 - e^{-\nu R_M}}{\nu}\right] < M,$$

and

(1.2b)
$$F^-(M) \left[\frac{1 - e^{-\nu R_N}}{\nu}\right] < N.$$

If $\nu = 0$, then the term in brackets in (1.2a) ((1.2b)) is understood to be R_M (R_N).

Note that (B) is a restriction on ν, τ and f. We point out some more easily verified sufficient conditions for (B) to hold. If $\nu = 0$, then sufficient conditions for (B) to hold were given in [23]. Here, we will assume $\nu > 0$. If

(1.3) $F_\nu(x) \equiv -\nu^{-1} f(x)$ satisfies $F_\nu([-M, N]) \subset [-M, N]$

then $\nu^{-1} F^+(N) \leq M$ and $\nu^{-1} F^-(M) \leq N$ implying that (B) holds. Also, (1.3) has the advantage of not involving the delay τ and being relatively easy to verify.

For the same values of M and N as in (B), we also assume:

(D4) If $\phi \in C_B$ is Lipschitz, takes values in $[-M, N]$ and satisfies $\phi(-\tau(\phi)) = 0$, then for any $\epsilon > 0$ and any extension $y : (-\infty, \epsilon) \to \mathbf{R}$ of ϕ, $y_0 = \phi$, where $y|_{[0,\epsilon)}$ is continuously differentiable and $y'(0) + \nu\phi(0) = 0$, there exists $\delta \in (0, \epsilon)$ such that

$$\tau(y_t) - \tau(\phi) < t, \qquad 0 < t < \delta.$$

If, in addition, $\phi'(0)$ exists and $\phi'(0) + \nu\phi(0) = 0$, then

$$\tau(\phi_s) - \tau(\phi) > s, \qquad -\delta < s < 0.$$

For the $M, N > 0$ such that (B) and (D4) hold, let

$$\tilde{K}(M, N) = \{\phi \in C_B : \phi(s) = 0,\ s \leq -R,\ \phi(s)e^{\nu s} \text{ is nondecreasing}$$

$$\text{on } [-R, 0],\ \phi(s) \leq N \text{ for } s \leq 0 \text{ and } \phi \text{ is Lipschitz}\}.$$

Note that as $\phi(s)e^{\nu s}$ is nondecreasing on $[-R, 0]$, it follows that either $\phi = 0$ or $s_\phi \equiv \sup\{s \leq 0 : \phi(s) = 0\}$ satisfies $-R \leq s_\phi < 0$ and $0 < \phi(s) \leq N$ holds for $s_\phi < s \leq 0$. Hereafter, we write $\tilde{K}$ for $\tilde{K}(M, N)$.

For the remainder of this section, we assume that (B) and (D4) hold and we only consider initial data $\phi \in \tilde{K}$. For such ϕ, let $\tilde{t} = \tilde{t}(\phi) = \sup\{t \geq 0 : -M \leq x(s, \phi) \leq N \text{ for } 0 \leq s \leq t\}$. If $\phi \in \tilde{K}\backslash 0$, then $\phi(0) \leq N$ and $x'(0) < 0$ so $0 < \tilde{t} \leq \omega$. If $\tilde{t} < \infty$ then by Lemma 1.1, $x(\tilde{t}) \in \{-M, N\}$. Observe that $\tau(x_t) \leq R$ for $0 \leq t \leq \tilde{t}$ if $\phi \in \tilde{K}$.

LEMMA 1.2. *Let* $\phi \in \tilde{K}$ *and* $x(t) = x(t, \phi)$. *If* $\frac{d}{dt}(x(t)e^{\nu t}) = 0$ *for some* $t_0 \in [0, \tilde{t}]$, *then*

$$t - \tau(x_t) > t_0 - \tau(x_{t_0}) \text{ for } t_0 < t \leq \tilde{t}$$

and

$$t - \tau(x_t) < t_0 - \tau(x_{t_0}) \text{ for } 0 \leq t < t_0.$$

PROOF. By (1.1),

$$\frac{d}{dt}\Big|_{t=t_0} e^{\nu t} x(t) = -e^{\nu t_0 - \eta \tau(x_{t_0})} f(x(t_0 - \tau(x_{t_0}))) = 0$$

implies that $x_{t_0}(-\tau(x_{t_0})) = 0$. By (D4) with $y(t) = x(t_0 + t)$, there exists $\delta > 0$ such that $\tau(x_t) - \tau(x_{t_0}) < t - t_0$ for $0 < t - t_0 < \delta$, if $t_0 < \tilde{t}$, and $\tau(x_t) - \tau(x_{t_0}) > t - t_0$ for $-\delta < t - t_0 < \delta$, if $t_0 > 0$. Thus the inequality assertions of the lemma hold locally near t_0. If the first inequality above does not hold for all $t \in (t_0, \tilde{t}]$, then there exists a point $\bar{t} \in (t_0, \tilde{t}]$ such that

$$\bar{t} - \tau(x_{\bar{t}}) = t_0 - \tau(x_{t_0}) < t - \tau(x_t)$$

for $t_0 < t < \bar{t}$. As $x_{\bar{t}}(-\tau(x_{\bar{t}})) = x_{t_0}(-\tau(x_{t_0})) = 0$ we may argue a contradiction to the last inequality above using (D4) again at $\bar{t}$. A similar argument applies to show that the second inequality must hold for all $t < t_0$. This completes our proof.

PROPOSITION 1.3. *In addition to the assumptions above, assume that* $f'(0)e^{-\eta} > 1$. *If* $\phi \in \tilde{K}\backslash 0$ *and* $x(t) = x(t, \phi)$, *then*

(i) *There exists* $z_1 = z_1(\phi) > 0$ *such that* $x(t) > 0$ *and* $x'(t) \leq 0$ *on* $[0, z_1)$, $x(z_1) = 0$ *and* $x'(z_1) < 0$. *Moreover, there exists* $P = P(M, N) > 0$ *such that* $z_1(\phi) \leq P$ *for all* $\phi \in \tilde{K}\backslash 0$.

(ii) *There exists* $t_1 = t_1(\phi) \in (z_1, z_1 + R]$ *such that* $t_1 - \tau(x_{t_1}) = z_1 > t - \tau(x_t)$ *for* $z_1 \leq t < t_1$. *Consequently,* $\frac{d}{dt}(e^{\nu t} x(t))$ *is negative on* $[z_1, t_1)$ *and vanishes at* t_1. *Moreover,* $0 \geq x(t) > -M$ *on* $[z_1, t_1]$.

(iii) *There exists* $z_2 = z_2(\phi) > t_1$ *such that* $-M < x(t) < 0$ *on* (z_1, z_2), $x(z_2) = 0$, $x'(z_2) > 0$. *There exists* $Q = Q(M, N) > 0$ *such that* $z_2(\phi) \leq Q$ *for all* $\phi \in \tilde{K}\backslash 0$.

(iv) *There exists* $t_2 = t_2(\phi) \in (z_2, z_2 + R]$ *such that* $t_2 - \tau(x_{t_2}) = z_2$. *For* $t_1 < t < t_2$, $z_1 < t - \tau(x_t) < z_2$ *and consequently* $\frac{d}{dt}[e^{\nu t} x(t)] > 0$ *and vanishes at both endpoints. Finally,* $0 < x(t) < N$ *on* $(z_2, t_1]$.

PROOF. Clearly $x'(t) \leq 0$ for all $t \in [0, z_1]$ where $z_1 = \inf\{t > 0 : x(t) = 0\} \leq +\infty$. Define $F(x) = \min\{f(u) : x \leq u \leq N\}$ for $0 \leq x \leq N$ so that F is Lipschitz, $F(x) > 0$ for $0 < x \leq N$, F is nondecreasing and $F(0) = 0$. If $z_1 > R$, then for $R \leq t \leq z_1$, $t - \tau(x_t) \geq 0$ since $\tau(x_t) \leq R$, so

$$x'(t) = -\nu x(t) - e^{-\eta \tau(x_t)} f(x(t - \tau(x_t)))$$
$$\leq -\nu x(t) - e^{-\eta R} F(x(t - \tau(x_t))) \leq -\nu x(t) - e^{-\eta R} F(x(t))$$

where we have used the fact that F is nondecreasing and $x(t)$ is decreasing. Standard differential inequality arguments imply that $x(t) \leq u(t)$, $R \leq t \leq z_1$, where

$$u'(t) = -\nu u(t) - e^{-\eta R} F(u(t)), \qquad u(R) = N.$$

Observe that $u(t)$ is independent of $\phi \in \tilde{K}\backslash 0$. Obviously $u(t) \to 0$ as $t \to +\infty$.

Let k be such that $f'(0)e^{-\eta} > k > 1$ and choose $\gamma \in (0, 1)$ and $\delta > 0$ such that $e^{-\eta(1+\gamma)} f(x)/x > k > 1$ if $0 < x \leq \delta$. If $0 \leq \phi \leq N$ then by (D3), $|\tau(\phi) - 1| \leq V \max_{-R \leq s \leq 0} |\phi(s)|$. Choose δ smaller if necessary so that $V\delta < \min\{\gamma, 1 - k^{-1}\}$. Now there exists $T > 0$ such that $u(t) \leq \delta$ for $t \geq T$. Suppose $z_1 > 2R + T + k^{-1}$. Then $0 < x(t) < u(t) \leq \delta$ for $T \leq t \leq 2R + T + k^{-1}$ and $|\tau(x_t) - 1| \leq V \max_{-R \leq s \leq 0} |x(t+s)| \leq V\delta$ for $T + R \leq t \leq 2R + T + k^{-1}$. Thus $1 + \gamma \geq \tau(x_t) > k^{-1}$ on the latter interval. For $2R + T \leq t \leq 2R + T + k^{-1}$, we have $R + T \leq t - R \leq t - \tau(x_t) < t - k^{-1} \leq 2R + T$ and hence $0 < x(t - \tau(x_t)) < \delta$ and $\tau(x_t) < 1 + \gamma$. Thus

$$x'(t) = -\nu x(t) - e^{-\eta \tau(x_t)} f(x(t - \tau(x_t))) \leq -e^{-\eta(1+\gamma)} f(x(t - \tau(x_t)))$$
$$< -kx(t - \tau(x_t)) \leq -kx(2R + T)$$

for $2R + T \leq t \leq 2R + T + k^{-1}$. This implies that

$$x(2R + T + k^{-1}) < x(2R + T) - kx(2R + T)k^{-1} = 0,$$

a contradiction to our assumption that $z_1 > 2R + T + k^{-1}$. Hence, $z_1 \leq 2R + T + k^{-1} \equiv P(M, N)$.

Note that for $0 \leq t \leq z_1$, $t - \tau(x_t) \geq t - R \geq -R$. If $t - \tau(x_t) \leq -s_\phi$ for $0 \leq t \leq z_1$, then $x'(t) = -\nu x(t)$ on this interval so $x(z_1) = e^{-\nu z_1}\phi(0) > 0$, a contradiction to $x(z_1) = 0$. Hence, there exists $t_0 \in (0, z_1)$ such that $t_0 - \tau(x_{t_0}) > -s_\phi$. Lemma 1.2 implies that $t - \tau(x_t) > -s_\phi$ for $z_1 \geq t \geq t_0$. It follows that $x'(z_1) = x'(z_1) + \nu x(z_1) < 0$.

Suppose that there does not exist $t_1 \in [z_1, \tilde{t}]$ such that $t_1 - \tau(x_{t_1}) = z_1$. Then $t - \tau(x_t) < z_1$ for all $t \in [z_1, \tilde{t}]$ and (1.1) implies $\frac{d}{dt}(e^{\nu t}x(t)) < 0$ for all $t \in [z_1, \tilde{t}]$. This implies that $x(t) < 0$ for $z_1 < t \leq \tilde{t}$. If $z_1 + R \leq \tilde{t}$, then $z_1 + R - \tau(x_{z_1+R}) \geq z_1$ contradicting our assumption that t_1 does not exist. Consequently, $\tilde{t} < z_1 + R$ so $x(\tilde{t}) = -M$. Then, using (1.1), and considering the case $\nu > 0$ ($\nu = 0$ proceeds similarly, see [23]), we conclude that

$$
\begin{aligned}
-x(\tilde{t}) = M &= \int_{z_1}^{\tilde{t}} e^{-\nu(\tilde{t}-t)} e^{-\eta \tau(x_t)} f(x(t - \tau(x_t)))\, dt \\
&\leq F^+(N) \int_{z_1}^{\tilde{t}} e^{-\nu(\tilde{t}-t)}\, dt = F^+(N) \left(\frac{1 - e^{-\nu(\tilde{t}-z_1)}}{\nu} \right) \\
&\leq F^+(N) \left(\frac{1 - e^{-\nu \tau(x_t)}}{\nu} \right) \\
&\leq F^+(N) \left(\frac{1 - e^{-\nu R_M}}{\nu} \right)
\end{aligned}
$$

where we have used $\tilde{t} - \tau(x_{\tilde{t}}) < z_1$ and $\tau(x_{\tilde{t}}) \leq R_M$. As this contradicts (1.2a) of (B), our assumption that t_1 does not exist must be discarded. We denote by t_1 the smallest t for which $t - \tau(x_t) = z_1$. Then $x(t) < 0$ on $(z_1, t_1]$ since $\frac{d}{dt}(x(t)e^{\nu t}) < 0$ and the argument above implies that $x(t) > -M$ on $[z_1, t_1]$. This in turn implies that $t_1 \leq z_1 + R$ as noted above.

By Lemma 1.2, $t - \tau(x_t) > z_1 = t_1 - \tau(x_{t_1})$ for $t_1 < t \leq \tilde{t}$. Thus for $t_1 < t < z_2 \equiv \inf\{t > t_1 : x(t) = 0\}$, $x(t) < 0$, $x(t - \tau(x_t)) < 0$ so $x' > 0$. The argument that $z_2 < \infty$ is entirely symmetrical to the argument that $z_1 < \infty$. Indeed, this analogous argument establishes the existence of $Q > 0$, independent of $\phi \in \tilde{K}\backslash 0$, such that $z_2(\phi) \leq Q$.

If there does not exist $t_2 \in (z_2, \tilde{t}]$ such that $t_2 - \tau(x_{t_2}) = z_2$ then $z_1 < t - \tau(x_t) < z_2$ for $z_2 \leq t \leq \tilde{t}$. Consequently, $\frac{d}{dt}(e^{\nu t}x(t)) > 0$ and $x(t) > 0$ on $z_2 < t \leq \tilde{t}$. If $z_2 + R \leq \tilde{t}$, then $z_2 + R - \tau(x_{z_2+R}) \geq z_2$ in contradiction to our assumption that t_2 doesn't exist. Hence $\tilde{t} < z_2 + R$ which implies that $x(\tilde{t}) = N$.

Then, using (1.1) and considering only the case $v > 0$, we conclude that

$$N = \int_{z_2}^{\tilde{t}} -e^{-\nu(\tilde{t}-t)}e^{-\eta\tau(x_t)}f(x(t-\tau(x_t)))\,dt$$

$$\leq F^-(M) \int_{z_2}^{\tilde{t}} e^{-\nu(\tilde{t}-t)}\,dt = F^-(M)\left[\frac{1-e^{-\nu(\tilde{t}-z_2)}}{\nu}\right]$$

$$\leq F^-(M)\left[\frac{1-e^{-\nu\tau(x_{\tilde{t}})}}{\nu}\right] \leq F^-(M)\left[\frac{1-e^{-\nu R_N}}{\nu}\right],$$

since $\tilde{t} - \tau(x_{\tilde{t}}) < z_2$ and $\tau(x_{\tilde{t}}) \leq R_N$. As this violates (1.2b) we have a contradiction to our assumption that there exists no $t_2 \in (z_2, \tilde{t}]$ such that $t_2 - \tau(x_{t_2}) = z_2$. Consequently, such a t_2 exists, $x(t) < N$ on $(z_2, t_2]$ and $t_2 \leq z_2 + R$ as the previous arguments show.

COROLLARY 1.4. *Let the hypotheses of Proposition 1.3 hold. For each $\phi \in \tilde{K}\backslash 0$, $x(t) = x(t, \phi)$ satisfies the following properties.*

(i) *$x(t)$ is defined for all $t > 0$ and $-M < x(t) < N$ for $t > 0$.*

(ii) *There exist sequences $\{z_n\}$ and $\{t_n\}$ such that*

$$z_1 < t_1 < z_2 < t_2 < \cdots < z_n < t_n < z_{n+1} < \cdots$$

satisfying $z_n \to +\infty$ as $n \to \infty$ with the properties:

(a) *$\{z_n\} = \{t \geq 0 : x(t) = 0\}$, $x'(z_{2k-1}) < 0$, $x'(z_{2k}) > 0$,*

(b) *t_n is the unique solution t of $t - \tau(x_t) = z_n$ in (z_n, z_{n+1}),*

(c) *$t - \tau(x_t) \in (z_n, z_{n+1})$ for $n \geq 1$ if and only if $t \in (t_n, t_{n+1})$.*

(iii) *$e^{\nu t}x(t)$ is increasing on (t_{2k-1}, t_{2k}) and decreasing on (t_{2k}, t_{2k+1}).*

PROOF. The arguments in the proof of Proposition 1.3 can be continued beyond t_2, by an induction argument, in order to obtain the sequences $\{z_n\}$ and $\{t_n\}$ where $x(z_n) = 0$, $x'(z_{2n-1}) < 0$, $x'(z_{2n}) > 0$, $t_n - \tau(x_{t_n}) = z_n$, $t - \tau(x_t) \lessgtr z_n$ as $t \lessgtr t_n$, and $-M < x(t) < N$. The keys to extending the results of Proposition 1.3 are Lemma 1.2 and (D2). Lemma 1.2 asserts that for $t > t_2$, $t - \tau(x_t) > z_2$, thus only $x_{t_2}|_{[z_2-t_2,0]} = x_{t_2}|_{[-\tau(x_{t_2}),0]}$ is relevant for extending $x(t,\phi)$ to the right of t_2. By Lemma 1.2 and (D2), we may as well replace x_{t_2} by $\psi \in C_B$ defined by $\psi|_{[-\tau(x_{t_2}),0]} = x_{t_2}|_{[-\tau(x_{t_2}),0]}$ and $\psi \equiv 0$ on $(-\infty, -\tau(x_{t_2}))$, in order to extend $x(t,\phi)$ to the right of t_2. As $\psi \in \tilde{K}$, we see that Proposition 1.3 itself applies to extend $x(t,\phi)$ to $[t_2, t_4]$. In order to see that $z_n, t_n \to +\infty$ we show that $z_{n+1} - t_n$ is bounded below, independent of n. Using the differential equation and the fact that $z_n < t - \tau(x_t) < z_{n+1}$ for $t_n < t < z_{n+1}$ we have

$$x(t_n) = \int_{t_n}^{z_{n+1}} e^{\nu(t-t_n)}e^{-\eta\tau(x_t)}f(x(t-\tau(x_t)))\,dt.$$

For definiteness, assume n is even so $x(t_n) > 0$ and $x(t - \tau(x_t)) > 0$ on $t_n < t <$ z_{n+1}; the other case is similar. If p is a Lipschitz constant for $f|_{[-M,N]}$, then

$$x(t_n) \le p \int_{t_n}^{z_{n+1}} e^{\nu(t-t_n)} x(t - \tau(x_t)) \, dt$$

$$\le p \int_{t_n}^{z_{n+1}} e^{\nu(\tau(x_t)-t_n)} e^{\nu(t-\tau(x_t))} x(t - \tau(x_t)) \, dt$$

$$\le p \int_{t_n}^{z_{n+1}} e^{\nu(\tau(x_t)-t_n)} e^{\nu t_n} x(t_n) \, dt$$

$$\le p e^{\nu R} x(t_n)(z_{n+1} - t_n)$$

where we have used the fact that $e^{\nu t} x(t)$ achieves its maximum on $[z_n, z_{n+1}]$ at t_n and $z_n < t - \tau(x_t) < z_{n+1}$. Hence,

$$(1.4) \qquad\qquad z_{n+1} - t_n \ge p^{-1} e^{-\nu R}$$

for all even n. It is now clear that $z_n, t_n \to +\infty$ as $n \to \infty$.

Let $L = \nu \max\{M, N\} + \max\{|f(x)| : -M \le x \le N\}$ and define

$$K = \{\phi \in \tilde{K} : \operatorname{lip} \phi \le L\}.$$

K is a compact, convex subset of C_B containing the zero function. Define $T : K \to C_B$ by

$$(T\phi)(\theta) = \begin{cases} x(t_2(\phi) + \theta, \phi), & z_2 - t_2 \le \theta \le 0 \\ 0, & \theta < z_2 - t_2 \end{cases}$$

if $\phi \in K\backslash 0$ and $T0 = 0$. Our choice of L insures that $\operatorname{lip} x(\cdot, \phi) \le L$ for all $\phi \in K$.

PROPOSITION 1.5. T maps K into K. If $T\phi = \phi$ for some $\phi \in K\backslash 0$, then $x(t, \phi)$ is a nonconstant periodic solution of (1.1) with minimal period $t_2(\phi)$.

PROOF. If $\phi \in K\backslash 0$, then $0 < x(t) < N$ on $(z_2, t_2]$ so $0 < (T\phi)(\theta) < N$ on $z_2 - t_2 < \theta \le 0$ and $(T\phi)(\theta) = 0$ for $\theta \le z_2 - t_2$. As $t_2 - \tau(x_{t_2}) = z_2$, $z_2 - t_2 \ge -R$ and hence $(T\phi)(\theta) = 0$ for $\theta \le -R$. Since $e^{\nu t} x(t)$ is increasing on $[z_2, t_2]$, $e^{\nu\theta} x(t_2 + \theta)$ is increasing on $z_2 - t_2 \le \theta \le 0$ and so $e^{\nu\theta}(T\phi)(\theta)$ is nondecreasing on $-R \le \theta \le 0$. Obviously, $T\phi$ is Lipschitz and $\operatorname{lip}(T\phi) \le L$. It follows that $T\phi \in K$.

If $T\phi = \phi \in K\backslash 0$, then $s_\phi = z_2 - t_2$ and $\phi(\theta) = x(t_2 + \theta)$ for $z_2 - t_2 \le \theta \le 0$. As $t_2 - \tau(x_{t_2}) = z_2$ and $\phi|_{[-\tau(x_{t_2}),0]} = x_{t_2}|_{[-\tau(x_{t_2}),0]}$ it follows from (D2) that $\tau(x_{t_2}) = \tau(\phi)$ and hence $-\tau(\phi) = s_\phi$. By Lemma 1.2 and the fact that $\phi'(0) + \nu\phi(0) = x'(t_2) + \nu x(t_2) = 0$, $t - \tau(x_t) > s_\phi$ for $0 < t \le t_2$. We can now verify that the function $y : \mathbf{R} \to \mathbf{R}$ defined by $y_0 = \phi$, $y(t) = x(s)$ for $t \ge 0$ where $t = nt_2 + s$, $0 \le s < t_2$, $n \ge 0$ an integer, is a t_2-periodic solution of (1.1). Note that $y_{t_2} = x_{t_2}$. Thus, for $0 \le s \le t_2$, $x_s|_{[-\tau(x_s),0]} = y_s|_{[-\tau(x_s),0]}$ so by (D2) $\tau(x_s) = \tau(y_s)$. Then $y_s|_{[-\tau(x_s),0]} = y_{nt_2+s}|_{[-\tau(x_s),0]}$ for $n \ge 1$ so

$\tau(x_s) = \tau(y_s) = \tau(y_t)$ by (D2), where $t = nt_2 + s$. But this implies that

$$y'(t) = x'(s) = -\nu x(s) - e^{-\eta\tau(x_s)}f(x(s - \tau(x_s)))$$
$$= -\nu y(t) - e^{-\eta\tau(y_t)}f(x_s(-\tau(x_s)))$$
$$= -\nu y(t) - e^{-\eta\tau(y_t)}f(y_t(-\tau(y_t))).$$

Thus $y(t)$ satisfies (1.1) for all $t \geq 0$. Now $y(t) = x(t, \phi)$ for $t \geq 0$ by the uniqueness of solutions of initial value problems asserted in Lemma 1.1.

PROPOSITION 1.6. $T : K \to K$ is continuous.

PROOF. We begin by establishing an estimate on $|x(t, \phi) - y(t, \phi)|$ for $\phi, \psi \in K$. Let $m = \max\{|f(x)| : -M \leq x \leq N\}$ and $|f(x) - f(y)| \leq p|x - y|$ for all $x, y \in [-M, N]$. If $\phi, \psi \in K$, put $Z(t) = \sup\{e^{\nu s}|x(s) - y(s)| : -\infty \leq s \leq t\}$, where $x(t) = x(t, \phi)$ and $y(t) = x(t, \psi)$. From (1.1) we have

$$|e^{\nu t}(x(t) - y(t))| \leq |x(0) - y(0)| + \int_0^t e^{\nu s}|e^{-\eta\tau(x_s)}f(x(s - \tau(x_s)))$$
$$- e^{-\eta\tau(y_s)}f(y(s - \tau(y_s)))|\, ds$$

$$\leq Z(0) + \int_0^t e^{\nu s}|f(x(s - \tau(x_s)))|\,|e^{-\eta\tau(x_s)} - e^{-\eta\tau(y_s)}|\, ds$$

$$+ \int_0^t e^{\nu s}e^{-\eta\tau(y_s)}|f(x(s - \tau(x_s))) - f(y(s - \tau(y_s)))|\, ds$$

$$\leq Z(0) + m\eta \int_0^t e^{\nu s}|\tau(x_s) - \tau(y_s)|\, ds + \int_0^t e^{\nu s}p|x(s - \tau(x_s)) - y(s - \tau(y_s))|\, ds$$

$$\leq Z(0) + Vm\eta \int_0^t e^{\nu s} \max_{-R \leq r \leq 0} |x(s + r) - y(s + r)|\, ds$$

$$+ p \int_0^t e^{\nu s}|x(s - \tau(x_s)) - x(s - \tau(y_s))|\, ds$$

$$+ p \int_0^t e^{\nu s}|x(s - \tau(y_s)) - y(s - \tau(y_s))|\, ds$$

$$\leq Z(0) + Vm\eta e^{\nu R} \int_0^t Z(s)\, ds + p \int_0^t Le^{\nu s}|\tau(x_s) - \tau(y_s)|\, ds$$

$$+ pe^{\nu R} \int_0^t Z(s)\, ds$$

$$\leq Z(0) + [Vm\eta e^{\nu R} + pe^{\nu R} + pLVe^{\nu R}] \int_0^t Z(s)\, ds.$$

Hence,

$$Z(t) \leq Z(0) + A \int_0^t Z(s)\, ds$$

and Gronwall's inequality yields $Z(t) \leq Z(0)e^{At}$ where $A = e^{\nu R}[Vm\eta + p + pLV]$. This implies

$$|x(t) - y(t)| \leq \max_{-R \leq s \leq 0} e^{\nu s}|\phi(s) - \psi(s)|e^{(A - \nu)t} \text{ for } t \geq 0.$$

Hence, for each $\bar{t} > 0$, $x(t, \phi)$ is uniformly close to $x(t, \psi)$ on $[0, \bar{t}]$ if ϕ is uniformly close to ψ on $[-R, 0]$ and $\phi, \psi \in K$.

It is now easy to establish the continuity of the maps $\phi \rightarrow z_i(\phi)$, $i = 1, 2$, at each $\phi \in K\backslash 0$, using the fact that these are simple zeros of $x(t, \phi)$ and that the z_i are uniformly bounded from above. One can then use the fact that $t_2(\phi)$ is the unique solution of $t - \tau(x_t(\phi)) = z_2(\phi)$ together with the continuity of τ to show that $\phi \rightarrow t_2(\phi)$ is also continuous at each $\phi \in K\backslash 0$. The continuity of T on K follows easily from this.

PROPOSITION 1.7. *If* $T\phi = \phi \in K\backslash 0$, *then the* $t_2(\phi)$-*periodic solution* $x(t, \phi)$ *has the following properties:*

   (i) $x(t)$ *has precisely two zeros in* $[0, t_2]$, *namely* z_1 *and* z_2, $0 < z_1 < z_2 < t_2$ *with* $x'(z_1) < 0$ *and* $x'(z_2) > 0$,
   (ii) $e^{\nu t}x(t)$ *has two extrema in* $(0, t_2]$, *namely* t_1 *and* t_2, $0 < z_1 < t_1 < z_2 < t_2$ *with* $e^{\nu t}x(t)$ *increasing on* (t_1, t_2) *and decreasing on* $(0, t_1)$,
   (iii) $-M < x(t) < N$ *for all* t,
   (iv) *If* $\tau(x_t) = \tau(x(t))$, *then* $z_2 - z_1 > 1$ *and* $t_2 > 2$.

PROOF. (i), (ii) and (iii) follow from Proposition 1.3 and the periodicity of $x(t)$. If $\tau(x_t) = \tau(x(t))$, then since $\tau(0) = 1$ and $z_2 - \tau(x(z_2)) = z_2 - 1 \in (z_1, z_2)$, we have $z_2 - z_1 > 1$. Hence $t_2 = z_2 - s_\phi = z_2 - z_1 + z_1 - s_\phi > 1 + 1 = 2$ as $z_1 - s_\phi > 1$ by the same argument as above.

Finally, we assume

   (H) $e^{-\eta}f'(0) > \alpha_\nu$ where α_ν is the smallest positive solution of

$$\nu + \alpha \cos \sqrt{\alpha^2 - \nu^2} = 0.$$

As noted in [15], $\alpha_\nu \geq \pi/2 = \alpha_0$ and α_ν increases with ν. With this assumption we can prove the following Lemma, using classical ideas of ([35], [29], [30], [15]), with slight modifications.

LEMMA 1.8. *Assume that* (H) *holds. Then there exists a positive number* a, *independent of* $\phi \in K\backslash 0$ *such that*

$$\sup_{t \geq z_n} |x(t, \phi)| \geq a$$

for all positive integers n *where* z_n *is the* n'*th positive zero of* $x(t, \phi)$.

PROOF. Let $\alpha = e^{-\eta}f'(0)$. Hypothesis (H) implies, see [15], that the characteristic equation $\nu + \lambda + \alpha e^{-\lambda} = 0$ has a root $\lambda = \mu + i\gamma$ satisfying $\mu > 0$ and $\frac{\pi}{2} < \gamma < \pi$. Express (1.1) as

$$x'(t) = -\nu x(t) - \alpha x(t-1) + g(x_t) + h(x_t) + k(x_t)$$

where $g(x_t) = \alpha x(t-1) - e^{-\eta}f(x(t-1))$, $h(x_t) = [e^{-\eta} - e^{-\eta\tau(x_t)}]f(x(t-1))$ and $k(x_t) = e^{-\eta\tau(x_t)}[f(x(t-1)) - f(x(t-\tau(x_t)))]$. Choose $\epsilon > 0$ such that $\epsilon < \frac{1}{2}\mu \cos\frac{\alpha}{2}e^{-3\nu/2}$ and choose a such that $|\alpha x - e^{-\eta}f(x)| \leq \frac{\epsilon}{3}|x|$ if $|x| \leq a$. We will choose a still smaller below. Let p be a Lipschitz constant for $f|_{[-M,N]}$.

If $\phi \in K \backslash 0$, then $-M \leq x(t, \phi) \leq N$ so we can estimate the functions $h(x_t), k(x_t)$ along the solution by

$$|h(x_t)| \leq p|x(t-1)| |e^{-\eta\tau(0)} - e^{-\eta\tau(x_t)}| \leq p|x(t-1)| |\eta| |\tau(x_t) - \tau(0)|$$
$$\leq p\eta V |x(t-1)| \max_{-Q \leq s \leq 0} |x(t+s)|,$$

and

$$|k(x_t)| \leq p|x(t-1) - x(t-\tau(x_t))| \leq p|x'(\xi)| |\tau(x_t) - \tau(0)|$$
$$\leq p|x'(\xi)| V \max_{-Q \leq s \leq 0} |x(t+s)|$$
$$\leq pV \max_{-Q \leq s \leq 0} |x(t+s)| [\nu|x(\xi)| + p|x(\xi - \tau(x_\xi))|]$$
$$\leq pV(\nu + p) \max_{-Q \leq s \leq 0} |x(t+s)| \max\{|x(\xi)|, |x(\xi - \tau(x_\xi))|\},$$

where ξ is a point between $t-1$ and $t-\tau(x_t)$ and $Q = \max\{1, \tau(x_t)\} \leq R$. Choose a smaller if necessary so that the following hold:

$$(1.5) \qquad\qquad p\eta Va \leq \frac{\epsilon}{3},$$

$$(1.6) \qquad\qquad pV(\nu + p)a \leq \frac{\epsilon}{3},$$

$$(1.7) \qquad\qquad pVae^{3\nu/2} \leq \frac{1}{3}\cos\frac{\gamma}{2}e^{-\mu/2},$$

$$(1.8) \qquad\qquad aV \leq \frac{1}{2},$$

$$(1.9) \qquad\qquad apV < e^{-\nu R},$$

$$(1.10) \qquad\qquad (\nu + p)aVe^{3\nu/2} \leq \frac{1}{3},$$

and

$$(1.11) \qquad\qquad \gamma(\frac{1}{2} + aV) < \frac{\pi}{2}.$$

Using (1.5) and (1.6) and the estimates above, we have

$$(1.12) \quad |g(x_t) + h(x_t) + k(x_t)| \leq \frac{\epsilon}{3}|x(t-1)| + \frac{\epsilon}{3}|x(t-1)| + \frac{\epsilon}{3}\max_{-Q \leq s \leq 0}|x(t+s)|$$

provided $\max_{2R \leq s \leq 0} |x(t+s)| \leq a$ since ξ and $\xi - \tau(x_\xi) \geq t - 2R$.

If $\sup_{t \geq z_n} |x(t, \phi)| \geq a$ fails to hold for some integer, then clearly it fails to hold for all larger integers so we may choose an n such that $\sup_{t \geq z_n} |x(t, \phi)| = \delta < a$ and such that $|x(t)| \leq a$ for $t \geq z_n - 2R$ and such that there exists $t_0 \in (z_n, z_{n+1})$ for which $|x(t_0)| = \max_{z_n < t < z_{n+1}} |x(t)| \geq 3\delta/4$ and $x'(t_0) = 0$. We assume $x(t_0) > 0$ for definiteness; the other case is handled similarly. As $x'(t) < 0$ on $[t_n, z_{n+1}]$, it follows that $t_0 < t_n$.

We first argue that $z_{n+1} - z_n > 1$. From (1.4), $z_{n+1} - t_n \geq p^{-1}e^{-\nu R}$. As $t_n - z_n = \tau(x_{t_n}) \geq 1 - aV$ by (D3), we have

$$z_{n+1} - z_n \geq p^{-1}e^{-\nu R} + 1 - aV > 1$$

by (1.9).

Now put $T = 1 + z_n \in (z_n, z_{n+1})$. The estimate

(1.13) $|T - t_n| = |\tau(0) - \tau(x_{t_n})| \leq Va$

will be used routinely.

We estimate $x(t_n)$ by noting that $e^{\nu t}x(t)$ is increasing on $[z_n, t_n]$ and t_0 is a point in this interval. Hence,

$$x(t_n) \geq x(t_0)e^{-\nu(t_n - t_0)} \geq x(t_0)e^{-\nu(t_n - z_n)} = x(t_0)e^{-\nu\tau(x_{t_n})}$$

(1.14)

$$\geq x(t_0)e^{-\nu(1+Va)} \geq x(t_0)e^{-3\nu/2} \geq \frac{3\delta}{4}e^{-3\nu/2},$$

where (1.8) was used.

Now we estimate $x(T)$ as follows

$$S = \frac{|x(T) - x(t_n)|}{x(t_n)} = \frac{|x'(s)| \, |T - t_n|}{x(t_n)}$$

$$= |x'(s)|\frac{|1 + z_n - t_n|}{x(t_n)} = |x'(s)|\frac{|1 - \tau(x_{t_n})|}{x(t_n)}$$

$$\leq (\nu + p)\max\{|x(s)|, |x(s - \tau(x_s))|\}\frac{|1 - \tau(x_{t_n})|}{x(t_n)}$$

where s lies between t_n and $1 + z_n$. If $t_n < 1 + z_n$, then $s > t_n$ so $s - \tau(x_s) > z_n$ by Corollary 1.4 and $0 < x(s - \tau(x_s)) < x(t_0)$. Hence

$$S \leq (\nu + p)x(t_0)\frac{Va}{x(t_n)} \leq (\nu + p)Vae^{3\nu/2} \leq \frac{1}{3}$$

where we have used (1.14). If $t_n \geq 1 + z_n$, then $\tau(x_{t_n}) = t_n - z_n \geq 1$ so $|\tau(x_{t_n}) - 1| \leq V\max_{-\tau(x_{t_n}) \leq r \leq 0}|x(t_n + r)| \leq Vx(t_0)$. Hence, again

$$S \leq (\nu + p)a\frac{Vx(t_0)}{x(t_n)} \leq (\nu + p)Vae^{3\nu/2} \leq \frac{1}{3}.$$

The last inequality follows from (1.10). Finally, we have

$$|x(T)| \geq |x(t_n)|[1 - S] \geq \frac{2}{3}|x(t_n)| \geq \frac{\delta}{2}e^{-3\nu/2}.$$

Now we proceed as in [15]. If λ is the root of the characteristic equation as described above, then an integration by parts yields

$$\int_T^\infty x'(t)e^{-\lambda t}\,dt = -x(T)e^{-\lambda T} + \lambda\int_T^\infty x(t)e^{-\lambda t}\,dt.$$

Using the differential equation in the left side, we get

$$\int_T^\infty x'(t)e^{-\lambda t}\,dt = \int_T^\infty [-\nu x(t) - \alpha x(t-1) + g(x_t) + h(x_t) + k(x_t)]e^{-\lambda t}\,dt$$

$$= -\nu \int_T^\infty x(t)e^{-\lambda t}\,dt - \alpha e^{-\lambda}\int_{T-1}^\infty x(t)e^{-\lambda t}\,dt + \int_T^\infty [g+h+k]e^{-\lambda t}\,dt$$

$$= \nu \int_{T-1}^T x(t)e^{-\lambda t}\,dt + \lambda \int_{T-1}^\infty x(t)e^{-\lambda t}\,dt + \int_T^\infty [g+h+k]e^{-\lambda t}\,dt$$

where we have used the characteristic equation $\nu + \alpha e^{-\lambda} = -\lambda$. Putting this in the equality above, we have

$$-x(T)e^{-\lambda T} - (\lambda + \nu)\int_{T-1}^T x(t)e^{-\lambda t}\,dt = \int_T^\infty [g+h+k]e^{-\lambda t}\,dt.$$

Writing the integrand in the integral on the left as $[x(t)e^{\nu t}]e^{-(\lambda+\nu)t}$ and integrating by parts, using $x(T-1) = 0$, gives

$$-\int_{T-1}^T y'(t)e^{-(\lambda+\nu)t}\,dt = \int_T^\infty [g+h+k]e^{-\lambda t}\,dt.$$

where $y(t) = x(t)e^{\nu t}$. Finally, multiply through by $e^{\lambda(T-1/2)}$ to get

$$(1.15) \qquad \int_{T-1}^T y'(t)e^{-\lambda(t-T+1/2)}e^{-\nu t}\,dt = -\int_T^\infty [g+h+k]e^{-\lambda(t-T+1/2)}\,dt.$$

There are two cases to consider. If $T = 1 + z_n \le t_n$, then $y' \ge 0$ on $[T-1, T] = [z_n, 1 + z_n]$ so

$$\mathrm{Re}\int_{T-1}^T y'(t)e^{-\lambda(t-T+1/2)}e^{-\nu t}\,dt$$

$$= \int_{T-1}^T y'(t)e^{-\mu(t-T+1/2)}\cos\gamma(t-T+1/2)e^{-\nu t}\,dt$$

$$\ge e^{-\mu/2}\cos\frac{\gamma}{2}e^{-\nu T}[y(T) - y(T-1)]$$

$$\ge e^{-\mu/2}\cos\frac{\gamma}{2}x(T)$$

$$\ge e^{-\mu/2}\cos\frac{\gamma}{2}\frac{\delta}{2}e^{-3\nu/2}.$$

On the other hand, estimating the real part of the right side of (1.15),

$$\mathrm{Re}\left[-\int_T^\infty [g+h+k]e^{-\lambda(t-T+1/2)}\,dt\right]$$

$$\le \int_T^\infty [|g| + |h|]e^{-\mu(t-T+1/2)}\,dt - \int_T^{t_n} ke^{-\mu(t-T+1/2)}\cos\gamma(t-T+1/2)\,dt$$

$$+ \int_{t_n}^\infty |k|e^{-\mu(t-T+1/2)}\,dt.$$

The integrand in the second integral is nonnegative on $[T, t_n]$ since on that interval $t - \tau(x_t) < z_n < t - 1$ so $f(x(t-1)) > 0$ and $f(x(t - \tau(x_t))) < 0$ (see the definition of k). Also note that $\cos \gamma(t - T + 1/2) > 0$ on $[T, t_n]$ by virtue of (1.11), $\pi/2 < \gamma < \pi$ and (1.13). Using (1.12) in the above estimate yields

$$\text{Re}\left[-\int_T^\infty [g + h + k]e^{-\lambda(t - T + 1/2)}\, dt\right] \leq \frac{\epsilon}{3}\delta\frac{e^{-\mu/2}}{\mu} + \frac{\epsilon}{3}\delta\frac{e^{-\mu/2}}{\mu} + \frac{\epsilon}{3}\delta\frac{e^{-\mu/2}}{\mu}$$

$$= \frac{\epsilon\delta e^{-\mu/2}}{\mu},$$

where we have used the fact that $t - \tau(x_t) > z_n$ for $t > t_n$ and thus $|k| \leq pV(\nu + p)a \max_{-\tau(x_t) \leq s \leq 0} |x(t + s)| \leq \frac{\epsilon}{3}\delta$ by (1.6). But these two estimates imply that

$$\frac{\delta}{2}e^{-\mu/2}\cos\frac{\gamma}{2}e^{-3\nu/2} \leq \epsilon\delta e^{-\mu/2}\mu^{-1}$$

contradicting our choice of ϵ.

If $T > t_n$, then $y' \geq 0$ on $[T - 1, t_n]$ and $y' \leq 0$ on $[t_n, T]$ so we must estimate the two sides of (1.15) as follows.

$$I = \text{Re}\int_{T-1}^T y'(t)e^{-\lambda(t - T + 1/2)}e^{-\nu t}\, dt$$

$$= \int_{T-1}^{t_n} y'(t)e^{-\mu(t - T + 1/2)}\cos\gamma(t - T + 1/2)e^{-\nu t}\, dt$$

$$+ \int_{t_n}^T y'(t)e^{-\mu(t - T + 1/2)}\cos\gamma(t - T + 1/2)e^{-\nu t}\, dt$$

$$= I_1 + I_2.$$

We may estimate I_1 as above,

$$I_1 \geq \cos\frac{\gamma}{2}e^{-\mu/2}e^{-\nu t_n}y(t_n) = \cos\frac{\gamma}{2}e^{-\mu/2}x(t_n).$$

To estimate I_2, note that by (1.13),

$$e^{-\mu(t - T + 1/2)}\cos\gamma(t - T + 1/2) \leq e^{-\mu(t_n - T + 1/2)} \leq 1.$$

Therefore

$$I_2 \geq \int_{t_n}^T y'(t)e^{-\nu t}\, dt = \int_{t_n}^T -e^{-\eta\tau(x_t)}f(x(t - \tau(x_t)))\, dt$$

$$\geq \int_{t_n}^T -px(t - \tau(x_t))\, dt \geq -px(t_0)[T - t_n]$$

$$\geq -pe^{3\nu/2}x(t_n)[T - t_n] \geq -paVe^{3\nu/2}x(t_n),$$

where we used (1.14) and $t - \tau(x_t) \geq z_n$ for $t_n \leq t \leq T$. Putting the estimates of I_1 and I_2 together and using (1.7) and (1.14), we have

$$
\begin{aligned}
I &\geq (\cos \frac{\gamma}{2} e^{-\mu/2} - paV e^{3\nu/2}) x(t_n) \\
&\geq \frac{2}{3} \cos \frac{\gamma}{2} e^{-\mu/2} x(t_n) \\
&\geq \frac{\delta}{2} \cos \frac{\gamma}{2} e^{-\mu/2} e^{-3\nu/2}.
\end{aligned}
$$

Note that this estimate is the same estimate obtained in the previous case.

We can estimate the real part of the right-hand side of (1.15), in this case, using (1.2) since $T - Q \geq z_n$

$$
\begin{aligned}
&\mathrm{Re} \left[-\int_T^\infty [g + h + k] e^{-\lambda(t-T+1/2)} \, dt \right] \\
&\leq \int_T^\infty [|g| + |h| + |k|] e^{-\mu(t-T+1/2)} \, dt \\
&\leq \int_T^\infty \left[\frac{2\epsilon}{3} |x(t-1)| + \frac{\epsilon}{3} \max_{-Q \leq s \leq 0} |x(t+s)| \right] e^{-\mu(t-T+1/2)} \, dt \\
&\leq \epsilon \delta \frac{e^{-\mu/2}}{\mu}.
\end{aligned}
$$

Thus we reach the same contradiction as before.

THEOREM 1.9. *If* (H) *holds, then* $T : K \to K$ *has a nonzero fixed point. In particular,* (1.1) *has a nonconstant periodic solution with the properties described in Proposition* 1.7.

PROOF. We use the theorem of Browder [5] (see also [18], Thm. 11.2.4) on the existence of a nonejective fixed point, showing that 0 is an ejective point of T and therefore there must exist a nonzero fixed point of T. The point 0 of K is an ejective point if there is an open neighborhood G of 0 in C_B such that for every $\phi \in G \cap K$, $\phi \neq 0$, there is an integer $m = m(\phi)$ such that $T^m \phi \notin G \cap K$. Recall that K is a compact and convex subset of C_B. If $\phi \in K \backslash 0$ we denote by r_n, a point of (z_n, z_{n+1}) at which $|x(t, \phi)|$ achieves its maximum value, $n \geq 1$. Then $\|T^n \phi\| = x(r_{2n}, \phi)$, $n \geq 1$. Lemma 1.8 implies that $|x(r_n, \phi)| \geq a/2$ for infinitely many values of n, if $\phi \in K \backslash 0$. As $r_n < t_n$, an estimate as in (1.14) gives

$$
|x(t_n, \phi)| \geq e^{-\nu R} |x(r_n, \phi)|.
$$

Arguing as in the estimates yielding (1.4),

$$
x(t_{2n+1}) = \int_{z_{2n+1}}^{t_{2n+1}} -e^{\nu(t-t_{2n+1})} e^{-\eta \tau(x_t)} f(x(t - \tau(x_t))) \, dt
$$

so that, as $z_{2n} < t - \tau(x_t) < z_{2n+1}$ for $z_{2n+1} < t < t_{2n+1}$,

$$|x(t_{2n+1})| \leq p \int_{t_{2n+1}}^{z_{2n+1}} e^{\nu(\tau(x_t)-t_{2n+1})} e^{\nu(t-\tau(x_t))} |x(t-\tau(x_t))| \, dt$$
$$\leq px(t_{2n}) e^{\nu R} e^{t_{2n}-t_{2n+1}} (z_{2n+1} - t_{2n+1})$$
$$\leq pRe^{\nu R} x(t_{2n}).$$

Hence,

$$x(r_{2n}) \geq x(t_{2n}) \geq (pRe^{\nu R})^{-1} |x(t_{2n+1})|$$
$$\geq (pRe^{\nu R})^{-1} e^{-\nu R} |x(r_{2n+1})|.$$

It follows that $x(r_{2n}) \geq (pRe^{\nu R})^{-1} e^{-\nu R} a/2$ for infinitely many values of n, proving that 0 is an ejective fixed point of T. This proves the Theorem.

2. Generalized Threshold Delays

In this section we show that our main result applies to (1.1) in the case that $\tau(x_t)$ is determined implicitly by

$$(2.1) \qquad \int_{t-\tau}^{t} k(x(t), x(s)) \, ds = m$$

where $k : \mathbf{R}^2 \to (0, \infty)$ is locally Lipschitz and $m > 0$. Dividing through by m, incorporating m into k and rescaling time, it may be assumed that

$$m = 1 \qquad \text{and} \qquad k(0,0) = 1.$$

We assume the above normalization throughout this section. The delay (2.1) was treated extensively in [23] and we will quote results from there as necessary.

The delay appearing in (2.1) can be viewed as a delay functional $\tau : C_B \to (0, \infty)$ defined implicitly by the formula

$$(2.2) \qquad \int_{-\tau(\phi)}^{0} k(\phi(0), \phi(s)) \, ds - 1 = 0.$$

LEMMA 2.1. *Given $L > 0$, let $p, q, m, M > 0$, depending on L, be such that*

$$|k(x,y) - k(u,v)| \leq p|x-u| + q|y-v|$$

and

$$m \leq k(x,y) \leq M$$

for all $(x,y), (u,v)$ such that $|x|, |y|, |u|, |v| \leq L$. Then (2.2) defines a functional $\tau : C_B \to (0, \infty)$ satisfying:

  (i) $\tau(0) = 1$,
  (ii) $\|\phi\| \leq L$ implies $M^{-1} \leq \tau(\phi) \leq m^{-1}$,
  (iii) *There exists $P = P(L) > 0$ such that if $\|\phi\|, \|\psi\| \leq L$, then*

$$|\tau(\phi) - \tau(\psi)| \leq P \max_{-Q \leq s \leq 0} |\phi(s) - \psi(s)|$$

where $Q = \max\{\tau(\phi), \tau(\psi)\}$ and $\phi, \psi \in C_B$.

PROOF. See ([23], Lemma 2.1 and its proof).

As a consequence of Lemma 2.1 and the definition of $\tau(\phi)$, the latter satisfies hypotheses (D1), (D2) and (D3).

Hypothesis (B) concerning the choice of $M, N > 0$ such that solutions $x(t, \phi)$ corresponding to $\phi \in \tilde{K}(M, N)$ satisfy $-M < x(t, \phi) < N$, involves τ only through the numbers R_M, R_N. It is easy to verify that

$$R_N = \left[\min_{-M \leq y \leq N} k(N, y) \right]^{-1}$$

$$R_M = \left[\min_{-M \leq y \leq N} k(-M, y) \right]^{-1}.$$

The next Lemma provides sufficient conditions for the delay defined by (2.1) to satisfy (D4). The sufficient condition, (2.3), may be easily derived from formally differentiating (2.1) with respect to t.

LEMMA 2.2. *In addition to the assumptions above, assume that $\frac{\partial k}{\partial x}(x, y)$ exists when $x \neq 0$, $x, y \in [-M, N]$ and is continuous. If*

$$(2.3) \qquad \nu x \frac{\partial k}{\partial x}(x, y) < k(x, x)k(x, y), \qquad -M \leq x, y \leq N, \ x \neq 0,$$

then τ satisfies (D4).

PROOF. The proof follows a similar one in [23]. Let $\phi \in C_B$ be Lipschitz, take values in $[-M, N]$ and satisfy $\phi(-\tau(\phi)) = 0$. Actually, we do not use the last assumption on ϕ nor that it is Lipschitz. Suppose that y is an extension of ϕ as in (D4) satisfying $y'(0) + \nu\phi(0) = 0$. Let $\tau_0 = \tau(\phi)$ and $\tau(y_t) = \tau_0 + r(t)$ for simplicity in notation. Then from

$$1 = \int_{-\tau_0}^{0} k(\phi(0), \phi(s)) \, ds = \int_{-\tau_0 - r(t)}^{0} k(y(t), y(t + s)) \, ds$$

we have

$$\int_{-\tau_0}^{0} k(\phi(0), \phi(s)) \, ds = \int_{-\tau_0 - r}^{-t} k(y(t), y(t + s)) \, ds + \int_{-t}^{0} k(y(t), y(t + s)) \, ds$$

$$= \int_{t-\tau_0 - r}^{0} k(y(t), \phi(s)) \, ds + \int_{-t}^{0} k(y(t), y(t + s)) \, ds.$$

This leads to

$$0 = \int_{-\tau_0}^{0} t^{-1}[k(y(t), \phi(s)) - k(\phi(0), \phi(s))] \, ds + t^{-1} \int_{t-\tau_0 - r}^{-\tau_0} k(y(t), \phi(s)) \, ds$$

$$+ t^{-1} \int_{-t}^{0} k(y(t), y(t + s)) \, ds.$$

Letting $t \to 0+$, we find, assuming $\phi(0) \neq 0$, that
(2.4)

$$\lim_{t \to 0+} t^{-1} \int_{t-\tau_0 - r}^{-\tau_0} k(y(t), \phi(s)) \, ds = \int_{-\tau_0}^{0} \nu\phi(0) \frac{\partial k}{\partial x}(\phi(0), \phi(s)) \, ds - k(\phi(0), \phi(0)).$$

If $\phi(0) = 0$, then as $y'(0) = 0$ and k is Lipschitz, the right side of (2.4) is $-k(\phi(0), \phi(0))$. If we can show that the right-hand side is negative then it follows that $t - r(t) > 0$ for small positive t and hence $\tau(y_t) - \tau(\phi) = r(t) < t$ for small positive t. Now, using (2.3)

$$\int_{-\tau_0}^0 \nu \phi(0) \frac{\partial k}{\partial x}(\phi(0), \phi(s)) \, ds$$

$$= \int_{-\tau_0}^0 \nu \phi(0) \left(\frac{\partial k}{\partial x}(\phi(0), \phi(s)) / k(\phi(0), \phi(s)) \right) k(\phi(0), \phi(s)) \, ds$$

$$< \int_{-\tau_0}^0 k(\phi(0), \phi(0)) k(\phi(0), \phi(s)) \, ds$$

$$= k(\phi(0), \phi(0)).$$

Hence the integral on the right side of (2.4) is negative. Thus the first requirement of (D4) holds. The second is established in a similar manner.

In the special case that

$$k(x, y) = \frac{1}{\sigma(x)} \qquad \text{i.e.} \qquad \tau(x_t) = \sigma(x(t))$$

then (2.3) becomes

$$\nu x \sigma'(x) > -1 \qquad \text{for} \qquad x \neq 0.$$

In case

$$k(x, y) = k(y) \qquad \text{so} \qquad \int_{-\tau(x_t)}^0 k(x(t+s)) \, ds = 1$$

then (2.3) is automatically satisfied since $\frac{\partial k}{\partial x} = 0$.

REFERENCES

1. W. Alt, *Some periodicity criteria for functional differential equations*, Manuscripta Math. **23** (1978), 295–318.
2. ———, *Periodic solutions of some autonomous differential equations with variable time delay*, Proc. Conf. on Functional Differential Equations and Approx. of Fixed Points (Bonn, 1978), vol. 730, Springer, Berlin, Heidelberg, and New York, 1979.
3. J. Bélair and M. C. Mackey, *Consumer memory and price fluctuations in commodity markets: an integro-differential model*, J. Dynam. Differential Equations **1** (1989), 299–325.
4. J. Bélair, *Population models with state-dependent delays*, Lecture Notes in Pure and Applied Mathematics **131** (1991) (O. Arino, D. Axelrod, M. Kimmel, ed.), Academic Press, Proc. 2nd International Conf. on Mathematical Population Dynamics.
5. F. E. Browder, *A further generalization of the Schauder fixed point theorem*, Duke Math. J. **32** (1965), 575–578.
6. S. N. Chow, *Existence of periodic solutions of autonomous functional differential equations*, J. Differential Equations **15** (1974), 350–378.
7. Y. Cao and H. I. Freedman, *Global attractivity of a model for stage-structured population growth with density dependent time delay*, preprint.
8. S. N. Chow and J. K. Hale, *Periodic solutions of autonomous equations*, J. Math. Anal. Appl. **66** (1978), 495–506.
9. R. D. Driver, *Ordinary and Delay Differential Equations*, Springer, New York, 1977.
10. R. B. Grafton, *A periodicity theorem for autonomous functional differential equations*, J. Differential Equations **6** (1969), 87–109.

11. J. A. Gatica and P. Waltman, *A threshold model of antigen antibody dynamics with fading memory*, Nonlinear Phenomena in Mathematical Sciences (V. Lakshmikantham, ed.), Academic Press, New York, 1982, pp. 425–439.

12. _____, *Existence and Uniqueness of Solutions of a functional differential equation modeling thresholds*, Nonlinear Anal. **8** (1984), 1215–1222.

13. _____, *A system of functional differential equations modeling threshold phenomena*, Appl. Anal. **28** (1988), 39–50.

14. S. Grossman and J. A. Yorke, *Asymptotic behavior and stability criteria for delay differential equations*, J. Differential Equations **12** (1972), 236–255.

15. K. P. Hadeler and J. Tomiuk, *Periodic solutions of difference-differential equations*, Arch. Rat. Mech. Anal. **65** (1977), 87–95.

16. F. C. Hoppensteadt and P. Waltman, *A flow mediated control model of respiration*, Lectures on Mathematics in the Life Sciences, vol. 12, Amer. Math. Soc., 1979.

17. A. Halanay and J. A. Yorke, *Some new results and problems in the theory of differential-delay equations*, SIAM Rev. **13** (1971), 55–80.

18. J. K. Hale, *Theory of Functional Differential Equations*, Springer-Verlag, New York, 1977.

19. G. S. Jones, *The existence of periodic solutions of* $f'(x) = -2f(x-1)(1+f(x))$, J. Math. Anal. Appl. **5** (1962), 435–450.

20. Z. Jackiewicz and E. Lo, *The numerical solution of neutral functional differential equations by Adams predictor corrector methods*, Tech. report #118, Arizona State University, 1988.

21. J. Kirk, J. S. Orr, J. Forrest, *The role of chalone in the control of the bone marrow stem cell population*, Math. Biosciences **6** (1970), 129–143.

22. Y. Kuang and H. L. Smith, *Slowly oscillating periodic solutions of autonomous state-dependent delay equations*, J. Nonlinear Analysis, to appear.

23. _____, *Periodic solutions of autonomous state-dependent threshold-delay equations*, preprint.

24. J. L. Kaplan and J. A. Yorke, *On the nonlinear differential delay equation* $x'(t) = -f(x(t), x(t-1))$, J. Differential Equations **23** (1977), 293–314.

25. G. Ladas, Y. G. Sficas and I. P. Stavroulakis, *Necessary and sufficient conditions for oscillations*, Amer. Math. Monthly **90** (1983), 637–640.

26. J. A. J. Metz and O. Diekmann, *The Dynamics of Physiologically Structured Populations*, Lecture Notes in Biomathematics **68** (1986), Springer-Verlag, New York.

27. J. Mallet-Paret and R. Nussbaum, *Periodic solutions of state-dependent differential delay equations: I*, preprint.

28. R. M. Nisbet and W. S. C. Gurney, *The systematic formulation of population models for insects with dynamically varying instar duration*, Theoret. Population Biol. **23** (1983), 114–135.

29. R. D. Nussbaum, *Periodic solutions of some nonlinear, autonomous functional differential equations II.*, J. Differential Equations **14** (1973), 360–394.

30. _____, *Periodic solutions of some nonlinear, autonomous functional differential equations*, Ann. Mat. Pura Appl. (4) **101 yr 1974**, 263–306.

31. H. L. Smith, *Hopf bifurcation in a system of functional equations modeling the spread of infectious disease*, SIAM J. Appl. Math. **43** (1983), 370–385.

32. H. O. Walther, *Existence of a nonconstant periodic solution of a non-linear autonomous functional differential equation representing the growth of a single species population*, J. Math. Biol. **1** (1975), 227–240.

33. _____, *On instability, ω-limit sets and periodic solutions of nonlinear autonomous delay equations.*, Functional Differential Equations and Approx. of Fixed Points (Proc. Bonn, 1978) (H. O. Pertgen and H. O. Walther, eds.), Lecture Notes in Math., Vol.730, Springer, 1979, pp. 489–503.

34. P. Waltman, *Deterministic threshold models in the theory of epidemics*, Lecture Notes in Biomathematics **1** (1974), Springer Verlag.

35. E. M. Wright, *A nonlinear differential-difference equation*, J. Reine Angew. Math. **194** (1955), 66–87.

DEPARTMENT OF MATHEMATICS, ARIZONA STATE UNIVERSITY, TEMPE, AZ 85287-1804

Department of Mathematics, Arizona State University, Tempe, AZ 85287–1804

Contemporary Mathematics
Volume **129**, 1992

Unstable Manifolds of Periodic Orbits
of a Differential Delay Equation

HANS-OTTO WALTHER

ABSTRACT. Unstable sets of periodic solutions to delay equations

$$x'(t) = -\mu x(t) + f(x(t-1))$$

with monotone nonlinearities are shown to be smooth 2-dimensional graphs whose boundaries are formed by further periodic orbits.

1. Introduction

Let a continuously differentiable function $f : \mathbb{R} \longrightarrow \mathbb{R}$ be given, with

$$f(0) = 0 \quad \text{and} \quad \xi f(\xi) < 0 \quad \text{for} \quad \xi \neq 0.$$

Let $\mu \geq 0$. The equation

$$(1.1) \qquad\qquad x'(t) = -\mu x(t) + f(x(t-1))$$

models a system governed by delayed negative feedback and decay. Initial data $\phi : [-1,0] \longrightarrow \mathbb{R}$ in the infinite–dimensional phase space $C = C([-1,0], \mathbb{R})$ define continuous solutions $x^\phi : [-1, \infty) \longrightarrow \mathbb{R}$ which satisfy eq. (1.1) for $t > 0$. The relations

$$F(t, \phi) = x_t^\phi, \quad x_t^\phi = x^\phi(t + \cdot) \in C \quad \text{for} \quad t \geq 0,$$

constitute a semiflow F on C whose dynamics appears in many cases to be structured by periodic orbits. The most accessible periodic solutions are those which are slowly oscillating in the sense that consecutive zeros are spaced at distances

$$z_{j+1} - z_j > 1$$

greater than the delay.

1991 *Mathematics Subject Classification.* 34K15, 58F09.

Results on existence, uniqueness and nonuniqueness, stability and instability, bifurcation and more have been obtained, for example, in [**4, 5, 9, 11, 12, 18–20, 22, 26–31, 33, 35, 38**].

The dominating role of slowly oscillating periodic solutions is to a certain extent explained by a result in [**34**] which guarantees that for $\mu = 0$ and for some f, initial values of eventually slowly oscillating solutions are an open and dense set, and by the Morse decomposition result from [**21**].

The simplest situations occur when f is monotone, and bounded from above or from below. We shall assume in this paper

$$f(0) = 0, \quad f'(\xi) < 0 \quad \text{for all} \quad \xi \in \mathbb{R}, \quad \sup f < \infty$$

and consider slowly oscillating periodic solutions y which are unstable and hyperbolic. Results in [**2, 3, 8, 32**] indicate that such solutions exist for suitable monotone nonlinearities as above. Examples will be given in [**17**].

The unstable set W of a periodic solution y consists of the phase curves $\mathbb{R} \ni t \longrightarrow x_t \in \mathbb{R}$ of solutions which converge to the orbit

$$|\eta| := \{y_s : s \in \mathbb{R}\}$$

as $t \longrightarrow -\infty$.

We are interested in the nature of W. An important intermediate result on the way to the description of W is Theorem 5.3 which asserts that, for y unstable and hyperbolic, there is precisely one Floquet multiplier outside the unit circle, necessarily situated on the ray $(1, \infty)$. This implies that local unstable manifolds $W_{loc} \subset W$ of $|\eta|$ are 2–dimensional and orientable (no Moebius strips).

The main result, Theorem 5.4, states that for y unstable and hyperbolic, W is a smooth, annulus–like graph of dimension two, bordered either by two other periodic orbits, or by one periodic orbit and by the stationary state $0 \in C$.

On the invariant set $\overline{W}$ the semiflow F can be represented by the flow of a planar vectorfield. Phase curves in W spiral away from the orbit of y, towards the closed orbits forming the boundary $\overline{W} \setminus W$.

The closure $\overline{W}$ should be viewed as a part of the global attractor of the semiflow F. Results in [**8, 18, 26, 32**] suggest that there are special cases, with a single unstable periodic orbit and a stable steady state, where $\overline{W}$ is the full attractor.

We briefly mention further aspects of this study. Section 4 deals with the linearization of eq. (1.1) at the zero solution,

$$(1.2) \qquad\qquad x'(t) = -\mu x(t) - \alpha x(t - 1)$$

where $\alpha = -f'(0) > 0$. The semigroup of eq. (1.2) determines an invariant decomposition

$$C = L \oplus Q, \qquad \dim L = 2$$

which serves as a coordinate system for the description of W as a graph in Theorem 5.4.

Let us point out that the set W may be far away from $0 \in C$. The relevant link between the nonlinear semiflow F and the semigroup of eq. (1.2) in our situation is not linearization, but the presence of a set which is invariant for both: This is the set S of nonzero data $\phi \in C$ with at most one change of sign in $[-1, 0]$. S contains segments $x_t \in C$ of slowly oscillating solutions; W and the space L (except 0) belong to S while

$$S \cap Q = \emptyset.$$

Theorem 5.1 and Corollary 5.1 show that all L–projections of orbits of slowly oscillating periodic solutions are simple closed curves which wind around $0 \in L$; they are nested in L.

Several ingredients for the proof of Theorem 5.4 are taken from [36] where a somewhat different, partly simpler situation is studied: If the zero solution of eq. (1.1) is linearly unstable then the global unstable set of the stationary point $0 \in C$ contains an invariant smooth 2-dimensional graph bordered by the orbit of a slowly oscillating periodic solution. Other constructions which have no counterpart in [36] employ winding numbers and homotopies.

All results in this paper are formulated and proved for eq. (1.1) with $\mu > 0$, i.e. for a system with decay. It is not difficult to extract the version for the case $\mu = 0$, which is technically simpler.

A general reference for most of the basic properties of differential delay equations which are used in the sequel is [13]. For calculus in Banach spaces and for properties of the winding number, see [7]. For submanifolds and transversality, see [1].

2. Notation, Preliminaries

$\mathbb{N}_0$ denotes the set of nonnegative integers. $\mathbb{R}^+$ stands for the interval $[0, \infty)$. Spectra of closed linear operators T in complex Banach spaces are denoted by $\operatorname{spec}(T)$, or spec; in case of a real Banach space, $\operatorname{spec}(T)$ is the spectrum of the complexification of T.

Consider a subset $X \subset E$ of a real Banach space E, and a point $x \in X$. The set $T_x X$ of tangents to X at x is defined to be the set of all vectors

$$v = Dc(0)1$$

where $c : (-1, 1) \longrightarrow E$ is a differentiable curve with $c(0) = x$ and $c((-1, 1)) \subset X$. Note $0 \in T_x X$. In general, $T_x X$ is *not* a vectorspace. For a differentiable map $g : U \longrightarrow G, U \supset X$,

$$Dg(x)T_x X \subset T_{g(x)}g(X).$$

The closure, the interior and the boundary of X are denoted by

$$\overline{X}, \quad X^\circ, \quad \partial X,$$

respectively.

Let Y denote a 2–dimensional normed vectorspace over $\mathbb{R}$. For a map

$$c : [a,b] \longrightarrow Y$$

we denote by

$$| c | := c([a,b])$$

its trace. We say c is (piecewise) smooth if c is (piecewise) of class C^1.

Fix an isomorphism ζ of Y onto the $\mathbb{R}$–vectorspace $\mathbb{C}$. Then the winding number of a closed, piecewise smooth curve c with respect to a point $y \in Y \setminus | c |$ is defined as

$$\text{wind}(y,c) := \text{wind}(\zeta y, \zeta \circ c) \in \mathbb{Z}$$

where on the right hand side we have the familiar winding number of $\mathbb{C}$-valued curves.

If c is a simple closed, piecewise smooth curve, $y \in Y \setminus | c |$, then

$$\text{wind}(y,c) \in \{-1,0,1\}.$$

For such c, the interior and exterior of c are defined by

$$\text{int}(c) := \{y \in Y \setminus | c | : \text{wind}(y,c) \neq 0\}$$

$$\text{ext}(c) := \{y \in Y \setminus | c | : \text{wind}(y,c) = 0\}.$$

Both sets are open and connected; $\text{int}(c)$ is bounded and $\text{ext}(c)$ is unbounded; the boundaries coincide with the trace $| c |$.

The winding number is a homotopy invariant: If

$$\text{hom} : [0,1] \times [a,b] \longrightarrow Y \quad \text{is continuous,} \quad y \in Y \setminus \text{hom}([0,1] \times [a,b]),$$

$$\text{hom}(\cdot, a) = \text{hom}(\cdot, b),$$

and if both $\text{hom}(0,\cdot)$ and $\text{hom}(1,\cdot)$ are piecewise smooth, then

$$\text{wind}(y, \text{hom}(0,\cdot)) = \text{wind}(y, \text{hom}(1,\cdot)).$$

(It is not necessary that the closed curves $\text{hom}(t,\cdot), 0 < t < 1$, are piecewise smooth.)

Let a function $g : \mathbb{R}^2 \longrightarrow \mathbb{R}$ be given. A solution of the differential delay equation

$$(2.1) \qquad\qquad x'(t) = g(x(t), x(t-1))$$

is either a differentiable function $x : \mathbb{R} \longrightarrow \mathbb{R}$ so that (2.1) is satisfied for all real t, or a continuous function $x : [t_0 - 1, \infty) \longrightarrow \mathbb{R}$, $t_0 \in R$, which is differentiable on (t_0, ∞) and satisfies (2.1) for all $t > t_0$.

Analogously one defines complex-valued solutions in case g is linear, and solutions of nonautonomous equations

$$x'(t) = g(t, x(t-1))$$

for functions $g : \mathbb{R}^2 \longrightarrow \mathbb{R}$ or $g : [t_0, \infty) \times \mathbb{R} \longrightarrow \mathbb{R}, t_0 \in R$.

The space C, and the complex vectorspace C' of continuous functions ϕ :
$[-1,0] \longrightarrow \mathbb{C}$, are equipped with the maximum-norm:

$$\|\phi\| = \max_{t \in [-1,0]} |\phi(t)|.$$

Solutions define phase curves $t \longrightarrow x_t$ with values in C or C' by

$$x_t(s) := x(t+s) \quad \text{for all} \quad s \in [-1,0],$$

provided the interval $[t-1,t]$ belongs to the domain of x.

The term trajectory is used in connection with maps: If P is a mapping, then
sequences (ϕ_n) with $\phi_{n+1} = P(\phi_n)$ are called trajectories.

3. Basic Properties of Solutions

Let a continuously differentiable function $f : \mathbb{R} \longrightarrow \mathbb{R}$ be given as in the
introduction:

$$f(0) = 0, \quad f'(\xi) < 0 \quad \text{for all} \quad \xi, \quad \sup f < \infty.$$

Let a constant $\mu > 0$ be given.

We collect a series of basic facts on solutions of equation (1.1). For proofs,
see e.g. [36].

The phase curves of solutions $x = x^\phi$ of initial value problems

$$x'(t) = -\mu x(t) + f(x(t-1)) \quad \text{for} \quad t > 0, \quad x|[-1,0] = \phi \in C$$

define a continuous semiflow

$$F : \mathbb{R}^+ \times C \ni (t, \phi) \longrightarrow x_t^\phi \in C.$$

We have continuous dependence on initial data also in the following sense:

For $\phi \in C, t \geq 0$, and $\epsilon > 0$ given, there exists $\delta > 0$ such that for all $\psi \in C$
with $|\psi - \phi| \leq \delta$ and for all $s \in [0, t]$,

$$|x^\psi(s) - x^\phi(s)| < \epsilon.$$

Each map $F(t, \cdot)$ is injective. Any two solutions $x : \mathbb{R} \longrightarrow \mathbb{R}$ and $x^* : \mathbb{R} \longrightarrow \mathbb{R}$
with $x_t = x_t^*$ for some $t \in \mathbb{R}$ coincide. Each map $F(t, \cdot), t \geq 1$, is compact.

F is of class C^1 on $(1, \infty) \times C$; each $F(t, \cdot), t \geq 0$, is of class C^1. Each $D_2 F(t, \phi)$
is injective, and $D_2 F$ is continuous on $\mathbb{R}^+ \times C$.

For $t > 1$ and $\phi \in C$,

$$D_1 F(t, \phi)1 = (x')_t \quad \text{where} \quad x = x^\phi,$$

and for $t \geq 0, \phi \in C, \psi \in C$,

$$D_2 F(t, \phi)\psi = v_t$$

where $v : [-1, \infty) \longrightarrow \mathbb{R}$ is the solution of the initial value problem for the linear
variational equation along $x := x^\phi$,

(3.1) $$v'(t) = -\mu v(t) + f'(x(t-1))v(t-1),$$

$$v_0 = \psi.$$

The maps $V_x(t,0), t \geq 0$, given by $\psi \longrightarrow v_t$ are linear, continuous, and compact for $t \geq 1$. For more on linear variational equations, see e.g. [6, 13].

Note that if $x : [-2, \infty) \longrightarrow \mathbb{R}$ is a solution of eq. (1.1), then

$$D_2 F(t, x_0)(x')_0 = (x')_t \quad \text{for all} \quad t \geq 0.$$

Among the most elementary properties of single solutions are the following: Every solution $x : [t_0 - 1, \infty) \longrightarrow \mathbb{R}$ is bounded. Its ω–limit set

$$\omega(x) = \{\phi \in C : \text{There is a sequence} \quad (t_n)_0^\infty$$

$$\text{with} \quad t_n \longrightarrow \infty, \quad F(t_n, x_{t_0}) \longrightarrow \phi\}$$

is nonempty, compact, connected. Each $\phi \in \omega(x)$ defines a unique solution $x^* : \mathbb{R} \longrightarrow \mathbb{R}, x_0^* = \phi$, with phase curve in $\omega(x)$. Every positive (negative) solution $x : [t_0 - 1, \infty) \longrightarrow \mathbb{R}$ tends to 0 as $t \longrightarrow \infty$. For every bounded solution $x : \mathbb{R} \longrightarrow \mathbb{R}$,

$$\inf x^{-1}(0) = -\infty.$$

PROOF. Suppose $x(t) \neq 0$ for all $t \leq t_0$. Set $\hat{x}(t) := e^{\mu t} x(t)$. The solution $\hat{x}$ of the equation

$$\hat{x}'(t) = e^{\mu t} f(e^{-\mu(t-1)} \hat{x}(t-1))$$

satisfies $\operatorname{sign}(\hat{x}(t)) = -\operatorname{sign}(\hat{x}'(t)) \neq 0$ for $t \leq t_0$ and $\hat{x}(t) \longrightarrow 0$ as $t \longrightarrow -\infty$, a contradiction. $\square$

An important observation is that initial values in the convex cone

$$K := \{\phi \in C : \phi(-1) = 0, \quad 0 < \phi \quad \text{in} \quad (-1, 0]\}$$

define solutions $x = x^\phi$ which are slowly oscillating functions (with respect to the delay 1) in the sense that any pair of zeros $z' > z$ satisfies

$$z' - z > 1.$$

Moreover, for $\phi \in K$, the zeros of $x := x^\phi$ form a sequence $(z_j)_0^J$, $J \in \mathbb{N}_0$ or $J = \infty$, such that

(3.2) $z_j + 1 < z_{j+1} \quad \text{and} \quad x'(z_{j+1}) \neq 0 \quad \text{for} \quad j < J.$

(It can be shown that the case $J < \infty$ occurs for

$$-f'(0) \cdot e^\mu < e^{-1}.$$

See also Section 4: Cases I and II provide solutions of the linearized equation without zeros.)

When convenient we shall also write $z_j(x), z_j(\phi), J(x), J(\phi)$.

Phase curves of solutions starting in K have values in the set S of nonzero functions with at most one change of sign; i.e.

$$\phi \in S \quad \text{if and only if} \quad \phi \neq 0, \quad \text{and there exists}$$

$$z \in [-1, 0] \quad \text{such that} \quad \phi \leq 0 \quad \text{in} \quad [-1, z], \quad 0 \leq \phi \quad \text{in} \quad [z, 0],$$

or there is $z \in [-1,0]$ with $0 \le \phi$ in $[-1,z]$, $\phi \le 0$ in $[z,0]$.

S is a cone (if $t > 0$ and $\phi \in S$, then $t\phi \in S$), but not convex. One can show that S is homotopy equivalent to a circle [10]. Elementary considerations yield

$$\overline{S} = S \cup \{0\}.$$

The set S is flow-invariant:

$$F(\mathbb{R}^+ \times S) \subset S,$$

and, more generally than above, every $\phi \in S$ defines a solution $x = x^\phi$ such that for t_0 sufficiently large, $x|[t_0, \infty)$ is slowly oscillating.

The monotonicity of f implies an even stronger statement.

PROPOSITION 3.1. 1. *For initial data* ϕ, ψ *with* $\phi - \psi \in S$, *there exists* $s \in [0,4]$ *such that* $F(s,\phi) - F(s,\psi)$ *has no zero;*

$$F(t,\phi) - F(t,\psi) \in S \quad \text{for all} \quad t \ge 0.$$

2. *Let a solution* $x : [-1,\infty) \longrightarrow \mathbb{R}$ *of eq.* (1.1) *be given. Every solution* $v : [-1,\infty) \longrightarrow \mathbb{R}$ *of eq.* (3.1), *with* $v_0 \in S$, *satisfies*

$$v_t \in S \quad \text{for all} \quad t \ge 0,$$

and there exists $s \in [0,4]$ *so that* v_s *has no zero. On* $[4,\infty)$, v *is slowly oscillating.*

PROOF. 1. Apply [36, **Remark 6.1**] and [36, **Proposition 6.1**] to the solution

$$t \longrightarrow e^{\mu t}(x^\phi(t) - x^\psi(t))$$

of the equation

$$x'(t) = g(t, x(t-1))$$

where

$$g(t,\xi) = e^{\mu t}[f(e^{-\mu(t-1)}(\xi + x^\psi(t-1))) - f(e^{-\mu(t-1)}x^\psi(t-1))]$$

$$= e^{\mu t} \int_{e^{-\mu(t-1)}x^\psi(t-1)}^{e^{-\mu(t-1)}(\xi + x^\psi(t-1))} f'.$$

2. Apply [36, **Remark 6.1, Proposition 6.1**] to the solution $t \longrightarrow e^{\mu t}v(t)$ of the equation

$$d'(t) = e^\mu f'(x(t-1))d(t-1). \quad \square$$

The zeros of any bounded solution $x : \mathbb{R} \longrightarrow \mathbb{R}$ with all values x_t of the phase curve in S form a sequence $(z_j)_{-\infty}^J$, $J \in \mathbb{Z}$ or $J = \infty$, with property (3.2).

PROPOSITION 3.2. 1. *Let* $x : \mathbb{R} \longrightarrow \mathbb{R}$ *be a solution of eq. (1.1) whose zeros form a sequence* $(z_j)_{-\infty}^{J}$, $J \in \mathbb{Z}$ *or* $J = \infty$, *with property (3.2). Then each local extremum of* x *belongs to some interval* $(z_j, z_j + 1]$.

2. *There exist positive constants* b_f, c_f, d_f *such that for every solution as in assertion 1,*

$$|x| \leq b_f; \quad |x(t) - x(s)| \leq c_f \cdot |t - s|$$

$$and \quad \|x_t - x_s\| \leq c_f \cdot |t - s| \quad for \ all \ real \quad t, s.$$

Furthermore,

$$\max_{[z_j, z_{j+1}]} |x| \leq d_f \max_{[z_{j-1}, z_j]} |x|$$

for all $j \leq J - 1$, *in case* $J \in \mathbb{Z}$, *and for all* $j \in \mathbb{Z}$, *in case* $J = \infty$; *and*

$$\max_{[z_J, \infty)} |x| \leq d_f \cdot \max_{[z_{J-1}, z_J]} |x| \quad in \ case \quad J \in \mathbb{Z}.$$

PROOF. 1. For $z_j + 1 \leq t < z_{j+1}$, or for $z_j + 1 \leq t < \infty$ in case $j = J < \infty$,

$$x'(t) = -\mu x(t) + f(x(t-1)) < f(x(t-1)) \leq 0 \quad \text{in case} \quad 0 < x(t),$$

$$x'(t) = -\mu x(t) + f(x(t-1)) > f(x(t-1)) \geq 0 \quad \text{in case} \quad x(t) < 0.$$

Consequently, each local extremum of x lies in an interval $(z_j, z_j + 1)$, and in case $J < \infty$,

$$|x(t)| \leq \max_{[z_J, z_J + 1]} |x| \quad \text{for} \quad t \geq z_J + 1.$$

2. Estimate of local maxima:

If $0 < x$ in (z_j, z_{j+1}) and $z_j < t \leq z_j + 1$, then

$$x'(t) = -\mu x(t) + f(x(t-1)) \leq f(x(t-1)) \leq \sup(f),$$

hence

$$0 < x(t) \leq \sup(f).$$

Estimate of local minima:

If $x < 0$ in (z_j, z_{j+1}) and $z_j < t \leq z_j + 1$, then

$$x'(t) \geq f(x(t-1)) \geq f(\max_{[z_{j-1}, z_j]} x) \geq f(\sup(f));$$

hence

$$x(t) \geq f(\sup(f)).$$

Set $b_f := \max\{\sup(f), -f(\sup(f))\}$. Lipschitz continuity follows from

$$|x'(t)| \leq \mu b_f + \max\{f(-b_f), -f(b_f)\}, \quad \text{for} \quad t \in \mathbb{R}.$$

There exists $d_f > 0$ such that

$$|f(\xi)| \leq d_f \cdot |\xi| \quad \text{for} \quad |\xi| \leq b_f.$$

Consider a zero z_j with, say, $x'(z_j) > 0$. Then, for every $t \in (z_j, z_j + 1)$,

$$x'(t) = -\mu x(t) + f(x(t-1)) \leq f(x(t-1)) \leq d_f \cdot |x(t-1)| \leq d_f \max_{[z_{j-1}, z_j]} |x|,$$

and consequently

$$0 < x(t) = \int_{z_j}^{t} x'(s)ds \le d_f \max_{[z_{j-1}, z_j]} |x|;$$

and it becomes obvious how to deduce the estimates involving d_f. $\square$

To investigate slowly oscillating solutions we shall make use of a globally defined return map. Consider the subset

$$K_P := \{\phi \in K \cup (-K) : J(\phi) \ge 2\}$$

of the double cone $K \cup (-K)$. Phase curves $t \longrightarrow x_t^\phi$ which start in K_P intersect at $t = z_2(\phi) + 1$ transversally with the hyperplane

$$H \subset C \quad \text{given by} \quad \phi(-1) = 0$$

since the tangent vector $\chi := D_1 F(z_2(\phi) + 1, \phi)1$ satisfies

$$\chi(-1) = x'(z_2) \ne 0.$$

Using the Implicit Function Theorem and continuous dependence on initial data, one constructs an open neighborhood N of K_P and a C^1-map

$$stop : N \longrightarrow (1, \infty)$$

(such that $F(stop(\phi), \phi) \in H$ for $\phi \in N$) with

$$stop(\phi) = z_2(\phi) + 1 \quad \text{for all} \quad \phi \in K_P.$$

The C^1-map

$$P : N \longrightarrow C, \quad P(\phi) = F(stop(\phi), \phi)$$

satisfies

$$P(\phi) = F(z_2(\phi) + 1, \phi) \in K \quad \text{for} \quad \phi \in K \quad \text{and} \quad J(\phi) \ge 2,$$

$$P(\phi) = F(z_2(\phi) + 1, \phi) \in -K \quad \text{for} \quad \phi \in -K \quad \text{and} \quad J(\phi) \ge 2.$$

Let $p_\chi : C \longrightarrow C$ denote the projection onto H, parallel to χ;

$$p_\chi(\psi) = \psi - \frac{\psi(-1)}{\chi(-1)} \cdot \chi.$$

Then one computes, for $\phi \in K_P$ and $\psi \in C$,

$$DP(\phi)\psi = p_\chi D_2 F(z_2(\phi) + 1, \phi)\psi.$$

The same constructions which lead to the return map P yield also a smooth stopping map $P_{1/2}$ on an open neighborhood $N_{1/2}$ of

$$K_{1/2} := \{\phi \in K \cup (-K) : J(\phi) \ge 1\}$$

such that

$$P_{1/2}(\phi) = F(z_1(\phi) + 1, \phi) \quad \text{on} \quad K_{1/2};$$

i.e.,

$$P_{1/2}(\phi) \in -K \quad \text{for} \quad \phi \in K \quad \text{and} \quad J(\phi) \ge 1,$$

$$P_{1/2}(\phi) \in K \quad \text{for} \quad \phi \in -K \quad \text{and} \quad J(\phi) \geq 1.$$

In the sequel we shall consider slowly oscillating periodic solutions

$$y : \mathbb{R} \longrightarrow \mathbb{R}$$

of eq. (1.1). We may, and will, assume

$$z_0(y) = -1 \quad \text{and} \quad y_0 \in K;$$

the minimal period $\tau = \tau(y)$ of y is then of the form

$$\tau = z_{2n}(y) + 1 \quad \text{for some} \quad n \in \mathbb{N}$$

(later we shall prove that $n = 1$), and

$$\eta : [0, \tau] \ni t \longrightarrow y_t \in C$$

is a simple closed smooth curve. When convenient we shall write $\eta(y)$ instead of η.

The local stability properties of y are governed by the Floquet multipliers, i.e. by the spectrum of the linear continuous, compact map $V = V_y(\tau, 0)$ given by the solutions of the linear variational equation

$$(3.3) \qquad v'(t) = -\mu v(t) + f'(y(t-1))v(t-1)$$

The number 1 is an eigenvalue of V, and $(y')_0$ is an eigenvector.

The solution y is called hyperbolic if 1 is simple and if there are no other eigenvalues on the unit circle.

If y is unstable, i.e. not stable, and hyperbolic, then necessarily

$$|\lambda| > 1$$

for at least one Floquet multiplier λ.

We end this section with basic facts about the unstable set

$$W := W(y) := \{\phi \in C : \text{There is a solution} \quad x : \mathbb{R} \longrightarrow \mathbb{R} \quad \text{of eq. (1.1)}$$

$$\text{such that} \quad x_0 \doteq \phi \quad \text{and} \quad x_t \longrightarrow |\eta| \quad \text{as} \quad t \longrightarrow -\infty\}.$$

Observe that

$$F(\mathbb{R}^+ \times W) = W.$$

PROPOSITION 3.3. $W \subset S$.

PROOF. Let $\phi \in W$. Consider the solution $x : \mathbb{R} \longrightarrow \mathbb{R}$ with $x_0 = \phi$. Use the compactness of $|\eta|$ to find a sequence $t_n \longrightarrow -\infty$ and $t \in \mathbb{R}$ such that $x_{t_n} \longrightarrow y_t$. For some $s > t$, y_s has no zero. By continuity of F, $x_{t_n + s}$ has no zero for $|n|$ sufficiently large; $x_{t_n + s} \in S$. Forward invariance of S yields $\phi = x_0 \in S$. $\square$

Every solution $x : \mathbb{R} \longrightarrow \mathbb{R}$ of eq. (1.1) with phase curve in W is bounded. Using Proposition 3.3, and previous remarks, we infer that the zeros of x form a sequence $(z_j)_{-\infty}^{J}$ where $J = J(x) \in \mathbb{Z}$ or $J = J(x) = \infty$, so that statement (3.2) holds. Proposition 3.2 yields

$$\|x_t\| \leq b_f \quad \text{on} \quad \mathbb{R}.$$

It follows that

$$W \subset \overline{F(\{1\} \times \{\phi \in C : \|\phi\| \leq b_f\})}$$

is compact. Also,

$$\overline{W} \subset \overline{S} = S \cup \{0\}.$$

PROPOSITION 3.4. *For every $\phi \in \overline{W} \setminus W$ there is a solution $x : \mathbb{R} \longrightarrow \mathbb{R}$ of eq. (1.1) such that $x_0 = \phi$ and $x_t \in \overline{W} \setminus W$ for all $t \in \mathbb{R}$.*

PROOF. See the proof of [36, Proposition 8.1] for the case $\phi \neq 0$. $\square$

We infer

$$F(\mathbb{R}^+ \times (\overline{W} \setminus W)) = \overline{W} \setminus W.$$

The set

$$(\overline{W} \setminus W) \setminus \{0\}$$

is formed by phase curves of bounded, slowly oscillating solutions as above.

4. On the Linear Equation

The solutions $x : [-1, \infty) \longrightarrow \mathbb{R}$ of eq. (1.2), $\alpha := -f'(0)$ as in the introduction, define a C_0–semigroup of operators $T(t) = D_2 F(t, 0)$, $t \geq 0$. The spectrum spec of its generator consists of complex conjugate pairs of eigenvalues in the double strips S_k given by

$$2k\pi < |\mathrm{Im}(\lambda)| < 2k\pi + \pi, \quad k \in \mathbb{N},$$

and by at most two eigenvalues in the strip S_0 given by

$$|\mathrm{Im}(\lambda)| < \pi;$$

the total multiplicity of spec in S_0 is 2.

We have

$$\max \mathrm{Re}(\cup_{\mathbb{N}}(\mathrm{spec} \cap S_k)) \quad < \quad \min \mathrm{Re}(\mathrm{spec} \cap S_0).$$

Spectral decomposition and reellification yield a $T(t)$–invariant decomposition

(4.1) $$C = L \oplus Q$$

into closed subspaces. The generator of the induced semigroup on L has spectrum spec $\cap S_0$;

$$\dim L = 2.$$

Three cases are possible for spec $\cap S_0$:

I. $\alpha e^\mu < \frac{1}{e}$. spec $\cap S_0$ consists of two simple, real eigenvalues $u_{00} < u_0 < 0$; a basis $\{\beta_1, \beta_2\}$ of L is given by the restrictions of the functions

$$t \longrightarrow e^{u_0 t}, \quad t \longrightarrow e^{u_{00} t}$$

to $[-1, 0]$.

II. $\alpha e^\mu = \frac{1}{e}$. spec $\cap S_0$ consists of a double eigenvalue

$$u_0 = -1 - \mu;$$

a basis $\{\beta_1, \beta_2\}$ of L is given by

$$t \longrightarrow e^{u_0 t}, \quad t \longrightarrow -t e^{u_0 t}.$$

III. $\alpha e^\mu > \frac{1}{e}$. spec $\cap S_0$ consists of a complex conjugate pair $\lambda_0 = u_0 + i v_0$, $\overline{\lambda_0}$ of simple eigenvalues; $0 < v_0$. A basis $\{\beta_1, \beta_2\}$ of L is given by

$$t \longrightarrow e^{u_0 t} \sin(v_0 t), \quad t \longrightarrow e^{u_0 t} \cos(v_0 t).$$

REMARK 4.1. 1. Slowly and rapidly oscillating solutions. L consists of the segments x_t of the linear combinations of the functions $\mathbb{R} \longrightarrow \mathbb{R}$ used to define the bases above. All of these linear combinations (except the trivial one) are slowly oscillating solutions of eq. (1.2); we have

$$L \subset S \cup \{0\}.$$

Real–valued solutions associated with eigenvalues $\lambda = u + iv$ outside S_0 have zeros spaced at distances

$$\frac{\pi}{v} < \frac{1}{2};$$

hence they are not slowly oscillating.

2. Stability. Consider

$$v(\mu) \in \left(\frac{\pi}{2}, \pi \right) \quad \text{defined by} \quad v(\mu) = -\mu \tan(v(\mu)).$$

For

$$\alpha < -\frac{\mu}{\cos(v(\mu))} \quad \left(= \frac{v(\mu}{\sin(v(\mu))} > \frac{\pi}{2} \right),$$

$$\text{Re}(\text{spec}) < 0;$$

the zero solution of eq. (1.2) is exponentially attractive. It follows that there exists a neighborhood U of 0 and a positive constant c such that

$$\|F(t, \phi)\| \leq c e^{\frac{u_0 t}{2}} \quad \text{for all} \quad \phi \in U, t \geq 0.$$

At $\alpha = -\frac{\mu}{\cos(v(\mu))}$, $u_0 = 0$, and for $\alpha > -\frac{\mu}{\cos(v(\mu))}$, $u_0 > 0$.

3. For $\alpha e^\mu > 1$, every solution $x : [-1, \infty) \longrightarrow \mathbb{R}$ of the nonlinear equation (1.1) has an unbounded zeroset (see [**36, Proposition 6.3, proof, part a**]). We conclude that in case there exists a solution x of eq. (1.1) with $x^{-1}(0)$ bounded from above,

$$\alpha \leq 1, \quad \text{and} \quad \text{Re}(\text{spec}) < 0.$$

Let $p : C \longrightarrow C$ and $q : C \longrightarrow C$ denote the projections onto L and Q given by (4.1).

LEMMA 4.1. $0 \notin pS$.

PROOF. In case III, see [**36, Lemma 6.3, proof**]. Analogous arguments work in the cases I and II where nontrivial solutions with phase curve in L have at most one zero. $\square$

In other words,

$$S \cap Q = \emptyset,$$

the cone S contains $L \setminus \{0\}$ and stays away from the complementary subspace Q.

Observe that

(4.2) $$pK \cap (-pK) = \emptyset$$

since $p\phi = p\psi$ with $\phi \in K$, $\quad \psi \in -K$ would imply

$$0 = p(\phi - \psi) \in p(K + K) \subset pK \subset pS,$$

a contradiction.

The remainder of this section is devoted to simple closed curves in L which wind around the origin. The subsequent constructions will later be used to define homotopies with values in the set S.

We begin with certain subcones of S. It will become important that these subcones are convex (recall that S is not).

For every integer $k \geq 3$, define

$$K_{k0} := \{\phi \in C : 0 < \phi(t) \quad \text{in} \quad (-1, 0)\},$$

$$K_{k1} := \left\{ \phi \in C : 0 < \phi(t) \quad \text{in} \quad \left[-1, -\frac{1}{k} \right), \quad \phi'(t) \quad \text{exists} \right.$$

$$\left. \text{and is negative in} \quad \left[-\frac{1}{k}, 0 \right] \right\},$$

$$K_{k\kappa} := \left\{ \phi \in C : 0 < \phi(t) \quad \text{in} \quad \left[-1, -\frac{\kappa}{k} \right), \quad \phi'(t) \quad \text{exists and is negative} \right.$$

$$\left. \text{in} \quad \left[-\frac{\kappa}{k}, -\frac{\kappa - 1}{k} \right], \quad \phi(t) < 0 \quad \text{in} \quad \left(-\frac{\kappa - 1}{k}, 0 \right) \right\}$$

$$\text{for} \quad \kappa = 2, \ldots, k - 1,$$

$$K_{kk} := \left\{ \phi \in C : \phi'(t) \quad \text{exists and is negative in} \quad \left[-1, -1 + \frac{1}{k} \right], \right.$$

$$\left. \phi(t) < 0 \quad \text{in} \quad \left(-1 + \frac{1}{k}, 0 \right] \right\}.$$

Our aim is to find, in each of the cases I, II, III, a simple closed piecewise smooth curve

$$c : [0, b] \longrightarrow C, \quad |c| \subset L,$$

with

$$1 = \text{wind}(0, c)$$

and an integer $k(c) \geq 3$ such that for every integer $k \geq k(c)$, there is a subdivision

$$0 = t_0 < t_1 < \cdots < t_{2k+2} = b$$

so that

$$c(t) \in K_{k\kappa} \quad \text{for} \quad t_\kappa \leq t \leq t_{\kappa+1}, \quad \kappa = 0, \ldots, k;$$

$$c(t) \in -K_{k\kappa} \quad \text{for} \quad t_{k+1+\kappa} \leq t \leq t_{k+1+\kappa+1}, \quad \kappa = 0, \ldots, k.$$

We begin with case III. The slowly oscillating solution

$$x : t \longrightarrow e^{u_0(t+1)} \sin(v_0(t+1))$$

of eq. (1.2) has segments $x_t \in L$; its zeros are given by

$$z_j = \frac{\pi j}{v_0} - 1, \quad j \in \mathbb{Z}.$$

Note

$$z_0 = -1, \quad x_0 \in K, \quad z_2 + 1 = \frac{2\pi}{v_0}.$$

Except for $u_0 = 0$, x is not periodic. For $0 \leq t \leq \frac{2\pi}{v_0}$, we set

$$c(t) := e^{-u_0 t} x_t$$

$$= e^{u_0} \cos(v_0(t+1)) \cdot \beta_1 + e^{u_0} \sin(v_0(t+1)) \cdot \beta_2.$$

This defines a simple closed smooth curve in L. Employing the isomorphism $\zeta : L \longrightarrow \mathbb{C}$ with $\beta_1 \longrightarrow 1, \quad \beta_2 \longrightarrow i$, we have

$$1 = \text{wind}(0, c).$$

Choose an integer $k(c) \geq 3$ so large that

$$x'(t) \neq 0 \quad \text{for} \quad |t - z_j| \leq \frac{1}{k(c)}, \quad j \in \{0, 1, 2\}.$$

One verifies easily that for every integer $k \geq k(c)$,

$$x_t \in K_{k0} \quad \text{for} \quad 0 \leq t \leq z_1,$$

$$x_t \in K_{k\kappa} \quad \text{for} \quad z_1 + \frac{\kappa}{k} \leq t \leq z_1 + \frac{\kappa+1}{k}, \quad \kappa = 0, \ldots, k-1;$$

$$x_t \in -K_{k0} \quad \text{for} \quad z_1 + 1 \leq t \leq z_2;$$

$$x_t \in -K_{k\kappa} \quad \text{for} \quad z_2 + \frac{\kappa}{k} \leq t \leq z_2 + \frac{\kappa+1}{k}, \quad \kappa = 0, \ldots, k-1.$$

The values $c(t)$ have the same property. A subdivision as required is given by

$$0 = t_0 < t_1 = z_1 < t_2 = z_1 + \frac{1}{k} < \cdots < t_{k+1} = z_1 + 1$$

$$< t_{k+2} = z_2 < t_{k+3} = z_2 + \frac{1}{k} < \cdots < t_{2k+2} = z_2 + 1.$$

In cases I and II, the construction of c is different. We shall obtain c as a parameterization of the boundary of the quadrangle

$$\{r \cdot \beta_1 + s \cdot \beta_2 \in L : |r| \le 1, |s| \le 1\};$$

in these cases, points on $|c|$ are not multiples of segments of one solution of eq. (1.2).

Case I. Consider the straight line parameterizations

$$c_1 : [-1, 1] \ni s \longrightarrow \beta_1 + s \cdot \beta_2 \in L,$$

$$c_2 : [-1, 1] \ni r \longrightarrow -r \cdot \beta_1 + \beta_2 \in L.$$

It is an elementary exercise to prove the following result.

PROPOSITION 4.1. *Let real numbers r and s be given with* $0 < |r| + |s|$. *The function*

$$g = g_{rs}, \quad g(t) = re^{u_0 t} + se^{u_{00} t} \quad \text{for} \quad t \in \mathbb{R},$$

has at most one zero. A zero $z = z(r, s)$ exists if and only if $r \ne 0 \ne s$ and $sign(s) = -sign(r)$. In this case,

$$z = -\frac{1}{u_{00} - u_0} \log\left(-\frac{s}{r}\right), \quad sign(g'(z)) = sign(-s) \ne 0.$$

If g and g' have zeros z and z', respectively, then the distance

$$(4.3) \qquad z - z' = \frac{\log(u_0) - \log(u_{00})}{u_0 - u_{00}} > 0$$

does not depend on r, s.

For c_1, we have

$$c_1(s)(t) = g_{1s}(t) \quad (|s| \le 1, \quad -1 \le t \le 0).$$

Consequently, each $c_1(s)$ has at most one zero; $c_1(s)$ has a zero (in $[-1, 0]$) if and only if

$$-1 \le s < 0 \quad \text{and} -1 \le z(1, s) \le 0;$$

in this case, the zero of $c_1(s)$ equals

$$z(1, s) = \frac{1}{u_0 - u_{00}} \log(-s).$$

Define $s_1 \in [-1, 0)$ by

$$-1 = \frac{1}{u_0 - u_{00}} \log(-s_1) \quad (= z(1, s_1)).$$

Then

$$c_1(s) \quad \text{has a zero} \quad z(1, s) \quad \text{for} \quad -1 \le s \le s_1,$$
$$z(1, -1) = 0,$$

$$(4.4) \qquad \frac{\partial z}{\partial s}(1, s) = \frac{1}{u_0 - u_{00}} \cdot \frac{1}{s} < 0 \quad \text{for} \quad -1 \le s \le s_1 \quad (< 0),$$

$$z(1, s_1) = -1,$$

and

$$\text{sign}(c_1(s)'(z(1,s)) = \text{sign}(g'_{1s}(z(1,s)) = \text{sign}(-s) > 0 \quad \text{for} \quad -1 \le s \le s_1.$$

Choose $k \in \mathbb{N}$, $k \ge 3$ so large that

(4.5) $$\frac{1}{k} < \frac{\log(u_0) - \log(u_{00})}{u_0 - u_{00}}.$$

Using (4.4) we find a subdivision

$$-1 = \sigma_0 < \sigma_1 < \cdots < \sigma_k = s_1$$

such that for $\sigma_{\kappa-1} \le s \le \sigma_\kappa$, $\kappa = 1, \ldots, k$, we have

$$-\frac{\kappa}{k} \le z(1,s) \le -\frac{\kappa - 1}{k}.$$

Using (4.3) and (4.5) we infer that for such s and κ,

$$c_1(s)'(t) > 0 \quad \text{for} \quad -\frac{\kappa}{k} \le t \le -\frac{\kappa - 1}{k},$$

$$c_1(s)(t) < 0 \quad \text{for} \quad t \in [-1, 0] \cap \left(-\infty, -\frac{\kappa}{k}\right],$$

$$0 < c_1(s)(t) \quad \text{for} \quad t \in [-1, 0] \cap \left[-\frac{\kappa - 1}{k}, \infty\right),$$

or

$$c_1(s) \in -K_{k\kappa}.$$

Next, we see that for $s_1 < s < 0$ and for $0 \le s \le 1$, the function $c_1(s)$ is positive on $(-1, 0)$; i.e.,

$$c_1(s) \in K_{k0}.$$

We turn to c_2. From the definition, $c_2(r)$ is positive on $(-1, 0)$ for $|r| \le 1$, hence

$$c_2(r) \in K_{k0} \quad \text{for} \quad -1 \le r \le 1.$$

Set $c_3 := -c_1$, $c_4 := -c_2$. Now it becomes obvious how to complete the construction.

Case II. Consider c_1 and c_2 defined as in Case I.

PROPOSITION 4.2. *Let real numbers r and s with $0 < |r| + |s|$ be given. The function*

$$g = g_{rs}, \quad g(t) = e^{u_0 t}(r + st) \quad \text{for} \quad t \in \mathbb{R},$$

has at most one zero. A zero $z = z(r, s)$ exists if and only if $s \neq 0$. In this case,

$$z = -\frac{r}{s}, \quad \text{sign}(g'(z)) = \text{sign}(s).$$

If g and g' have zeros z and z', respectively, then

(4.6) $$z' - z = -\frac{1}{u_0}.$$

Note that for $|s| \leq 1$ and $-1 \leq t \leq 0$,

$$c_1(s)(t) = \beta_1(t) + s \cdot \beta_2(t) = e^{u_0 t}(1 + (-s) \cdot t) = g_{1,-s}(t).$$

The function $c_1(s)$ is strictly positive on $(-1, 0)$ for $|s| \leq 1$, so that $c_1(s) \in K_{k0}$ for all integers $k \geq 3$, $|s| \leq 1$.

We turn to c_2. For $|r| \leq 1$ and $-1 \leq t \leq 0$,

$$c_2(r)(t) = -r \cdot \beta_1(t) + \beta_2(t) = e^{u_0 t}(-r + (-1) \cdot t) = g_{-r,-1}(t).$$

It follows that $c_2(r)$ has a zero (in $[-1, 0]$) if and only if $0 \leq r \leq 1$; in this case, the zero equals

$$z(-r, -1) = -r;$$
$$z(0, -1) = 0,$$
$$\frac{\partial}{\partial r}(r \longrightarrow z(-r, -1))(r) = -1 < 0 \quad \text{for} \quad 0 \leq t \leq 1,$$
$$z(-1, 1) = -1,$$

and for $0 \leq r \leq 1$,

(4.7) $\text{sign}(c_2(r)'(z(-r, -1))) = \text{sign}(g'_{-r,-1}(z(-r, -1))) = \text{sign}(-1) = -1.$

For $-1 \leq r \leq 0$, we see that

$$c_2(r) \quad \text{is strictly positive on} \quad (-1, 0).$$

Choose an integer $k \geq 3$ with

(4.8) $\dfrac{1}{k} < -\dfrac{1}{u_0}.$

For $\dfrac{\kappa - 1}{k} \leq r \leq \dfrac{\kappa}{k}$, $\kappa = 1, \ldots, k$, we have

$$-\frac{\kappa}{k} \leq z(-r, -1) \leq -\frac{\kappa - 1}{k}.$$

By (4.8) and (4.6), and by (4.5),

$$c_2(r)'(t) < 0 \quad \text{for} \quad -\frac{\kappa}{k} \leq t \leq -\frac{\kappa - 1}{k};$$

and

$$0 < c_2(r)(t) \quad \text{for} \quad t \in [-1, 0] \cap \left(-\infty, -\frac{\kappa}{k}\right],$$
$$c_2(r)(t) < 0 \quad \text{for} \quad t \in [-1, 0] \cap \left[-\frac{\kappa - 1}{k}, \infty\right);$$

or

$$c_2(r) \in K_{k\kappa}.$$

Now the construction is completed as in case I.

5. Results

THEOREM 5.1. *For every slowly oscillating periodic solution y of eq. (1.1) with $y_0 \in K$ and $z_0 = -1$,*

$$\tau = z_2(y) + 1 \quad and \quad 0 \in int(p \circ \eta).$$

COROLLARY 5.1. *For any two slowly oscillating periodic solutions y^1 and y^2 with initial values in K and $z_0(y^i) = -1$, the simple closed curves*

$$\eta(y^i) : [0, \tau(y^i)] \ni t \longrightarrow y_t^i \in C$$

have nested projections: Either $y^1 = y^2$, or $|p \circ \eta(y^i)| \subset int(p \circ \eta(y^j))$, $i \neq j$.

THEOREM 5.2. *For every slowly oscillating periodic solution y of eq. (1.1), y' is also slowly oscillating and has minimal period $\tau(y)$. The zeros of y and y' alternate.*

REMARK 5.1. We shall use the projection p to prove these results. Alternatively, one can work with evaluation maps

$$x_t \longrightarrow (x(t), x'(t)) \quad or \quad x_t \longrightarrow (x(t), x(t-1))$$

into the plane $\mathbb{R}^2$ (one leaves the phase space of eq. (1.1)), derive the statements about minimal periods and about y', and obtain analogues of the others. See [23, 24].

THEOREM 5.3. *(Floquet multipliers) Let a slowly oscillating periodic solution y of eq. (1.1) be given. If y is unstable and hyperbolic, then there exists exactly one Floquet multiplier u with $|u| > 1$. Also, u is simple, and $u \in (1, \infty)$. If y is unstable but not hyperbolic, then the multiplier 1 has multiplicity 2, and $|\lambda| < 1$ for all multipliers $\lambda \neq 1$.*

THEOREM 5.4. *(On the unstable set of y) Suppose y is a slowly oscillating periodic solution of eq. (1.1) which is unstable and hyperbolic.*
1. *The closure $\overline{W}$ of the unstable set $W = W(y)$ is a graph $\overline{w} : \overline{pW} \longrightarrow Q$. Moreover, $\overline{w}$ is Lipschitz continuous. The set pW is an open subset of L, and $w := \overline{w}|pW$ is C^1.*
2. *$E := (\overline{W} \setminus W) \cap p^{-1}(ext(p \circ \eta))$ is the orbit of a slowly oscillating periodic solution of eq. (1.1).*
3. *$I := (\overline{W} \setminus W) \cap p^{-1}(int(p \circ \eta))$ is either the orbit of a slowly oscillating periodic solution of eq. (1.1), or $I = \{0\}$.*

6. Proofs of Theorem 5.1, Corollary 5.1 and Theorem 5.2

PROPOSITION 6.1. 1. *If y is a slowly oscillating periodic solution of eq. (1.1), $y_0 \in K$ and $z_0 = -1$, then*

$$y_t - y_s \in S \quad for \quad 0 \le s < t < \tau.$$

2. *For any two slowly oscillating periodic solutions* y^i, $i = 1, 2$, *with disjoint orbits in* C,

$$y^1_t - y^2_s \in S \quad \text{for all real} \quad t, s.$$

PROOF. 1. Fix $t_0 \in (0, \tau)$ and $t_1 \in (t_0, t_0 + \tau)$ so that $y_{t_0} > 0 > y_{t_1}$. By continuity,

$$M := \{t \in (t_0, t_1) : y_{t_0} - y_s \in S \quad \text{for all} \quad s \in [t, t_1]\} \neq \emptyset,$$

and $t_- := \inf(M) \in [t_0, t_1)$. Suppose $t_0 < t_-$. By continuity, $y_{t_0} - y_{t_-} \in \overline{S} = S \cup \{0\}$. Minimality of τ yields

$$y_{t_0} \neq y_{t_-}, \quad \text{so} \quad y_{t_0} - y_{t_-} \in S.$$

By Proposition 3.1, there exists $t \in [0, 4]$ such that $y_{t_0 + t} - y_{t_- + t}$ has zero. By continuity, there exists $\epsilon > 0$ such that for $|s| < \epsilon$

$$y_{t_0 + t} - y_{t_- + t + s} \quad \text{has no zero, and is in} \quad S.$$

Fix $j \in \mathbb{N}$ so large that

$$t_0 + t + \epsilon - j\tau < t_-.$$

By periodicity,

$$y_{t_0 + t - j\tau} - y_{t_- + t - j\tau + s} \in S \quad \text{for} \quad |s| < \epsilon.$$

Proposition 3.1 gives

$$y_{t_0} - y_{t_- + s} \in S \quad \text{for} \quad |s| < \epsilon,$$

which implies a contradiction to the definition of t_-. We have shown that

$$[y_{t_0} - y_t \in S \quad \text{for all} \quad t \in (t_0, t_1].$$

Analogously one finds

$$y_{t_0} - y_t \in S \quad \text{for all} \quad t \in (t_1, t_0 + \tau).$$

For $0 \le s < t < \tau$, we have $t_0 + s - t + \tau \in (t_0, t_0 + \tau)$; the relations

$$y_t - y_s = F(t - (t_0 - \tau), y_{t_0} - \tau) \quad - \quad F(t - (t_0 - \tau), y_s + (t_0 - \tau) - t)$$

$$= F(\ldots, y_{t_0}) \quad - \quad F(\ldots, y_{t_0 + s - t + \tau})$$

and Proposition 3.1, applied to $t - (t_0 - \tau) > 0$, give the assertion. 2. Fix points t_i such that $y^1_{t_1} > 0 > y^2_{t_2}$. Then

$$M := \{t < t_2 : \quad y^1_{t_1} - y^2_s \in S \quad \text{for all} \quad s \in [t, t_2]\} \neq \emptyset.$$

Suppose $t_- := \inf(M) > -\infty$. By continuity,

$$y^1_{t_1} - y^2_{t_-} \in \overline{S} = S \cup \{0\}.$$

As the orbits are disjoint,

$$y^1_{t_1} - y^2_{t_-} \in S.$$

For some $t_0 \in [0, 4]$,

$$y^1_{t_1 + t_0} - y^2_{t_- + t_0} \quad \text{has no zero}$$

(Proposition 3.1), and there exists $\epsilon > 0$ such that

$$y^1_{t_1 + t_0} - y^2_{t_- + t_0 + t} \in S \quad \text{for} \quad |t| < 2\epsilon.$$

Let $\tau_i = \tau(y^i)$ for $i = 1, 2$. There exist positive integers $k > \frac{t_0}{\tau_1}$ and l such that

$$|k\tau_1 - l\tau_2| < \epsilon;$$

or $k\tau_1 = l\tau_2 + \hat{t}$ where $|\hat{t}| < \epsilon$. Hence, for $|t| < \epsilon$,

$$y^1_{t_1 + t_0 - k\tau_1} - y^2_{t_- + t_0 - k\tau_1 + t} = y^1_{t_1 + t_0} - y^2_{t_- + t_0 - l\tau_2 - \hat{t} + t}$$

$$= y^1_{t_1 + t_0} - y^2_{t_- + t_0 - \hat{t} + t} \in S.$$

Proposition 3.1 yields

$$y^1_{t_1} - y^2_{t_- + t} \in S \quad \text{for} \quad |t| < \epsilon,$$

which implies a contradiction to the definition of t_-. We have shown that

$$y^1_{t_1} - y^2_t \in S \quad \text{for all} \quad t \leq t_2.$$

Analogously, one obtains

$$y^1_{t_1} - y^2_t \in S \quad \text{for all} \quad t > t_2.$$

Now let real numbers s and t be given. Choose $k \in \mathbb{N}$ so that $s - t_1 + n\tau_1 > 0$. Then

$$y^1_s - y^2_t = F(s - t_1 + n\tau_1, y^1_{t_1 - n\tau_1}) - F(\ldots, y^2_{t - (s - t_1 + n\tau_1)})$$

$$= F(\ldots, y^1_{t_1}) - F(\ldots, y^2_{\ldots}),$$

and Proposition 3.1 implies the assertion.   □

PROOF OF THEOREM 5.1. 1. Recall the closed curves $c : [0, b] \longrightarrow C$, $|c| \subset L$, and the integers $k(c)$ from Section 4.

2. Consider the simple closed curve $\eta : [0, \tau] \longrightarrow C$, $\tau = z_{2n} + 1$. As in Section 4, one obtains, for each subinterval

$$[z_{2\nu} + 1, z_{2\nu+2} + 1], \quad \nu = 0, \ldots, n - 1$$

an integer $k(\nu)$ so that for each integer $k \geq k(\nu)$ there is a subdivision

$$z_{2\nu} + 1 = t^\nu_0 < t^\nu_1 < \cdots < t^\nu_{2k+2} = z_{2\nu+2} + 1$$

with

$$y_t \in K_{k\kappa} \quad \text{for} \quad t^\nu_\kappa \leq t \leq t^\nu_{\kappa+1}, \quad \kappa = 0, \ldots, k,$$

$$y_t \in -K_{k\kappa} \quad \text{for} \quad t^\nu_{k+1+\kappa} \leq t \leq t^\nu_{k+1+\kappa+1}, \quad \kappa = 0, \ldots, k.$$

Fix an integer

$$k \geq k(c) + \max_{\nu=0,\ldots,n-1} k(\nu).$$

3. Reparameterize η so that one obtains a simple closed piecewise smooth curve $\eta_r : [0,n] \longrightarrow C$ such that for $\nu = 0,\ldots,n-1$ and for $\kappa = 0,\ldots,2k+1$,

$$\eta_r \left(\left[\nu + \frac{\kappa}{2k+2}, \nu + \frac{\kappa+1}{2k+2} \right] \right) = \eta([t_\kappa^\nu, t_{\kappa+1}^\nu]),$$

i.e., the interval $[z_{2\nu} + 1, z_{2\nu+2} + 1]$ corresponds to $[\nu, \nu+1]$ and t_κ^ν corresponds to $\nu + \frac{\kappa}{2k+2}$.

4. Fix a subdivision

$$0 = t_0 < t_1 < \cdots < t_{2k+2} = b$$

as in Section 4, i.e.

$$c(t) \in K_{k\kappa} \quad \text{for} \quad t_\kappa \leq t \leq t_{\kappa+1}, \quad \kappa = 0,\ldots,k,$$

$$c(t) \in -K_{k\kappa} \quad \text{for} \quad t_{k+1+\kappa} \leq t \leq t_{k+1+\kappa+1}, \quad \kappa = 0,\ldots,k.$$

Reparameterize c so that one obtains a simple closed piecewise smooth curve $c_r : [0,1] \longrightarrow C$ with

$$c_r \left(\left[\frac{\kappa}{2k+2}, \frac{\kappa+1}{2k+2} \right] \right) = c([t_\kappa, t_{\kappa+1}]) \quad \text{for} \quad \kappa = 0,\ldots,2k+2.$$

Extend c_r by periodicity to a curve $n \cdot c_r : [0,n] \longrightarrow C$. Note that $|c| \subset L$ gives $p \circ n \cdot c_r = n \cdot c_r$.

5. Define a homotopy of closed curves as follows. For

$$0 \leq \nu \leq n-1, \quad 0 \leq \kappa \leq 2k+1, \quad \nu + \frac{\kappa}{2k+2} \leq t \leq \nu + \frac{\kappa+1}{2k+2}, \quad s \in [0,1],$$

set

$$\hom(s,t) := s \cdot n \cdot c_r(t) + (1-s) \cdot \eta_r(t).$$

Convexity of the subcones $K_{k\kappa}, -K_{k\kappa}$ of S now yields that

$$\hom([0,1] \times [0,n]) \subset S.$$

In particular, $0 \notin p \circ \hom(\ldots)$, by Lemma 4.1. It follows that

$$\text{wind}(0, p \circ \eta) = \text{wind}(0, p \circ \eta_r) = \text{wind}(0, p \circ \hom(0,\cdot)) = \text{wind}(0, p \circ \hom(1,\cdot))$$

$$= \text{wind}(0, p \circ n \cdot c_r) = \text{wind}(0, n \cdot c_r) = n \cdot \text{wind}(0, c_r) = n \cdot \text{wind}(0, c) = n;$$

hence $0 \in \text{int}(p \circ \eta)$.

6. Proposition 6.1 and Lemma 4.1 yield

$$0 \neq p(y_s - y_t) = py_s - py_t \quad \text{for} \quad 0 \leq s < t < \tau;$$

p is injective on $|\eta|$. It follows that $p \circ \eta$ is a simple closed curve; therefore $n = 1$, and $\tau = z_2 + 1$. $\square$

PROOF OF COROLLARY 5.1. Theorem 5.1 implies that the minimal period of y^i is $z_2(y^i) + 1$. Set $\eta_i := \eta(y^i)$. It follows that in case $y_0^1 \neq y_0^2$ the orbits $|\eta_i|$ are disjoint. Proposition 6.1 and Lemma 4.1 yield

$$(6.1) \qquad\qquad |p \circ \eta_1| \cap |p \circ \eta_2| = \emptyset.$$

Connect $0 \in \text{int}(p \circ \eta_2) \cap \text{int}(p \circ \eta_1)$ by a straight line to a point $\chi \in |p \circ \eta_1|$ with minimal norm. Then $[0,1)\chi \subset \text{int}(p \circ \eta_1)$. In case $[0,1)\chi \cap |p \circ \eta_2| = \emptyset$, we conclude that $\chi \in \text{int}(p \circ \eta_2)$, and using (6.1), that $|p \circ \eta_1| \subset \text{int}(p \circ \eta_2)$. Otherwise,

$$\emptyset \neq [0,1)\chi \cap |p \circ \eta_2| \subset [0,1)\chi \subset \text{int}(p \circ \eta_1),$$

and (6.1) implies $|p \circ \eta_2| \subset \text{int}(p \circ \eta_1)$. $\square$

PROOF OF THEOREM 5.2. Proposition 6.1 yields

$$(y')_0 = \lim_{t \searrow 0} \frac{1}{t}(y_t - y_0) \in \overline{S} = S \cup \{0\}.$$

Since $y'(-1) > 0$, $(y')_0 \in S$. Proposition 3.1 and periodicity imply that y' is slowly oscillating. As in the proof of Proposition 3.2,

$$y'(t) < 0 \quad \text{on} \quad [0, z_1].$$

There exists a zero of y' in $[-1, z_1]$. As y' is slowly oscillating, there is exactly one zero t_0 of y' in $[-1, 0)$, and $-1 < t_0$. Analogously one sees that there is precisely one zero t_1 of y' in $[z_1, z_2]$, and $z_1 < t_1 < z_1 + 1$. Now it is obvious how to complete the proof. $\square$

7. Proof of Theorem 5.3

We have to generalize results from [5, 37] where eq.(1.1) was studied with $\mu = 0$. Note first that for $\mu > 0$ and for every continuous function $b : \mathbb{R} \longrightarrow \mathbb{R}$, a function $x : \mathbb{R} \longrightarrow \mathbb{R}$ is a solution of the equation

$$(7.1) \qquad\qquad x'(t) = -\mu x(t) + b(t)x(t - 1)$$

if and only if

$$d : t \longrightarrow e^{\mu t}x(t)$$

is a solution of the equation

$$(7.2) \qquad\qquad d'(t) = e^{\mu}b(t)d(t - 1).$$

A differentiable function $x : \mathbb{R} \longrightarrow \mathbb{R}$ is called slowly oscillating *at* t (with respect to the delay 1), if either x has no zero in $[t - 1, t]$, or if x has precisely one zero in $[t - 1, t]$ and this zero is simple.

COROLLARY 7.1. *Consider solutions x of eq. (7.1) and d of eq. (7.2) as above; x is slowly oscillating at t if and only if d is slowly oscillating at t.*

PROOF. The zeros of x and d coincide. $x(z) = 0 \neq x'(z)$ implies $d(z) = 0$ and $d'(z) = \mu d(z) + e^{\mu z} x'(z) \neq 0$; $d(z) = 0 \neq d'(z)$ implies $x(z) = 0 \neq x'(z)$. $\square$

Assume

$$b(t) < 0 \quad \text{for all} \quad t \in \mathbb{R}$$

from now on. Using [5, **Lemma 1**] and Corollary 7.1 we infer:

LEMMA 7.1. *A solution $x : \mathbb{R} \longrightarrow \mathbb{R}$ of eq. (7.1) which is slowly oscillating at $t \in \mathbb{R}$, is slowly oscillating at every $s \geq t$.*

Note, also, that every slowly oscillating solution $x : \mathbb{R} \longrightarrow \mathbb{R}$ of eq. (7.1) is necessarily slowly oscillating at every $t \in \mathbb{R}$; i.e., has only simple zeros. (Proof: The function $d : t \longrightarrow e^{\mu t} x(t)$ is a slowly oscillating solution of eq. (7.2). $d(t) = 0$ and $d'(t) = 0$ would imply $d(t - 1) = 0$, a contradiction. So, d has only simple zeros. Apply Corollary 7.1.)

It is easy to see that the set

$$\Sigma \quad \text{of slowly oscillating solutions} \quad x : \mathbb{R} \longrightarrow \mathbb{R} \quad \text{of eq. (7.1)}$$

equals

$$e^{-\mu \cdot} \Sigma_2$$

where Σ_2 denotes the set of slowly oscillating solutions $d : \mathbb{R} \longrightarrow \mathbb{R}$ of eq. (7.2).

Let $C(\mathbb{R})$ denote the real vectorspace of continuous functions $x : \mathbb{R} \longrightarrow \mathbb{R}$, equipped with the topology of uniform convergence on compact sets.

LEMMA 7.2. 1. $\overline{\Sigma} \subset \Sigma \cup \{0\}$
2. *For every linear space $Z \subset \Sigma \cup \{0\}$, $\dim Z \leq 2$.*

PROOF. 1. Apply [5, **Lemma 2**] to Σ_2, multiply by the function $\mathbb{R} \ni t \longrightarrow e^{-\mu t} \in \mathbb{R}$, use $\Sigma = e^{-\mu \cdot} \Sigma_2$.
2. $Z \subset \Sigma \cup \{0\} = e^{-\mu \cdot}(\Sigma_2 \cup \{0\})$ implies $e^{\mu \cdot} Z \subset \Sigma_2 \cup \{0\}$. By [5, **Lemma 3**], $\dim e^{\mu \cdot} Z \leq 2$. Hence $\dim Z \leq 2$. $\square$

The results of [5, **Section 3**] hold for eq. (7.1) without change; the proofs of [5, **Lemmas 4, 5, 6, 7**] and of [5, **Corollary 1**] remain valid.

We consider the linear variational equation (3.3). Recall that with every Floquet multiplier $\lambda \in \text{spec}(V) \setminus \{0\}$, $\text{Im}(\lambda) \geq 0$, is associated a linear space $\mathcal{G}(\lambda) \subset C(\mathbb{R})$ of solutions of eq. (3.3) which are real parts of complex–valued solutions whose segments at $t = 0$ belong to the eigenspace of the spectral set $\{\lambda, \overline{\lambda}\}$. This is as in [5]. We have

$$\dim \mathcal{G}(\lambda) = \text{m}(\lambda)$$

where $\text{m}(\lambda)$ denotes the multiplicity of λ as an eigenvalue of V. The slowly oscillating periodic solution $y' : \mathbb{R} \longrightarrow \mathbb{R}$ of eq. (3.3) belongs to the space $\mathcal{G}(1)$.

It follows from [5, **Lemmas 5, 6, 7**] that every nonzero element of the space

$$\bigoplus_{\lambda \in \text{spec}(V), \text{Im}(\lambda) \geq 0, |\lambda| \geq 1} \mathcal{G}(\lambda)$$

is slowly oscillating. By [**5, Corollary 1**],

$$\dim \oplus_{\lambda \in \operatorname{spec}(V), \operatorname{Im}(\lambda) \geq 0, |\lambda| \geq 1} \mathcal{G}(\lambda) \leq 2.$$

We infer that

$$\text{either} \quad m(1) = 1, \quad |\lambda| < 1 \quad \text{for all} \quad \lambda \in \operatorname{spec}(V) \setminus \{1\},$$

$$\text{or} \quad m(1) = 2, \quad |\lambda| < 1 \quad \text{for all} \quad \lambda \in \operatorname{spec}(V) \setminus \{1\},$$

$$\text{or} \quad m(1) = 1, \quad \text{and there exists} \quad u \in \operatorname{spec}(V) \quad \text{in} \quad (-\infty, -1] \cup (1, \infty),$$

$$m(u) = 1, \quad |\lambda| < 1 \quad \text{for all} \quad \lambda \in \operatorname{spec}(V) \setminus \{1, u\}.$$

In order to prove Theorem 5.3, it remains to exclude in the last case that

$$u \leq -1.$$

This can be done as in [**37**]. We begin with the following lemma.

LEMMA 7.3. *Let* $Z \subset C(\mathbb{R})$ *be a linear space of solutions of eq. (7.1) such that every* $x \in Z \setminus \{0\}$ *is slowly oscillating. Suppose there exists* $\underline{x} \in Z \setminus \{0\}$ *such that its zeroset is neither bounded from below nor from above.*
1. *Then the same holds true for every* $x \neq 0$ *in* Z, *and* $x^{-1}(0)$ *is given by a strictly increasing sequence of simple zeros*

$$t_{x,n+1} > t_{x,n} + 1, \quad n \in \mathbb{Z}.$$

2. *For* x, v *in* $Z \setminus \{0\}$ *and* $t_{x,n} < t_{v,n} < t_{x,n+1}$,

$$t_{x,n+1} < t_{v,n+1}.$$

PROOF. 1. The zeros of $\underline{x}$ form a sequence of points $t_n, n \in \mathbb{Z}$, with $t_n + 1 < t_{n+1}$. Suppose the zeros of some $x \in Z$, $x \neq 0$, are bounded below by some real a. Then $x(t) \neq 0$ for $t < a$, hence

$$\operatorname{sign}(x'(t)) = \operatorname{sign}(-\mu x(t) + b(t)x(t-1)) = -\operatorname{sign}(x(t))$$

for all $t < a$. Observe that x is not a real multiple of $\underline{x}$. Choose $t_{n+1} < a$ so that

$$\operatorname{sign}(\underline{x}) = \operatorname{sign}(x) \quad \text{on} \quad (t_n, t_{n+1}).$$

There exist $c > 0$, $t \in (t_n, t_{n+1})$ such that

$$|c\underline{x}| \leq |x| \quad \text{on} \quad (t_n, t_{n+1}) \quad \text{and} \quad c\underline{x}(t) = x(t).$$

Consequently, $c\underline{x}'(t) = x'(t)$; t is a double zero of $c\underline{x} - x \in Z$, so $c\underline{x} - x = 0$, a contradiction. If one assumes an upper bound a for the zeros of x then

$$\operatorname{sign}(x'(t)) = -\operatorname{sign}(x(t)) \quad \text{for} \quad t > a + 1,$$

and one can argue as above.

2. Assume

$$t_{x,n} < t_{v,n} < t_{x,n+1} \quad \text{and} \quad t_{v,n+1} \le t_{x,n+1}.$$

In case $t_{v,n+1} = t_{x,n+1}$, define

$$c := \frac{v'(t_{x,n+1})}{x'(t_{x,n+1})}.$$

Then $cx - v \in Z$ has a double zero at $t_{x,n+1}$, hence $cx - v = 0$, a contradiction to $v(t_{v,n}) = 0 \ne x(t_{v,n})$. In case $t_{v,n+1} < t_{x,n+1}$ there exist $c \in \mathbb{R}$ and $t \in (t_{v,n}, t_{x,n+1})$ such that

$$|cv| \le |x| \quad \text{and} \quad \text{sign}(cv) = \text{sign}(x) \quad \text{in} \quad (t_{v,n}, t_{x,n+1}), \quad cv(t) = x(t).$$

It follows that $cv'(t) = x'(t)$; t is a double zero of $cv - x \in Z$, hence $cv - x = 0$, a contradiction as before. $\square$

REMARK 7.1. Assertion 2 above expresses a synchronization among solutions in Z.

Now assume that there exists $u \le -1$ in $\text{spec}(V)$. Set

$$b(t) := f'(y(t-1)), \quad \text{for} \quad t \in \mathbb{R}.$$

The space $Z := \mathcal{G}(1) \oplus \mathcal{G}(u)$ satisfies the hypotheses in Lemma 7.3. Choose $x \in \mathcal{G}(u) \setminus \{0\}$. $m(u) = 1$ implies

$$x_\tau = V x_0 = u \cdot x_0.$$

There are consecutive zeros

$$-1 < t_0 < 0, \quad t_1, \quad \tau - 1 < t_2 < \tau$$

of y', and

$$\tau = t_2 - t_0,$$

by Theorem 5.2 and its proof. It follows that

$$\text{sign}(x(t)) = \text{sign}(u \cdot x(t-\tau)) = -\text{sign}(x(t-\tau)) \quad \text{for} \quad \tau - 1 \le t \le \tau;$$

hence

(7.3) $$\text{sign}(x(t_2)) = -\text{sign}(x(t_0))$$

and

(7.4) $$\text{sign}(x'(t_2)) = -\text{sign}(x'(t_0)).$$

We may assume that in the sequence of zeros $t_{x,n}$ of x, given by Lemma 7.1,

$$t_{x,0} \le t_0 < t_{x,1}.$$

The case $t_{x,0} < t_0$. Repeated applications of Lemma 7.3 yield

$$t_{x,1} < t_1 < t_{x,2} < t_2 < t_{x,3}.$$

As zeros are simple, we arrive at a contradiction to (7.3).

The case $t_{x,0} = t_0$. By (7.3), $x(t_2) = 0$. Hence $t_{x,1} \leq t_2$. If $(t_0 <)\ \ t_{x,1} < t_1$, or if $t_1 < t_{x,1} < t_2$, then repeated applications of Lemma 7.3 show that t_2 is not a zero of x. This contradicts (7.3). Similarly,

$$t_2 = t_{x,1} \quad (> t_1 > t_0 = t_{x,0})$$

is not possible, due to Lemma 7.1. We have shown

$$t_{x,1} = t_1.$$

Arguments as before yield

$$t_{x,2} = t_2.$$

The simplicity of the zeros now gives

$$\text{sign}(x'(t_0)) = \text{sign}(x'(t_{x,0})) = \text{sign}(x'(t_{x,2})) = \text{sign}(x'(t_2)),$$

a contradiction to (7.4). □

It is convenient to state the following results at this point.

PROPOSITION 7.1. *If y is unstable and hyperbolic, and if u is the Floquet multiplier in $(1, \infty)$, then the set $\mathcal{G}(1) \oplus \mathcal{G}(u) \setminus \{0\}$ consists of slowly oscillating solutions $\mathbb{R} \longrightarrow \mathbb{R}$ of eq. (3.3). We have $\mathcal{G}(1) = \mathbb{R} \cdot y'$, $\mathcal{G}(u) = \mathbb{R} \cdot v$ for some slowly oscillating solution v of eq. (3.3) with $\|v_0\| = 1$, $V(y')_0 = (y')_0$ and $V v_0 = u \cdot v_0$.*

The proof should be obvious from preceding arguments.

8. Graph Representation of W

From now on we assume that the hypothesis of Theorem 5.4 is satisfied: y is a slowly oscillating periodic solution of eq. (1.1) with $y_0 \in K$ and $z_0 = -1$; y is unstable and hyperbolic. Then 1 is a simple Floquet multiplier. According to Theorem 5.3, there is a single Floquet multiplier u outside the unit circle; we have

$$u \in (1, \infty), \quad \text{m}(u) = 1.$$

It is convenient to begin with the return map

$$P_H : N \cap H \ni \phi \longrightarrow P(\phi) \in H.$$

Also, y_0 is a hyperbolic fixed point of P_H. u is a simple eigenvalue of the compact map $D P_H(y_0)$. All other eigenvalues satisfy $|\lambda| < 1$.

This is essentially as in case of a periodic orbit of a vectorfield on a finite–dimensional space [16]. For a proof in our situation, see the forthcoming book [6], or compare [15].

Now, [14, Theorem 3.1] (or [25, Theorem 2.7]) yields a one–dimensional local unstable manifold W_H of P_H at y_0:

Let G_H and R_H denote the eigenspaces given by the spectral sets $\{u\}$ and $\operatorname{spec}(DP_H(y_0)) \setminus \{u\}$, respectively;

$$H = G_H \oplus R_H; \quad \dim(G_H) = 1.$$

Fix

$$u_H \in (1, u).$$

There exist convex open neighborhoods N_G of 0 in G_H, N_R of 0 in R_H,

$$y_0 + N_G + N_R \subset N \cap H,$$

and a C^1–map

$$w_H : N_G \longrightarrow N_R, \quad w_H(0) = 0 \quad \text{and} \quad Dw_H(0) = 0,$$

such that the shifted graph

$$y_0 + \{\phi + w_H(\phi) : \phi \in N_G\} =: W_H$$

coincides with the set $W(y_0 + N_G + N_R)$ of $\phi \in y_0 + N_G + N_R$ such that there is a trajectory $(\phi_j)^0_{-\infty}$ of P_H with $\phi_0 = \phi$, $\phi_j \in y_0 + N_G + N_R$ for all $j \in -\mathbb{N}_0$; there exists a constant $c_H \geq 0$ such that every trajectory $(\phi_j)^0_{-\infty}$ of P in $y_0 + N_G + N_R$ satisfies

$$\|\phi_j - y_0\| \leq c_H \cdot \|\phi_0 - y_0\| \cdot u_H^j \quad \text{for all} \quad j \in -\mathbb{N}_0.$$

REMARK 8.1. $W_H \subset W$.

PROOF. Let $\phi \in W_H$. There is a trajectory $(\phi_j)^0_{-\infty}$ of P with $\phi = \phi_0$ and $\phi_j \longrightarrow y_0$ as $j \longrightarrow -\infty$. This defines a solution $x : \mathbb{R} \longrightarrow \mathbb{R}$ of eq. (1.1) whose zeros are given by a sequence $(z_j)^J_{-\infty}, J \in \mathbb{N} \cup \{\infty\}, J \geq 2, \quad z_0 = -1,$ with property (3.2) such that

$$\phi_j = x_{z_{2j}+1} \quad \text{for} \quad j \leq 0.$$

Continuity of $stop$ implies $z_2(\phi_j) + 1 \longrightarrow \tau$ as $j \longrightarrow -\infty$. Now it is easy to conclude that

$$x_t \longrightarrow |\eta| \quad \text{as} \quad t \longrightarrow -\infty,$$

or, $\phi = \phi_0 = x_0 \in W$. $\square$

REMARK 8.2. $P(W_H) \subset W \cap H$.

PROOF. Use $P(W_H) \subset F(\mathbb{R}^+ \times W_H) \subset F(\mathbb{R}^+ \times W) \subset W$. $\square$

We need exponential convergence and an asymptotic phase on the unstable set W. This can be achieved by standard arguments which use the return map P and its local unstable manifold W_H. However, a citeable treatment in the case of semiflows does not seem to be available in the literature, so we include a detailed proof.

Define $\rho > 0$ by

$$e^{\rho \tau} = u_H.$$

PROPOSITION 8.1. *For every solution* $x : \mathbb{R} \longrightarrow \mathbb{R}$ *of eq.* (1.1) *with phase curve in* W, *there are* $t = t(x) \leq 0$, $\gamma = \gamma(x) \in [0, \tau)$ *and a constant* $c = c(x) > 0$ *such that for all* $s \leq t$,

$$\|x_s - y_{s+\gamma}\| \leq c \cdot e^{\rho s}.$$

1. . Choose a neighborhood N_0 of y_0 and a constant $c > 0$ such that

$$\|Dstop(\phi)\| \leq c \quad \text{on} \quad N_0, \quad \|D_2 F(s, \phi)\| \leq c \quad \text{for} \quad 0 \leq s \leq \tau, \quad \phi \in N_0$$

(for the last property, use compactness of $[0, \tau]$ and continuity of $D_2 F$ on $[0, \infty) \times C$.)

2. Let a solution $\hat{x} : \mathbb{R} \longrightarrow \mathbb{R}$ of eq. (1.1) with phase curve in W be given. There exists $\hat{t} \geq 0$ so that for $x := \hat{x}(\cdot - \hat{t})$ we have

$$x_t \in N_0 \quad \text{for all} \quad t \leq 0, \quad J(x) \in \mathbb{N} \quad \text{or} \quad J(x) = \infty,$$

and for the zeros z_j of x

$$z_0 = -1 \quad \text{and} \quad x_{z_{2j}+1} \in K \cap (y_0 + N_G + N_R) \quad \text{for} \quad j \leq 0.$$

Set

$$\phi_j := x_{z_{2j}+1} \quad \text{for} \quad j \leq 0,$$

$$\Delta_j := z_{2j+2} - z_{2j} \quad (= stop(\phi_j)) \quad \text{for} \quad j \leq -1,$$

so that

$$\phi_{j+1} = P(\phi_j) = F(\Delta_j, \phi_j) \quad \text{for} \quad j \leq -1.$$

The estimate

$$|\tau - \Delta_j| = |stop(y_0) - stop(\phi_j)| \leq c \cdot \|y_0 - \phi_j\|$$

$$\leq c \cdot c_H \cdot \|y_0 - \phi_j\| \cdot u_H{}^j \quad \text{for} \quad j \leq 0$$

implies that

$$\gamma := -\sum_1^\infty (\tau - \Delta_{-j})$$

is well–defined. For some $l \in \mathbb{Z}$, $l\tau \leq \gamma < (l+1)\tau$.

3. For $k \in \mathbb{N}$, set $t_k := \gamma + \sum_1^k (\tau - \Delta_{-j})$. Then

$$\|x_{t_k - k\tau - l\tau} - x_{-k\tau - l\tau}\| \leq c_f \cdot |t_k| \quad \text{(Proposition 3.2)}$$

$$\leq c_f \cdot c \cdot c_H \cdot \|y_0 - \phi_0\| \cdot u_H^{-k} \cdot \frac{1}{u_H - 1}.$$

4. For all $k \in \mathbb{N}$ we obtain

$$\|x_{-k\tau - l\tau} - y_{\gamma - l\tau}\| \leq$$

$$\|x_{t_k - k\tau - l\tau} - x_{-k\tau - l\tau}\| + \|x_{t_k - k\tau - l\tau} - y_\gamma - l\tau\|$$

$$\leq c_f \cdot c \cdot c_H \cdot \|y_0 - \phi_0\| \cdot \frac{1}{u_H - 1} \cdot u_H^{-k} \quad + \quad \|x_{t_k - k\tau - l\tau} - y_\gamma - l\tau\|$$

$$= \quad \ldots \quad + \|F(\gamma - l\tau, \phi_{-k}) - F(\gamma - l\tau, y_0)\|$$

$$\left(\text{with} \quad t_k - k\tau = \gamma + \sum_1^k (-\Delta_{-j}) \quad \right)$$

$$\leq \quad \cdots + c \cdot \|\phi_{-k} - y_0\| \leq \hat{c} \cdot \|\phi_0 - y_0\| \cdot e^{-\gamma(k+l)\tau},$$

$$\text{where} \quad \hat{c} := \left(\frac{c_f}{u_H - 1} + 1\right) \cdot c_H \cdot c \cdot e^{\rho l\tau},$$

or

$$\|x_{-k\tau} - y_{\gamma - l\tau}\| \leq \hat{c} \cdot \|\phi_0 - y_0\| \cdot e^{-\rho k\tau} \quad \text{for integers} \quad k \geq |l| + 1.$$

Let $t \leq -(|l| + 1)\tau$ be given. For some integer $k \geq |l| + 2$,

$$-k\tau < t \leq -k\tau + \tau.$$

Hence,

$$\|x_t - y_{t+(\gamma - l\tau)}\| = \|F(t + k\tau, x_{-k\tau}) - F(t + k\tau, y_{\gamma - l\tau - k\tau})\|$$

$$= \| \quad \ldots \quad - F(t + k\tau, y_{\gamma - l\tau})\| \quad \text{(periodicity)}$$

$$\leq c \cdot \|x_{-k\tau} - y_{\gamma - l\tau}\| \leq c \cdot \hat{c} \cdot \|\phi_0 - y_0\| \cdot e^{-\rho k\tau}$$

$$= c \cdot \hat{c} \cdot \|\phi_0 - y_0\| \cdot e^{-\rho t - \rho k\tau} e^{\rho t} \leq c \cdot \hat{c} \cdot \|\phi_0 - y_0\| \cdot e^{\rho t}.$$

Finally, for $t \leq -\hat{t} - (|l| + 1)\tau,$

$$\|\hat{x}_t - y_{t+\hat{t}+\gamma - l\tau}\| = \|x_{t+\hat{t}} - y_{t+\hat{t}+\gamma - l\tau}\|$$

$$\leq c \cdot \hat{c} \cdot \|\phi_0 - y_0\| \cdot e^{\rho(t+\hat{t})} = c \cdot \hat{c} \cdot \|\phi_0 - y_0\| \cdot e^{\rho\hat{t}} e^{\rho t}.$$

Define the asymptotic phase of $\hat{x}$ by

$$\hat{\gamma} := \hat{t} + \gamma - l\tau - j\tau \in [0, \tau)$$

where $j \in \mathbb{Z}$. $\square$

We turn to the period map $F(\tau, \cdot) : C \longrightarrow C$. As before, there exists a local unstable manifold at y_0. More precisely, there is a smooth map

$$w_\tau : N_\tau \longrightarrow R'_\tau,$$

where N_τ is an open convex neighborhood of 0 in the one–dimensional (reellified) linear unstable eigenspace

$$G_\tau \subset C \quad \text{of} \quad V = D_2 F(\tau, y_0),$$

and R'_τ is an open convex neighborhood of 0 in the (reellified) eigenspace R_τ associated with the complementary spectral set

$$\text{spec}(V) \setminus \{u\},$$

such that we have

$$w_\tau(0) = 0, \quad Dw_\tau(0) = 0;$$

and the shifted graph

$$W_\tau := y_0 + \{\chi + w_\tau(\chi) : \chi \in N_\tau\}$$

coincides with the set of all $\phi \in y_0 + N_\tau + R'_\tau$ such that there is a trajectory $(\phi_j)^0_{-\infty}$ of $F(\tau, \cdot)$ with $\phi_0 = \phi$ and

$$(\phi_j - y_0) \cdot u_H^{-j} \in N_\tau + R'_\tau \quad \text{for all} \quad j \in -\mathbb{N}_0.$$

The map $\Phi : W_\tau \longrightarrow W_\tau$, $\phi \longrightarrow \phi_{-1}$, is a C^1–diffeomorphism onto an open neighborhood W'_τ of y_0 in W_τ. For a proof, we refer to [14, Theorem 3.1] on local unstable manifolds of hyperbolic fixed points. Of course, the present situation is non–hyperbolic, but as indicated in [36, Section 5], one can easily modify the proof from [14] in order to derive the result described above. A very detailed version of the proof of [14, Theorem 3.1] may be found in [25].

Proposition 7.1 implies that there is a slowly oscillating solution $v : \mathbb{R} \longrightarrow \mathbb{R}$ of eq. (3.3), $\|v_0\| = 1$, so that

$$T_{y_0} W_\tau = \quad (G_\tau =) \quad \mathbb{R} \cdot v_0.$$

Now we are in a position to prove a first part of Theorem 5.4, namely that W is given by a map w with domain $pW \subset L$ and range in Q. The assertion is equivalent to the injectivity of the projection p on W. The latter follows from $0 \notin pS$ (Lemma 4.1), provided we can show that nontrivial differences

(8.1) $$0 \neq \phi - \overline{\phi}$$

of points $\phi, \overline{\phi}$ in W belong to S.

So let $\phi \neq \overline{\phi}$ in W be given. There are solutions $x : \mathbb{R} \longrightarrow \mathbb{R}$, $\overline{x} : \mathbb{R} \longrightarrow \mathbb{R}$ of eq. (1.1) with $x_0 = \phi$, $\overline{x}_0 = \overline{\phi}$, respectively. Let γ and $\overline{\gamma}$ denote the asymptotic phases of x and $\overline{x}$.

Case 1: $\gamma \neq \overline{\gamma}$. Then

$$x_{-j\tau} \longrightarrow y_\gamma, \quad \overline{x}_{-j\tau} \longrightarrow y_{\overline{\gamma}} \quad \text{as} \quad j \longrightarrow \infty.$$

Now, γ and $\overline{\gamma}$ belong to $[0, \tau)$, and Proposition 6.1 gives

$$y_\gamma - y_{\overline{\gamma}} \in S.$$

According to Proposition 3.1, there exists $s \in [0, 4]$ such that

$$y_{s+\gamma} - y_{s+\overline{\gamma}} \quad \text{has no zero.}$$

The continuity of $F(s, \cdot)$ implies that for all $j \geq \frac{4}{\tau}$ sufficiently large,

$$x_{s-j\tau} - \overline{x}_{s-j\tau} \quad \text{has no zero.}$$

Again by Proposition 3.1,

$$\phi - \overline{\phi} = x_0 - \overline{x}_0 = F(j\tau - s, x_{s-j\tau}) - F(\ldots, \overline{x}_{s-j\tau}) \in S.$$

Case 2: $\gamma = \overline{\gamma}$. Now

$$\|x_{-j\tau+(\tau-\gamma)} - y_0\| \leq c(x) \cdot e^{-j\rho\tau} \quad \text{for all} \quad j \in \mathbb{N},$$

analogously for $\overline{x}$, with positive constants $c(x)$ and $c(\overline{x})$. It follows that there exists $j_0 \in \mathbb{N}$ such that all

$$x_{-j\tau+(\tau-\gamma)}, \quad \overline{x}_{-j\tau+(\tau-\gamma)} \quad \text{with} \quad j \geq j_0$$

belong to the local unstable manifold W_τ; the sequence of normed distances

$$\frac{1}{\|x_{-j\tau+(\tau-\gamma)} - \overline{x}_{-j\tau+(\tau-\gamma)}\|}(x_{-j\tau+(\tau-\gamma)} - \overline{x}_{-j\tau+(\tau-\gamma)}) =: d_j$$

is well–defined (since $\phi \neq \overline{\phi}$) and converges to the unit sphere

$$\{v_0, -v_0\} \subset T_{y_0} W_\tau$$

as $j \longrightarrow \infty$. For some $s \in [0, 4]$, $v_s = V(s, 0)v_0$ has no zero (Proposition 3.1). There exist $\epsilon > 0$ and $\delta > 0$ such that for

$$\text{dist}(\chi, \{v_0, -v_0\}) < \delta,$$

$$|V(s, 0)\chi(t)| > \epsilon \quad \text{on} \quad [-1, 0].$$

Further, there exists $\delta_1 > 0$ such that for $\|\psi\| \leq \delta_1$, $\|\overline{\psi}\| \leq \delta_1$,

$$\|F(s, y_0 + \psi) - F(s, y_0 + \overline{\psi}) - V(s, 0)[\psi - \overline{\psi}]\|$$

$$= \left\|\int_0^1 \{D_2 F(s, y_0 + \overline{\psi} + \theta(\psi - \overline{\psi})) - D_2 F(s, y_0)\}[\psi - \overline{\psi}]d\theta\right\|$$

$$\leq \epsilon \cdot \|\psi - \overline{\psi}\|.$$

Choose an integer $j \geq \frac{4 + \tau}{\tau}$ so large that

$$\text{dist}(d_j, \{v_0, -v_0\}) < \delta$$

and

$$\|x_{-j\tau+(\tau-\gamma)} - y_0\| \leq \delta_1, \quad \|\overline{x}_{-j\tau+(\tau-\gamma)} - y_0\| \leq \delta_1.$$

It follows that

$$\left\|\frac{1}{\|x_{-j\tau+(\tau-\gamma)} - \overline{x}_{-j\tau+(\tau-\gamma)}\|}\{x_{s-j\tau+\tau-\gamma} - \overline{x}_{s-j\tau+\tau-\gamma}\} - V(s, 0)d_j\right\| \leq \epsilon,$$

and

$$\frac{1}{\|\cdots\|}\{\ldots\} = d_j$$

has no zero. Consequently, $\{\ldots\} \in S$, and by Proposition 3.1 once again,

$$\phi - \overline{\phi} = x_0 - \overline{x}_0 \in S. \quad \square$$

COROLLARY 8.1. *For every $\phi \in W$,*

$$T_\phi W \in S \cup \{0\}.$$

PROOF. We saw that for any two points $\psi \neq \phi$ in W, $\psi - \phi \in S$. Use $\overline{S} = S \cup \{0\}$, $\mathbb{R} \cdot \overline{S} \subset \overline{S}$. $\square$

9. Smoothness of W

In this section we show that the domain pW of the map w obtained in Section 8 is open, and that w is C^1.

Let p_τ denote the eigenprojection onto G_τ, associated with the spectral set $\{u\}$ of V. The fact that W_τ consists of initial values of backward trajectories of $F(\tau, \cdot)$ which tend to y_0 implies

$$(9.1) \qquad\qquad\qquad\qquad W_\tau \subset W$$

PROPOSITION 9.1. *For every neighborhood N_0 of y_0 in C and for every $t_0 \geq 0$,*

$$W = F((t_0, \infty) \times (W_\tau \cap N_0)).$$

PROOF. 1. Positive invariance of W and (9.1) yield $W \supset F(\dots)$.
2. Let a neighborhood N_0 of y_0 in C be given. Let $\phi \in W$. There exist $\gamma \in [0, \tau)$, $t \leq 0$ and $c > 0$ such that the solution $x : \mathbb{R} \longrightarrow \mathbb{R}$ of eq. (1.1) with $x_0 = \phi$ satisfies

$$\|x_s - y_{s+\gamma}\| \leq c \cdot e^{\rho s} \quad \text{for} \quad s \leq t.$$

In particular,

$$\|x_{-k\tau - j\tau - \gamma} - y_0\| \leq (c \cdot e^{-\rho\gamma} \cdot u_H^{-j}) \cdot u_H^{-k}$$

for all integers $j \geq -\dfrac{t+\gamma}{\tau}$, and for all $k \in \mathbb{N}_0$. For j sufficiently large we obtain

$$x_{-j\tau - \gamma} \in W_\tau, \quad x_{-j\tau - \gamma} \in N_0, \quad j\tau + \gamma \geq t_0.$$

Hence,

$$\phi = x_0 = F(j\tau + \gamma, x_{-j\tau - \gamma}) \in F((t_0, \infty) \times (W_\tau \cap N_0)). \quad \square$$

COROLLARY 9.1. *W is pathwise connected.*

PROOF. For $\phi \in W = F(\mathbb{R}^+ \times W_\tau)$, $\phi = F(t, \psi)$ with $t \geq 0$, $\psi \in W_\tau$. For $0 \leq s \leq t$, $F(s, \psi) \in W$. The C^1-graph W_τ over the open convex set $N_\tau \subset G_\tau$ connects ψ to y_0 in W. $\square$

PROPOSITION 9.2. *There exist an open neighborhood N_1 of y_0 in C and $\epsilon_1 \in (0, \tau - 1)$ with*

$$F(t, \phi) \notin W_\tau \quad \text{for} \quad \phi \in N_1 \cap W_\tau, \quad 0 < t < 2\epsilon_1.$$

PROOF. 1. Set $\psi := (y')_0$. Choose $\epsilon \in (0, 1)$, $\epsilon < \|\psi\|$, so small that

$$\psi \notin \left[-\frac{\|\psi\| + \epsilon}{1 - \epsilon}, \frac{\|\psi\| + \epsilon}{1 - \epsilon} \right] \cdot (v_0 + U) \quad + \quad U$$

where $U := \{\phi \in C : \|\phi\| \le \epsilon\}$. This is possible since ψ and v_0 are linearly independent. We have

$$(0, \infty) \cdot (\psi + U) \cap \mathbb{R} \cdot (v_0 + U) = \emptyset$$

since otherwise,

$$\psi = r \cdot (v_0 + \phi) + \chi \quad \text{with} \quad r \in \mathbb{R}, \phi \in U, \chi \in U;$$

consequently

$$|r|(1 - \epsilon) \le \|\psi\| + \epsilon$$

which implies a contradiction to the choice of ϵ.

2. Choose an open convex neighborhood N'_τ of 0 in $N_\tau \subset G_\tau$ with

$$Dw_\tau(\chi)v_0 \in U \quad \text{for} \quad \chi \in N'_\tau.$$

For χ, χ' in N'_τ, the integral in

$$\chi_1 + w_\tau(\chi_1) = \chi + w_\tau(\chi) + \int_0^1 [(\chi_1 - \chi) + Dw_\tau(\chi + t(\chi_1 - \chi))(\chi_1 - \chi)]dt$$

has values in $\mathbb{R} \cdot (v_0 + U)$ (use $\chi_1 - \chi = rv_0$ for some $r \in \mathbb{R}$, and convexity). We get

$$y_0 + \chi_1 + w_\tau(\chi_1) \in y_0 + \chi + w_\tau(\chi) + \mathbb{R} \cdot (v_0 + U)$$

for all χ and χ_1 in N'_τ.

3. Recall

$$D_1 F(\tau, \Phi(y_0))1 = (y')_0 = \psi.$$

By continuity, there exist $\epsilon_1 \in (0, \tau - 1)$ and a neighborhood N_1 of y_0 in C with the following properties: For all $\phi \in N_1 \cap W_\tau$ and all $t \in [0, 2\epsilon_1]$,

$$D_1 F(t + \tau, \Phi(\phi))1 \in \psi + U$$

and

$$p_\tau(F(t, \phi) - y_0) \in N'_\tau.$$

Observe that for phase curves in W with segment $\phi \in W_\tau$ at time $t = 0$, the tangent vectors at times $t \ge 0$ equal

$$D_1 F(t + \tau, \Phi(\phi))1.$$

Now assume that for some $\phi \in N_1 \cap W_\tau$ and some $t \in (0, 2\epsilon_1]$,

$$F(t, \phi) \in W_\tau.$$

There are χ, χ_1 in N'_τ with

$$\phi = y_0 + \chi + w_\tau(\chi), \quad F(t, \phi) = y_0 + \chi_1 + w_\tau(\chi_1).$$

Hence,

$$\mathbb{R} \cdot (v_0 + U) \ni F(t, \phi) - \phi$$

$$= t \cdot \int_0^1 D_1 F(st + \tau, \Phi(\phi))1 ds \in (0, \infty) \cdot (\psi + U),$$

a contradiction to the result of part 1 above. $\square$

CORROLARY 9.2. *The map*

$$A_1 : (-\epsilon_1, \epsilon_1) \times (W_\tau \cap N_1) \ni (t, \phi) \longrightarrow F(t + \tau, \phi) \in C$$

is injective.

PROOF. Assume $F(t + \tau, \phi) = F(s + \tau, \psi)$, $-\epsilon_1 < s \leq t < \epsilon_1$, ϕ and ψ in $W_\tau \cap N_1$. Injectivity of all $F(t', \cdot), t' \geq 0$, yields $\psi = F(t - s, \phi)$. By Proposition 9.1, $t - s = 0$; hence $\psi = \phi$. $\square$

PROPOSITION 9.3. *There exist $\epsilon_2 \in (0, \epsilon_1)$ and an open neighborhood $N_2 \subset N_1$ of y_0 in C, $W_\tau \cap N_2 \subset \Phi(W_\tau)$, such that the C^1-map $p \circ A_2$,*

$$A_2 := A_1|(-\epsilon_2, \epsilon_2) \times (W_\tau \cap N_2),$$

has injective derivatives.

PROOF. 1. At a point $(t, \phi) \in (-\epsilon_1, \epsilon_1) \times (W_\tau \cap N_1)$, a basis of the tangent space to the C^1-manifold $(-\epsilon_1, \epsilon_1) \times (W_\tau \cap N_1)$ is given by the vectors $(1, 0)$ and $(0, v_0 + Dw_\tau(p_\tau(\phi - y_0))v_0)$ in $\mathbb{R} \times C$, and $D(p \circ A_1)(t, \phi)T_{(t,\phi)}(\ldots)$ consists of the linear combinations of the vectors

$$pD_1 F(t + \tau, \phi)1 \in C, \quad pD_2 F(t + \tau, \phi)[v_0 + Dw_\tau(p_\tau(\phi - y_0))v_0] \in C.$$

2. At $t = 0, \phi = y_0$, these vectors are

$$p(y')_\tau = p(y')_0 \quad \text{and} \quad pVv_0 = u \cdot pv_0, \quad \text{respectively.}$$

We show that $p(y')_0$, pv_0 are linearly independent:
The fact that all nonzero elements of $\mathcal{G}(1) \oplus \mathcal{G}(u) = \mathbb{R} \cdot y' \oplus \mathbb{R} \cdot v$ are slowly oscillating (Proposition 7.1) implies

$$\mathbb{R} \cdot (y')_0 \oplus \mathbb{R} \cdot v_0 \subset S \cup \{0\}.$$

By $0 \notin pS$,
$$0 \neq a_1 \cdot p(y')_0 + a_2 \cdot pv_0 = p(a_1 \cdot (y')_0 + a_2 \cdot v_0)$$

whenever $(a_1, a_2) \neq (0, 0)$.
3. Continuity permits us to find $\epsilon_2 \in (0, \epsilon_1)$ and an open neighborhood $N_2 \subset N_1$ of y_0 in C with $W_\tau \cap N_2 \subset \Phi(W_\tau)$ so that for $|t| < \epsilon_2$ and $\phi \in N_2$,

$$pD_1 F(t + \tau, \phi)1 \quad \text{and} \quad pD_2 F(t + \tau, \phi)[v_0 + Dw_\tau(p_\tau(\phi - y_0))v_0]$$

are linearly independent. $\square$
Define
$$W_A := A_2((-\epsilon_2, \epsilon_2) \times (W_\tau \cap N_2)) \subset W.$$

COROLLARY 9.3. pW_A *is an open subset of* L, *and the map* $w|pW_A$ *is* C^1; W_A *is a* C^1-*submanifold of* C.

PROOF. The map $id_{\mathbb{R}} \times \Phi$ defines a diffeomorphism A_0 of the C^1-manifold $(-\epsilon_2, \epsilon_2) \times \Phi^{-1}(W_\tau \cap N_2)$ onto the domain of A_2 (use $W_\tau \cap N_2 \subset \Phi(W_\tau)$). The map

$$B : (-\epsilon_2, \epsilon_2) \times \Phi^{-1}(W_\tau \cap N_2) \longrightarrow L$$

given by $id_{\mathbb{R}} \times \Phi$, A_2, p is a C^1- diffeomorphism onto the open subset pW_A of the 2-dimensional space L (Injectivity follows from Corollary 9.2 and from the injectivity of $p|W$; the injectivity of derivatives follows from Proposition 9.3.). For $\chi = p\phi$ and $\phi \in W_A$,

$$w(\chi) = (id - p)\phi \quad \text{and} \quad \phi = A_2((id_{\mathbb{R}} \times \Phi)(B^{-1}(p\phi))). \quad \square$$

We have

$$W = F(\mathbb{R}^+ \times W_A)$$

since $F(\mathbb{R}^+ \times W_A) \subset F(\mathbb{R}^+ \times W) \subset W$,

$$W = F((\tau, \infty) \times (W_\tau \cap N_2)) \quad \text{(Proposition 9.1)},$$

and for each $t \geq \tau$ and $\phi \in W_\tau \cap N_2$,

$$F(t, \phi) = F(t - \tau, F(\tau, \phi)) = F(t - \tau, A_2(0, \phi)).$$

In order to show that $pW \subset L$ is open, and that w is C^1, it is therefore enough to prove that for every $t \geq 0$,

$$pF(t, \cdot)(W_A) \subset L \quad \text{is open, and} \quad w|pF(t, \cdot)(W_A) \quad \text{is} \quad C^1.$$

So let $t \geq 0$ be given. The C^1-map $p \circ F(t, \cdot)|W_A$ is injective ($p|W$ is injective!); its derivatives are injective since

$$T_\phi W_A \subset T_\phi W \subset S \cup \{0\} \quad \text{(Corollary 8.1)},$$

$$D_2 F(t, \phi)S \subset S \quad \text{(Proposition 3.1.2)},$$

$$D_2 F(t, \phi) \quad \text{is injective, and} \quad 0 \notin pS.$$

It follows that $p \circ F(t, \cdot)$ defines a C^1-diffeomorphism B_t of W_A onto the open subset $B_t(W_A) = pF(t, \cdot)(W_A) \subset L$. For $\chi \in B_t(W_A)$ and $\chi = p\phi$, $\phi \in F(t, \cdot)(W_A) \subset W$, we obtain

$$w(\chi) = (id - p)\phi \quad \text{and} \quad \phi = F(t, \cdot) \circ B_t^{-1}(p\phi).$$

10. The Lipschitz Condition

We proceed as in [36, Section 7]. Lemma 4.1, i.e.

(10.1) $0 \notin pS,$

and [36, proof of Lemma 7.1] yield the following result.

LEMMA 10.1. *Let a subcone $S' \subset S$ be given. The following statements are equivalent.*
1. *There exists $c^* > 0$ with $c^* \|\phi\| \leq \|T(1)\phi\|$ for all $\phi \in S'$.*
2. *There exists $c > 0$ with $c \cdot \|\phi\| \leq \|p\phi\|$ for all $\phi \in S'$.*

In [36] we made the hypothesis (H2) that the zero solution of eq. (1.1) is linearly unstable. This is not required in the present paper.

Observe that all prerequisites for [36, Proposition 7.1] (with the exception of (10.1) above) and [36, Proposition 7.1] itself are derived without using (H2). Therefore we have our final result in this section.

PROPOSITION 10.1. *Let $r > 0$ be given. There exists a constant $c(r) > 0$ with the following property:*
If $x : [t_0 - 1, \infty) \longrightarrow \mathbb{R}$ and $x^* : [t_0 - 1, \infty) \longrightarrow \mathbb{R}$ are solutions of eq. (1.1) with $|x(t)| \leq r$ and $|x^*(t)| \leq r$ for all $t \geq t_0 - 1$, so that

$$d := x - x^* \quad \text{has no zero in} \quad [t_0 - 1, t_0]$$

then, for all $t \geq t_0 + 2$,

$$c(r)\|d_t\| \leq \|pd_t\|.$$

As $\overline{W}$ is compact there exists $r(W) > 0$ so that every solution of eq. (1.1) with phase curve in $\overline{W}$ is bounded by $r(W)$. Now it is easy to derive

$$\|w(\chi) - w(\chi^*)\| \leq \left(1 + \frac{1}{c(r(W))}\right) \|\chi - \chi^*\| \quad \text{for all} \quad \chi \in pW, \chi^* \in pW.$$

For $\chi \neq \chi^*$ in pW, consider the solutions $x : \mathbb{R} \longrightarrow \mathbb{R}$ and $x^* : \mathbb{R} \longrightarrow \mathbb{R}$ of eq. (1.1) with $x_0 = \chi + w(\chi)$, $x_0^* = \chi^* + w(\chi^*)$. For some $t \leq -6$, $d := x - x^*$ satisfies $d_t \in S$ (see (8.1)). There exists $t_0 \leq -2$ so that

$$d_{t_0} \quad \text{has no zero (Proposition 3.1.2).}$$

Proposition 10.1 yields

$$c(r(W)) \cdot \|\chi + w(\chi) - (\chi^* + w(\chi^*))\| = c(r(W)) \cdot \|d_0\| \leq \|pd_0\| = \|\chi - \chi^*\|.$$

Set

$$l_w := 1 + \frac{1}{c(r(W))}.$$

Finally, we obtain a Lipschitz continuous extension

$$\overline{w} : \overline{pW} \longrightarrow Q$$

of w, with Lipschitz constant l_w, so that

$$\overline{W} = \{\chi + \overline{w}(\chi) : \chi \in \overline{pW}\},$$

$$\overline{W} \setminus W = \{\chi + \overline{w}(\chi) : \chi \in \partial pW\}.$$

11. The Return Map on W

Phase curves on W intersect transversally with the hyperplane H. More precisely, for

$$X := W \cap H$$

we have the following proposition.

PROPOSITION 11.1. *For every solution* $x : \mathbb{R} \longrightarrow \mathbb{R}$ *of eq.* (1.1) *with* $x_t \in W$ *for all* t, *and* $x_0 \in W \cap H$,

$$(x')_0 = D(t \longrightarrow x_t)(0)1 \in (T_{x_0}W) \setminus H.$$

PROOF. (Compare the construction of the map *stop* in Section 3). We have $x(-1) = 0$. Zeros are simple. Hence $0 \neq x'(-1)$, or $(x')_0 \notin H$. $\quad\square$

Arguments from the proof of [**36, Proposition 9.1**] yield our next result.

COROLLARY 11.1. X *is a one-dimensional* C^1*-submanifold of* C.

Observe that X is the disjoint union of the open subsets

$$X^+ := W \cap K, \quad X^- := W \cap (-K)$$

(since $\phi \in W \cap H$ implies $\phi \in K \cup (-K)$).

The return map P may not be defined on all of X^+ and X^-. So consider

$$O^+ := \{\phi \in X^+ : J(\phi) \geq 2\},$$

$$O^- := \{\phi \in X^- : J(\phi) \geq 2\}.$$

The union of these sets forms the domain of P. They are open in X^+, X^-, respectively (this follows easily from the simplicity of zeros of solutions with phase curve in W, and from continuous dependence on initial data).

Note that the statement

(11.1) $J(x) = \infty$ for all solutions $x : \mathbb{R} \longrightarrow \mathbb{R}$ of eq. (1.1)

with phase curve in W

is equivalent to

$$O^+ = X^+ \quad \text{and} \quad O^- = X^-,$$

while in case

(11.2) $J(x) < \infty$ for at least one solution $x : \mathbb{R} \longrightarrow \mathbb{R}$ of eq. (1.1)

with phase curve in W

we have

$$O^+ \neq X^+ \quad \text{and} \quad O^- \neq X^-.$$

The arguments from [**36, Proposition 9.3, proof, parts 1, 3, 4**] imply our next result.

COROLLARY 11.2. *P maps O^+ diffeomorphically onto X^+ and O^- diffeomorphically onto X^-.*

We write

$$P^+ : O^+ \longrightarrow X^+, \quad P^- : O^- \longrightarrow X^-$$

for the C^1–diffeomorphisms given by P.

COROLLARY 11.3. *y_0 and $y_{z_1(y)} + 1$ are unstable hyperbolic fixed points of P^+ and P^-, respectively. For both maps, the eigenvalue of the linearization at the fixed point is*

$$u > 1.$$

PROOF. 1. Remarks 8.1 and 8.2 yield $W_H \subset X, P(W_H) \subset X$. Hence

$$T_{y_0} W_H \subset T_{y_0} X;$$

in fact,

$$T_{y_0} W_H = T_{y_0} X \quad (= T_{y_0} X^+ = T_{y_0} O^+)$$

since both spaces have dimension 1; and

$$DP(y_0) T_{y_0} W_H \subset T_{P(y_0)} X = T_{y_0} W_H.$$

Recall that $DP(y_0)$ expands the line $T_{y_0} W_H = G_H$ by the factor $u > 1$.

2. Let $z_j = z_j(y)$, for $j \in \mathbb{Z}$. It remains to consider P^- and $y_{z_1} + 1 \in O^-$. Note first that $P_{1/2}$ maps $O^+ \cup O^-$ into $K_{1/2}$, i.e. into the domain of $P_{1/2}$, and

$$P_{1/2}(O^+) \subset W \cap H = X.$$

Choose a basis vector ϕ^+ of $G_H = T_{y_0} O^+$. Then

$$\phi^- := DP_{1/2}(y_0)\phi^+ \in T_{y_{z_1} + 1} X = T_{y_{z_1} + 1} O^-.$$

We have $\phi^- \neq 0$ because

$$0 \neq u \cdot \phi^+ = DP(y_0)\phi^+ = DP_{1/2}(y_{z_1} + 1)(DP_{1/2}(y_0)\phi^+).$$

Finally,

$$DP^-(y_{z_1} + 1)\phi^- = DP(y_{z_1} + 1)\phi^-$$
$$= (DP_{1/2}(y_{z_2} + 1) \circ DP_{1/2}(y_{z_1} + 1) \circ DP_{1/2}(y_0))(\phi^+)$$
$$= DP_{1/2}[DP(y_0)\phi^+] = DP_{1/2}(y_0)[u \cdot \phi^+] = u \cdot [DP_{1/2}(y_0)\phi^+] = u \cdot \phi^-. \quad \square$$

It is convenient to introduce local coordinates at the point y_0 of the C^1–submanifold O^+ of C. Choose a C^1–diffeomorphism

$$h : (-1, 1) \longrightarrow N^+$$

onto an open neighborhood N^+ of y_0 in O^+, with

$$h(0) = y_0.$$

As y_0 is the only point on $|\eta|$ in O^+,

$$h(a) \notin |\eta| \quad \text{for} \quad 0 < |a| < 1.$$

Proposition 11.1 implies that the tangent vectors

$$(y')_0 \in (T_{y_0} W) \setminus H \quad \text{and} \quad Dh(0)1 \in T_{y_0} O^+ \subset (T_{y_0} W) \cap H$$

are linearly independent. As p defines a diffeomorphism of W onto pW, we obtain the next corollary.

COROLLARY 11.4. $D(p \circ h)(0)1 \in L$ and $p(y')_0 \in L$ are linearly independent.

So, for $\epsilon > 0$ sufficiently small, $p \circ h((-\epsilon, 0))$ and $p \circ h((0, \epsilon))$ belong to different components of $L \setminus |p \circ \eta|$. By a modification of h, we achieve

$$p \circ h((-\epsilon, 0)) \subset \text{int}(p \circ \eta) \quad \text{for some} \quad \epsilon \in (0, 1).$$

It follows that there exists $\underline{a} \in (0, 1)$ such that we have

$$p \circ h((-\underline{a}, 0)) \subset \text{int}(p \circ \eta),$$

$$P^+ \circ h((-\underline{a}, \underline{a})) \subset h((-1, 1)),$$

and the C^1-map

$$g : (-\underline{a}, \underline{a}) \longrightarrow (-1, 1), \quad g(a) := h^{-1}(P^+(h(a))) \quad \text{for} \quad |a| < \underline{a}$$

satisfies

$$g(0) = 0, \quad g'(0) = u > 1, \quad g'(a) > 1 \quad \text{for} \quad |a| < \underline{a}.$$

Each $a \in (-\underline{a}, \underline{a})$ is the endpoint $a = a_0$ of a backward trajectory $(a_j)^0_{-\infty}$ of g;

$$\text{sign}(a_j) = \text{sign}(a) \quad \text{for all} \quad j \quad \text{and} \quad a_j \longrightarrow 0 \quad \text{as} \quad j \longrightarrow -\infty.$$

The solution $x^a : \mathbb{R} \longrightarrow \mathbb{R}$ of eq. (1.1) given by

$$x_0^a = h(a) \in O^+$$

has segments

$$x_t^a \in W.$$

Its zeros form an increasing sequence of points $z_j = z_j(x^a)$ which has property (3.2). We have

$$x^a_{z_{2j}} + 1 = h(a_j) \quad \text{for all integers} \quad j \le 0.$$

12. Simple Closed Curves on W

Let $a \in (-\underline{a}, 0)$ be given. Set $x := x^a$ and consider the sequence of zeros $(z_j)^J_{-\infty}$ of x, $2 \le J \in \mathbb{Z} \cup \{\infty\}$, $z_0 = -1$. Now, a is the endpoint $a = a_0$ of a trajectory $(a_j)^0_{-\infty}$ of g in $(-\underline{a}, 0)$;

$$a_j \longrightarrow 0 \quad \text{as} \quad j \longrightarrow -\infty$$

and

(12.1) $$h(a_j) = x_{z_{2j}} + 1 \quad \text{for} \quad j \le 0.$$

We define piecewise smooth simple closed curves

$$\xi_j : [-1, 1] \longrightarrow C, \quad j \in -\mathbb{N}_0$$

with values in W:

$$\text{For} \quad -1 \le t \le 0, \quad \xi_j(t) := h(-t \cdot a_j + (1 + t) \cdot a_{j-1}).$$

I.e., this piece of ξ_j connects $h(a_j) = x_{z_{2j}} + 1$ to $h(a_{j-1}) = x_{z_{2j-2}} + 1$ in $O^+ \subset K$.

For $\quad 0 \leq t \leq 1, \quad \xi_j(t) := F(t \cdot (z_{2j} - z_{2j-2}), x_{z_{2j-2}} + 1)$.

Here we follow the phase curve from $x_{z_{2j-2}} + 1$ to $x_{z_{2j}} + 1 = \xi_j(-1)$.

In case (11.1) holds, i.e. $J = \infty$ and $O^+ = X^+$, we define ξ_j also for $j \in \mathbb{N}$, by

$$\xi_j(t) := (P^+)^j(\xi_0(t)) \in O^+ \subset K \quad \text{for} \quad -1 \leq t \leq 0,$$
$$\xi_j(t) := F(t \cdot (z_{2j} - z_{2j-2}), x_{z_{2j-2}} + 1) \quad \text{for} \quad 0 \leq t \leq 1.$$

Whenever ξ_j is defined,

$$\xi_j([-1, 0]) \subset W \cap K$$

while $\xi_j|[0, 1]$ is a reparameterization of the phase curve between $x_{z_{2j-2}} + 1$ and $x_{z_{2j}} + 1$.

PROPOSITION 12.1. *Let* $k \in \mathbb{N}_0$, $\quad j \in \mathbb{Z}$. *Assume* $j \leq 0$ *and* $k + j \leq 0$, *or assume that* (11.1) *holds. Then*

$$\xi_{k+j}([-1, 0]) = (P^+)^k(\xi_j([-1, 0])).$$

PROOF. In the first case, use $\xi_n([-1, 0]) = h([a_n, a_{n-1}])$ for all $n \leq 0$, and

$$g^k(a_j) = a_{k+j}, \quad g^k(a_{j-1}) = a_{k+j-1}, \quad \text{and}$$
$$h \circ g^k = (P^+)^k \circ h \quad \text{on} \quad [a_j, a_{j-1}].$$

In the second case, P^+ maps $O^+ = X^+$ one-to-one onto X^+, and it is easy to deduce the assertion. $\quad \square$

PROPOSITION 12.2. *Let* $j \in \mathbb{Z}$. *Assume* $j \leq 0$, *or assume that* (11.1) *holds. Then*

$$|p \circ \xi_j| \subset int(p \circ \eta).$$

PROOF. The case $j \leq 0$. Following ξ_j we see that each point on $|\xi_j|$ connects in $W \setminus |\eta|$ to, say, $h(a_j)$. Recall that

$$ph(a_j) \in \text{int}(p \circ \eta),$$

and use the fact that p maps W one-to-one onto $pW \subset L$. In case (11.1) holds and $j > 0$ we note that

$$\xi_j([-1, 0]) = (P^+)^j(\xi_0([-1, 0])) \subset K$$

is disjoint with $|\eta|$ since

$$P^+ : X^+ \longrightarrow X^+ \quad \text{is one-to-one}, \quad P^+(y_0) = y_0,$$
$$|\eta| \cap K = \{y_0\}, \quad \xi_0([-1, 0]) \cap |\eta| = \emptyset.$$

Also, $x_t \notin |\eta|$ for all $t \in \mathbb{R}$ (otherwise, we obtain a contradiction to (12.1) and $a_j \neq 0$ for all $j \in -\mathbb{N}_0$). Following ξ_j and the phase curve of x we see that each point of $|\xi_j|$ connects in $W \setminus |\eta|$ to $h(a_0)$. Continue as in the first case. $\quad \square$

PROPOSITION 12.3. *Let $j \in \mathbb{Z}$. Assume $j \leq 0$, or assume that (11.1) holds. Then*

$$|\xi_{j-1}| \cap |\xi_j| = \{x_{z_{2j-2}+1}\},$$

and for integers $k < j$,

$$|\xi_{k-1}| \cap |\xi_j| = \emptyset.$$

PROOF. It is obvious that the intersections of traces contain the sets on the right hand sides. We show that they are also subsets of the right hand side. Let an integer $k \leq j$ be given.

1. Proof of

$$\xi_{k-1}((0,1)) \cap |\xi_j| = \emptyset.$$

Otherwise, there exists $\xi_{k-1}(t) = x_s \notin K$ in $|\xi_j|$ where $t \in (0,1)$, $s < z_{2k-2}+1$. Necessarily,

$$x_s = \xi_j(\underline{t}) = x_{\underline{s}} \quad \text{where} \quad \underline{t} \in (0,1) \quad \text{and} \quad z_{2j-2}+1 < \underline{s}.$$

It follows that x is periodic, and we arrive at a contradiction to (12.1) and $a_j \nearrow 0$ as $j \longrightarrow -\infty$.

2. Proof that

$$\phi \in \xi_{k-1}([-1,0]) \cap |\xi_j| \quad \text{implies} \quad k = j \quad \text{and} \quad \phi = x_{z_{2j-2}+1}.$$

Such ϕ belong to $\xi_j([-1,0]) \subset K$. There exist $t, \underline{t}$ in $[-1,0]$ with

$$\xi_{k-1}(t) = \phi = \xi_j(\underline{t}).$$

Hence,

$$(P^+)^{-j}(\xi_{k-1}(t)) = (P^+)^{-j}(\xi_j(\underline{t})).$$

Proposition 12.1 implies that the left hand side belongs to

$$\xi_{k-j-1}([-1,0]) = h([a_{k-j-1}, a_{k-j-2}]);$$

the right hand side belongs to

$$\xi_0([-1,0]) = h([a_0, a_{-1}]).$$

Strict monotonicity of g yields $k = j$ and

$$(P^+)^{-j}(\phi) = h(a_{-1});$$

$$\phi = (P^+)^j(h(a_{-1})) = (P^+)^j(x_{z_{-2}+1}) = x_{z_{2j-2}+1}. \quad \square$$

Next we consider points on $|\xi_j|$ in $-K$ and their projections in L. Such points have the advantage that close to them ξ_j is smooth (given by the phase curve of x).

PROPOSITION 12.4. *Let $j \in \mathbb{Z}$. Assume $j \leq 1$, or assume that (11.1) holds. For every integer $k < j$,*

$$px_{z_{2j-1}+1} \in int(p \circ \xi_k).$$

PROOF. 1. In case $j \leq 1$,

$$x_{z_{2j-1}+1} = P_{1/2}(x_{z_{2j-2}+1}) = P_{1/2}(h(a_{j-1})).$$

Set $a^* := a_{j-1} \in (-\underline{a}, 0)$, and consider the map

$$\Pi : [a^*, 0] \ni \hat{a} \longrightarrow P_{1/2}(h(\hat{a})) \in C.$$

Π maps into $-K \subset H$.

In case that (11.1) holds, and $j > 1$,

$$x_{z_{2j-1}+1} = P_{1/2}(x_{z_{2j-2}+1}) = P_{1/2}((P^+)^{j-1}(x_0))$$

$$= P_{1/2}((P^+)^{j-1}(h(a_0))).$$

Set

$$a^* := a_0 \in (-\underline{a}, 0),$$

and define $\Pi : [a^*, 0] \longrightarrow C$ by

$$\Pi(\hat{a}) := P_{1/2}((P^+)^{j-1}(h(\hat{a}))) \in -K \subset H.$$

2. Π connects $x_{z_{2j-1}+1}$ in $W \cap H$ to $y_{z_1(y)+1} \in |\eta|$. Since all maps involved are given by diffeomorphisms,

$$0 \neq D\Pi(\hat{a})1 \quad \text{for} \quad a^* < \hat{a} < 0.$$

3. Proof of

$$|\xi_k| \cap |\Pi| = \{x_{z_{2k-1}+1}\}.$$

$|\xi_k| \cap |\Pi| \subset \{\ldots\}$ follows from $|\Pi| \subset -K$ and from $|\xi_k| \cap (-K) = \{\ldots\}$. In case (11.1) holds and $j > 1$ we deduce from $k < j$ that

$$|\xi_k| \ni x_{z_{2k-1}+1} = P_{1/2}(x_{z_{2k-2}+1}) = P_{1/2}((P^+)^{j-1}(x_{z_{2k-2-2(j-1)}+1}))$$

$$= P_{1/2}((P^+)^{j-1}(h(a_{k-j}))) \in |\Pi|,$$

since $a^* = a_0 < a_{k-j} < 0$.

In case $j \leq 1$ we obtain

$$|\xi_k| \ni x_{z_{2k-1}+1} = P_{1/2}(x_{z_{2k-2}+1}) = P_{1/2}(h(a_{k-1})) \in |\Pi|$$

since $a^* = a_{j-1} < a_{k-1} < 0$.

4. The nonzero tangent vectors $D\Pi(a)1 \in H \cap T_{\Pi(a)}W$ where

$$\Pi(a) = x_{z_{2k-1}+1},$$

and $D\xi_k(t)1$ where

$$t \in (0, 1), \quad \xi_k(t) = x_{z_{2k-1}+1},$$

are linearly independent since $D\xi_k(t)1$ is a multiple of

$$(x')_{z_{2k-1}} + 1 \in (T_{\xi_k(t)}W) \setminus H.$$

As p maps W diffeomorphically onto $pW \subset L$ we infer that the projected curves $p \circ \Pi$ and $p \circ \xi_k$ have linearly independent tangent vectors at their intersection point

$$px_{z_{2k-1}} + 1.$$

Now, $|p \circ \xi_k| \subset \text{int}(p \circ \eta)$ implies

$$p \circ \Pi(0) = py_{z_1(y)} + 1 \in |p \circ \eta| \subset \text{ext}(p \circ \xi_k).$$

It follows that the other endpoint of $|p \circ \Pi|$, namely,

$$px_{z_{2j-1}} + 1 = p \circ \Pi(a^*),$$

belongs to $\text{int}(p \circ \xi_k)$. $\square$

COROLLARY 12.1. *Let integers $k < j$ be given. Assume $j \leq 0$, or assume that (11.1) holds. Then 1. $|p \circ \xi_j| \setminus \{px_{z_{2j-2}} + 1\} \subset int(p \circ \xi_{j-1})$,*

2. $|p \circ \xi_j| \subset int(p \circ \xi_{k-1})$,
3. $int(p \circ \xi_j) \subset int(p \circ \xi_k)$,
4. $|p \circ \xi_k| \setminus |p \circ \xi_j| \subset ext(p \circ \xi_j)$.

PROOF. 1. Proposition 12.3 implies that each point in

$$|\xi_j| \setminus \{x_{z_{2j-2}} + 1\}$$

connects in the complement of $|\xi_{j-1}|$ to $x_{z_{2j-1}} + 1$. Apply p and use Proposition 12.4.

2. As before, with the last assertion of Proposition 12.3.

3. Assertions 1,2 and Proposition 12.3 yield

$$|p \circ \xi_j| \subset \text{int}(p \circ \xi_k) \cup |p \circ \xi_k|,$$

and a standard argument completes the proof.

4. This part is a routine consequence of the first assertions. $\square$

COROLLARY 12.2. $\omega(x) \subset \overline{W} \setminus W$.

PROOF. First, $\omega(x) \subset \overline{W}$ is obvious. Next, $px_0 \in \text{int}(p \circ \eta)$ (Proposition 12.2) and

$$x_t \notin |\eta| \quad \text{for all} \quad t$$

imply that $px_t \in \text{int}(p \circ \eta)$ for all $t \in \mathbb{R}$. It follows that

$$p\omega(x) \subset \overline{\text{int}(p \circ \eta)}.$$

Suppose $\omega(x) \cap W \neq \emptyset$. Then there is a solution $\underline{x} : \mathbb{R} \longrightarrow \mathbb{R}$ of eq. (1.1) with phase curve in $\omega(x) \cap W$. Convergence to $|\eta|$ as $t \longrightarrow -\infty$ implies that there exist $t \in \mathbb{R}$ with

$$p\underline{x}_t \in \text{ext}(p \circ \xi_0)$$

$(|p \circ \eta| \subset \text{ext}(p \circ \xi_0)$ since $|p \circ \xi_0| \subset \text{int}(p \circ \eta)$ (Proposition 12.2)). Using $\underline{x}_t \in \omega(x)$ we find $s > z_1 + 1$ such that

$$px_s \in \text{ext}(p \circ \xi_0).$$

Recall $px_{z_1 + 1} \in \text{int}(p \circ \xi_0)$ (Proposition 12.4). It follows that there exists $t > z_1 + 1$ so that

$$px_t \in |p \circ \xi_0| \quad (\text{in particular,} \quad x_t \in |\xi_0|),$$

$$px_v \in \text{int}(p \circ \xi_0) \quad \text{for} \quad z_1 + 1 \leq v < t.$$

In case $x_t \in \xi_0([0,1])$, $x_t = x_v$ for some $v \leq z_0 + 1$, and x is periodic which leads to a contradiction.

It remains to consider the case

$$x_t \in \xi_0((-1, 0)).$$

Then $x_t = h(a^*)$ where $a_0 < a^* < a_{-1}$. The backward trajectory $(a_j^*)^0_{-\infty}$ of g which ends at $a_0^* = a^*$ satisfies

$$a_j < a_j^* < a_{j-1}.$$

The solution

$$x^* := x(\cdot + t)$$

has zeros z_j^*, $j \in -\mathbb{N}_0$, such that

$$x^*_{z_{2j}^* + 1} = h(a_j^*),$$

and

$$x_t^* \notin K \quad \text{for all} \quad t \leq 0 \quad \text{not contained in} \quad \{z_{2j}^* + 1 : j \in -\mathbb{N}_0\}.$$

Therefore,

$$K \ni h(a_0) = x_0 = x^*_{-t} = x^*_{z_{2j}^* + 1} = h(a_j^*) \quad \text{for some} \quad j \in -\mathbb{N}_0$$

which is a contradiction to $a_0 < a_j^*$. $\square$

Next, we construct homotopies of closed curves in W which connect a reparameterization of η to a reparameterization of a curve ξ_j. We do this for $j = 0$, and in case (11.1) holds, also for every $j \in \mathbb{N}$.

The map

$$m : [a, 0) \times [-1, 1] \longrightarrow C$$

given by

$$m(s, t) = (P^+)^j \circ h((-t)s + (1 + t)g^{-1}(s)) \quad \text{for} \quad t \leq 0$$

$$m(s, t) = F(t \cdot (z_2((P^+)^j \circ h \circ g^{-1}(s)) + 1, (P^+)^j \circ h \circ g^{-1}(s)) \quad \text{for} \quad t \geq 0$$

is continuous since both formulae yield the same value $(P^+)^j \circ h \circ g^{-1}(s)$ at points $(s, 0)$. Note that

$$m(a, t) = \xi_j(t).$$

The closed curves $m(s, \cdot), a \leq s < 0$, are all analogues of ξ_j, composed of a piece

$$(P^+)^j \circ h([s, g^{-1}(s)]) \subset W \cap K,$$

and of a piece of the phase curve through

$$(P^+)^j \circ h(g^{-1}(s)) \quad \text{and} \quad (P^+)^j \circ h(s).$$

The number s serves as the homotopy parameter. The points

$$m(s, 1) = F(z_2((P^+)^j \circ h \circ g^{-1}(s)) + 1, (P^+)^j \circ h \circ g^{-1}(s))$$

$$= (P^+)^{j+1} \circ h \circ g^{-1}(s) = (P^+)^j \circ h(s)$$

on the curves $m(s, \cdot)$ fill the homeomorphic image of $h([a, 0))$ under the map $(P^+)^j$ (a connected set in the one–dimensional submanifold $X^+ = W \cap K$).

The transformation

$$\Psi : [a, 0) \times [0, 1] \longrightarrow [a, 0) \times [-1, 1]$$

given by

$$\Psi_1(s, t) = s$$

and

$$\Psi_2(s, t) = \frac{t}{s/2a} - 1 \quad \text{for} \quad t \leq \frac{s}{2a}$$

$$\Psi_2(s, t) = \frac{t - (s/2a)}{1 - (s/2a)} \quad \text{for} \quad \frac{s}{2a} \leq t$$

is continuous since both formulae for Ψ_2 yield the same value when

$$t = \frac{s}{2a};$$

$\Psi_2(s, \cdot)$ maps $\left[0, \frac{s}{2a}\right]$ onto $[-1, 0]$ and $\left[\frac{s}{2a}, 1\right]$ onto $[0, 1]$.

The interval $\left[0, \frac{s}{2a}\right]$ shrinks to $\{0\}$ as $s \nearrow 0$. Now define

$$M(s, t) := m \circ \Psi(s, t) \quad \text{on} \quad [0, a) \times [0, 1],$$

$$M(s, t) := \eta(t \cdot \tau) \quad \text{for} \quad s = 0, \quad 0 \leq t \leq 1.$$

We have that M is continuous on $[a, 0) \times [0, 1]$; $M(a, \cdot)$ and $M(0, \cdot)$ are reparameterizations of ξ_j and η since

$$M(a, t) = m(a, 2t - 1) = \xi_j(2t - 1) \quad \text{and} \quad M(0, t) = \eta(t \cdot \tau).$$

Each $M(s, \cdot)$ is a (simple) closed curve on W.

We next show the continuity of M at the points $(0, t)$:

Let a sequence of points $(s_n, t_n), n \in \mathbb{N}$, in the domain of M be given with

$$(s_n, t_n) \longrightarrow (0, t) \quad \text{as} \quad n \longrightarrow \infty.$$

Clearly, as $n \longrightarrow \infty$,

$$M\left(\{s_n\} \times \left[0, \frac{s_n}{2a}\right]\right) = (P^+)^j \circ h([s_n, g^{-1}(s_n)]) \longrightarrow (P^+)^j \circ h(0)$$

$$= y_0 = M(0,0).$$

For a subsequence of points $(s_{\nu(n)}, t_{\nu(n)})$, $\nu : \mathbb{N} \longrightarrow \mathbb{N}$ strictly increasing, such that

$$t_{\nu(n)} \leq \frac{s_{\nu(n)}}{2a} \quad \text{for all} \quad n \in \mathbb{N},$$

necessarily $t = 0$, and

$$M(s_{\nu(n)}, t_{\nu(n)}) \longrightarrow M(0,0).$$

For a subsequence of points $(s_{\nu(n)}, t_{\nu(n)})$, $\nu : \mathbb{N} \longrightarrow \mathbb{N}$ strictly increasing, such that

$$t_{\nu(n)} > \frac{s_{\nu(n)}}{2a} \quad (\leq 0) \quad \text{for all} \quad n \in \mathbb{N},$$

we obtain, in case $s_{\nu(n)} < 0$,

$$M(s_{\nu(n)}, t_{\nu(n)}) = m(s_{\nu(n)}, \Psi_2(s_{\nu(n)}, t_{\nu(n)}))$$

$$= F(\Psi_2(\ldots) \cdot (z_2((P^+)^j \circ h \circ g^{-1}(s_{\nu(n)})) + 1, (P^+)^j \circ h \circ g^{-1}(s_{\nu(n)})),$$

and in case $s_{\nu(n)} = 0$,

$$M(s_{\nu(n)}, t_{\nu(n)}) = \eta(t_{\nu(n)}\tau).$$

Observe that in case $s_{\nu(n)} < 0$,

$$\Psi_2(s_{\nu(n)}, t_{\nu(n)}) = \frac{t_{\nu(n)} - (s_{\nu(n)}/2a)}{1 - (s_{\nu(n)}/2a)}.$$

For every further subsequence of points $(s_{\nu \circ \kappa(n)}, t_{\nu \circ \kappa(n)})$, $\kappa : \mathbb{N} \longrightarrow \mathbb{N}$ strictly increasing, such that

$$s_{\nu \circ \kappa(n)} < 0 \quad \text{for all} \quad n \in \mathbb{N}$$

we infer

$$\Psi_2(s_{\nu \circ \kappa(n)}, t_{\nu \circ \kappa(n)}) \longrightarrow t \quad \text{as} \quad n \longrightarrow \infty.$$

Continuity of g and h and P^+, and the equations $g(0) = 0$, $h(0) = y_0$, $P^+(y_0) = y_0$ may now be used to derive

$$M(s_{\nu(n)}, t_{\nu(n)}) \longrightarrow F(t \cdot (z_2((P^+)^j \circ h(0)) + 1, (P^+)^j \circ h(0))$$

$$= \eta(t \cdot \tau) = M(0, t),$$

and it is routine to complete the proof.

PROPOSITION 12.5. $ext(p \circ \xi_0) \cap int(p \circ \eta) \subset pW$. If (11.1) holds, then

$$ext(p \circ \xi_j) \cap int(p \circ \eta) \subset pW \quad \text{for every} \quad j \in \mathbb{N}.$$

PROOF. Set $j := 0$, or assume that (11.1) holds, and let $j \in \mathbb{N}$ in this case. We have

$$|M(1, \cdot)| = |\xi_j|, \quad |M(0, \cdot)| = |\eta|.$$

Hence,

$$ext(p \circ \xi_j) = ext(p \circ M(1, \cdot)), \quad ext(p \circ M(0, \cdot)) = ext(p \circ \eta).$$

Suppose there exists

$$\chi \in ((\text{ext}(p \circ \xi_j) \cap \text{int}(p \circ \eta)) \setminus pW.$$

Then $\chi \in \text{ext}(p \circ M(1, \cdot))$, and

$$0 = \text{wind}(\chi, p \circ M(1, \cdot)) = \text{wind}(\chi, p \circ M(0, \cdot))$$

since $p \circ M$ is a continuous homotopy of closed curves in $pW, \chi \notin pW$. Therefore

$$\chi \in \text{ext}(p \circ M(0, \cdot)) = \text{ext}(p \circ \eta),$$

a contradiction to $\chi \in \text{int}(p \circ \eta)$. $\square$

Later on we shall sometimes write $\xi_{j,a}$ and $M_{j,a}$ instead of ξ_j and M.

All constructions in this section have counterparts if we begin at the other unstable fixed point of P on $|\eta|$, namely at the fixed point

$$y_{z_1}(y) + 1 \in O^- \subset -K$$

of the map P^-.

Let $\underline{a}^- < 0$, h^-, g^- denote the analogues of $\underline{a}$, h, g. Every $a^- \in (\underline{a}^-, 0)$ is the endpoint $a^- = a_0^-$ of a trajectory $(a_j^-)_{-\infty}^0$ of g^- in $(\underline{a}^-, 0)$;

$$a_j^- \longrightarrow 0 \quad \text{as} \quad j \longrightarrow -\infty.$$

a^- determines a solution $x^* : \mathbb{R} \longrightarrow \mathbb{R}$ of eq. (1.1) with phase curve in W, by

$$x_0^* = h^-(a^-) \in -K.$$

The zeros of x^* form a sequence $(z_j^*)_{-\infty}^J$, $3 \le J \in \mathbb{Z}$ or $J = \infty$, with property (3.2), and

$$z_1^* = -1.$$

For applications in the next section, it is convenient to perform a time shift. Set

$$x := x^{a^-} := x^*(\cdot + z_0^* + 1)$$

and

$$z_j := z_j^* - (z_0^* + 1) \quad \text{for all} \quad j.$$

The zeros of x are then given by the sequence $(z_j)_{-\infty}^J$; property (3.2) is satisfied, and

$$z_0 = -1,$$
$$x_0 = x_{z_0^*}^* + 1 \in K \quad (\text{since} \quad x_{z_1^*}^* + 1 \in -K),$$
$$x_{z_1} + 1 = x_{z_1^*}^* + 1 = x_0^* = h^-(a^-) \in -K.$$

As at the beginning of this section, we construct closed curves

$$\xi_j^- : [-1, 1] \longrightarrow C$$

for all $j \in -\mathbb{N}_0$, and in case (11.1) holds, for all integers j, so that

$$\xi_j^-([-1, 0]) \subset W \cap (-K),$$

and for $0 \leq t \leq 1$,

$$\xi_j^-(t) = F(t \cdot (z_{2j+1} - z_{2j-1}), x_{z_{2j-1}} + 1).$$

13. A Periodic Orbit in $\overline{W} \setminus W$

We assume in this section that (11.1) holds, i.e. $J(x) = \infty$ for all solutions $x : \mathbb{R} \longrightarrow \mathbb{R}$ of eq. (1.1) with phase curve in W. We look for a periodic orbit in $\overline{W} \setminus W$ which projects into $\text{int}(p \circ \eta) \subset L$.

Consider a solution $x^* = x^{a^*}, \underline{a} < a^* < 0$, as in the preceding section. Then $x_t^* \notin |\eta|$ for all $t \in \mathbb{R}$, and it follows that

$$(13.1) \qquad\qquad px_t^* \in \text{int}(p \circ \eta) \quad \text{for all} \quad t \in \mathbb{R}$$

since $px_0^* = ph(a^*) \in \text{int}(p \circ \eta)$. The zeros of x^* form a sequence $(z_j^*)_{-\infty}^{\infty}$ with property (3.2), and

$$x_{z_{2j}^*}^* + 1 \in K, \quad x_{z_{2j+1}^*}^* + 1 \in -K \quad \text{for all} \quad j \in \mathbb{Z}.$$

The relations

$$x_t \longrightarrow |\eta| \quad \text{as} \quad t \longrightarrow -\infty, \quad |\eta| \cap K = \{y_0\}, \quad |\eta| \cap (-K) = \{y_{z_1(y)} + 1\}$$

imply that the trajectories

$$(x_{z_{2j}^*}^* + 1)_{-\infty}^{\infty} \quad \text{of} \quad P^+ \quad \text{and} \quad (x_{z_{2j+1}^*}^* + 1)_{-\infty}^{\infty} \quad \text{of} \quad P^-$$

converge to y_0 and $y_{z_1(y)} + 1$, respectively, as $j \longrightarrow -\infty$. Using (13.1) we infer that there exists an integer k so that

$$x_{z_{2k}^*}^* + 1 \in h((\underline{a}, 0)),$$

$$x_{z_{2k+1}^*}^* + 1 \in h((\underline{a}^-, 0)).$$

Set

$$x := x^*(\cdot + z_{2k}^* + 1) , \quad \text{and} \quad z_j := z_{j+2k}^* - (z_{2k}^* + 1) \quad \text{for} \quad j \in \mathbb{Z}.$$

Observe that the sequence $(z_j)_{-\infty}^{\infty}$ has property (3.2);

$$x^{-1}(0) = \{z_j : j \in \mathbb{Z}\}$$

and

$$z_0 = -1,$$

$$x_0 = x_{z_{2k}^*}^* + 1 = h(a) \quad \text{for some} \quad a \in (\underline{a}, 0),$$

$$x_{z_1} + 1 = x_{z_{2k+1}^*}^* + 1 = h(a^-) \quad \text{for some} \quad a^- \in (\underline{a}^-, 0).$$

Therefore

$$x^a = x = x^{a^-}.$$

Let ξ_j and ξ_j^-, $\quad j \in \mathbb{Z}$, denote the closed curves, associated as in Section 12, with x^a and x^{a^-}, respectively. Then

$$\xi_j(1) = x_{z_{2j}} + 1 \in K, \quad \xi_j^-(1) = x_{z_{2j+1}} + 1 \in -K \quad \text{for all} \quad j \in \mathbb{Z}.$$

COROLLARY 13.1. *For integers* $k < j - 1$, $\quad ext(p \circ \xi_k^-) \subset ext(p \circ \xi_j)$.

PROOF. The analogue of Corollary 12.1.2 for the curves ξ_k^- yields

$$|p \circ \xi_j^-| \subset \text{int}(p \circ \xi_k).$$

In particular, the point $p x_{z_{2j}} + 1 \in |p \circ \xi_j|$ belongs to $\text{int}(p \circ \xi_k^-)$. We have

$$|\xi_j| \cap |\xi_k^-| = \emptyset$$

since

$$\xi_j([-1,0]) \cap \xi_k^-([-1,0]) \subset K \cap (-K) = \emptyset,$$
$$\xi_j([0,1]) = \{x_t : z_{2j-2} + 1 \le t \le z_{2j} + 1\},$$
$$\xi_k^-([0,1]) = \{x_t : z_{2k-1} + 1 \le t \le z_{2k+1} + 1\},$$
$$z_{2k+1} + 1 < z_{2j-2} + 1,$$

and x is not periodic. We conclude that

$$|p \circ \xi_j| \subset \text{int}(p \circ \xi_k^-)$$

from which the assertion follows.   □

There is a strictly increasing sequence

$$\iota : \mathbb{N} \longrightarrow \mathbb{N}$$

so that the subsequences $(x_{z_{2\iota(j)}})_1^\infty$ and $(x_{z_{2\iota(j)+1}})_1^\infty$ converge to points ϕ and ϕ^-, respectively, in $\omega(x) \subset \overline{W} \setminus W$ (see Corollary 12.2). These limit points determine solutions b and b^- of eq. (1.1), both defined on $\mathbb{R}$, so that

$$b_{-1} = \phi, \quad b_{-1}^- = \phi^-;$$

the phase curves of b and b^- belong to $\omega(x) \cap (\overline{W} \setminus W)$. On any compact interval $[-2, t], t \ge 0$, we have uniform convergence

(13.2) $x(s + z_{2\iota(j)} + 1) \longrightarrow b(s) \quad \text{as} \quad j \longrightarrow \infty,$

(13.3) $x(s + z_{2\iota(j)+1} + 1) \longrightarrow b^-(s) \quad \text{as} \quad j \longrightarrow \infty.$

(Working with phase points at $t = z_{2j}$ (and $t = z_{2j+1}$), $j \in \mathbb{Z}$, instead of $t = z_{2j} + 1$ also implies that derivatives $x'(t + z_{2\iota(j)} + 1), t \ge -1$, converge to $b'(t)$ as $j \longrightarrow \infty$. This will be shown and used in the proof of Proposition 13.4 below.) In particular,

$$x_{z_{2\iota(j)}} + 1 \longrightarrow b_0 \in \overline{K}$$

and

$$x_{z_{2\iota(j)+1}} + 1 \longrightarrow b_0^- \in \overline{-K}$$

as $j \longrightarrow \infty$. The case $b_0 = 0$ is equivalent to $b = 0$. In case $b_0 \neq 0$, the zeros of b are given by a sequence $(z_j(b))_{-\infty}^{J(b)}$ where $0 \leq J(b) \in \mathbb{Z}$ or $J(b) = \infty$, and $z_0(b) = -1$. Analogously, $b_0^- = 0$ if and only if $b^- = 0$; in case $b^- \neq 0$, the zeros of b^- are given by a sequence $(z_j(b^-))_{-\infty}^{J(b^-)}$ where $1 \leq J(b^-) \in \mathbb{Z}$ or $J(b^-) = \infty$, and $z_1(b^-) = -1$.

Incidentally, note that we can not immediately conclude that b and b^- have the same orbit since there is no estimate of the distances $z_{2\iota(j)+1} - z_{2\iota(j)}$. But we do have the following result.

COROLLARY 13.2. *If $b = 0$ then $b^- = 0$, and*

$$\xi_{\iota(j)+1}([0,1]) \longrightarrow 0, \quad \xi_{\iota(j)+1}^-([0,1]) \longrightarrow 0 \quad as \quad j \longrightarrow \infty.$$

PROOF. 1. Proposition 3.2 implies that for all $j \in \mathbb{N}$,

$$\|x_{z_{2\iota(j)+1}} + 1\| = \max_{[z_{2\iota(j)+1}, z_{2\iota(j)+2}]} |x| \leq d_J \cdot \max_{[z_{2\iota(j)}, z_{2\iota(j)+1}]} |x|$$

$$= d_J \cdot \|x_{z_{2\iota(j)}} + 1\|.$$

2. Also by Proposition 3.2, for all $j \in \mathbb{N}$,

$$\max_{[0,1]} \|\xi_{\iota(j)+1}(t)\| = \max_{[z_{2\iota(j)}, z_{2\iota(j)+2} + 1]} |x| \leq (d_J + 1)^2 \|x_{z_{2\iota(j)}} + 1\|.$$

This yields $\xi_{\iota(j)+1}([0,1]) \longrightarrow 0$. The argument for the curves $\xi_{\iota(j)+1}^-$ is analogous. $\square$

PROPOSITION 13.1. 1. *Let $t \in \mathbb{R}$. We have*

$$pb_t \in int(p \circ \eta) \ni pb_t^-,$$

and for all $j \in \mathbb{Z}$,

$$pb_t \in int(p \circ \xi_j), \quad pb_t^- \in int(p \circ \xi_j^-).$$

In particular,

$$b_t \notin |\eta| \cup |\xi_j| \quad and \quad b_t^- \notin |\eta| \cup |\xi_j^-|.$$

2. *In case $b \neq 0$, $J(b) = \infty$.*

PROOF. 1. The relations $|\eta| \cup |\xi_j| \subset W$, $b_t \in \overline{W} \setminus W$, imply that some convex neighborhood N_L of pb_t in L is disjoint with $|p \circ \eta| \cup |p \circ \xi_j|$. As $b_t \in \omega(x)$, there exists $s \geq z_{2j+2} + 1$ with $x_s \in p^{-1}(N_L)$. We have

$$x_s \in \xi_k([0,1]) \quad \text{for some} \quad k \geq j + 2,$$

and

$$|p \circ \xi_k| \subset int(p \circ \eta) \cap int(p \circ \xi_j)$$

(see Proposition 12.2 and Corollary 12.1.2). It follows that

$$b_t \in int(p \circ \eta) \cap int(p \circ \xi_j).$$

The proof for b_t^- and ξ_j^- is analogous.

2. Suppose $J(b) < \infty$. Then $b(t) \longrightarrow 0$ as $t \longrightarrow \infty$, and $0 \in C$ is an attractive stationary point of the semiflow F (Section 3, Remark 4.1, parts 2 and 3). For every ψ in some open neighborhood U of 0 in C,

$$F(t, \psi) \longrightarrow 0 \quad \text{as} \quad t \longrightarrow \infty.$$

Choose $t > 0$ with $b_t \in U$. There exists $s > 0$ with $x_s \in U$. Hence $x(t') \longrightarrow 0$ as $t' \longrightarrow \infty$, a contradiction to

$$x_{z_{2\iota(j)} + 1} \longrightarrow b_0 \neq 0 \quad \text{as} \quad j \longrightarrow \infty. \quad \square$$

Part 2 of Proposition 13.1, statement (13.2) and the simplicity of zeros yield our next result.

COROLLARY 13.3. *In case $b \neq 0$, for every $k \in \mathbb{N}_0$,*

$$z_{2\iota(j)+k} - z_{2\iota(j)} \longrightarrow z_k(b) + 1 \quad as \quad j \longrightarrow \infty,$$

and if $\epsilon = \epsilon(k) \in (0,1)$ satisfies $z_{k+2}(b) + 1 + \epsilon < z_{k+3}(b)$, then the set $\{x_t : z_{2\iota(j)+k} + 1 \leq t \leq z_{2\iota(j)+k+2} + 1\}$ converges to the set

$$\{b_s : z_k(b) + 1 - \epsilon \leq s \leq z_{k+2}(b) + 1 + \epsilon\}$$

as $j \longrightarrow \infty$.

COROLLARY 13.4. *In case $b \neq 0$,*

$$x_{z_{2\iota(j)+1} + 1} \longrightarrow b_{z_1(b) + 1} \quad as \quad j \longrightarrow \infty,$$

and

$$b^- = b(\cdot + z_1(b) + 1).$$

PROOF. The convergence

$$x_{z_{2\iota(j)} + 1} \longrightarrow b_0, \quad z_{2\iota(j)+1} - z_{2\iota(j)} \longrightarrow z_1(b) + 1 \quad \text{for} \quad j \longrightarrow \infty$$

and continuity of F imply

$$b_{z_1(b) + 1} = F(z_1(b) + 1, b_0) = \lim_{j \longrightarrow \infty} F(z_{2\iota(j)+1} - z_{2\iota(j)}, x_{z_{2\iota(j)} + 1})$$

$$= \lim_{j \longrightarrow \infty} x_{z_{2\iota(j)+1} + 1} = b_0^-. \quad \square$$

PROPOSITION 13.2. *In case $b \neq 0$, b is periodic with minimal period $z_2(b) + 1$.*

PROOF. 1. Suppose $b_0 \neq b_{z_2(b) + 1}$. Then

$$b_{z_1(b) + 1} \neq b_{z_3(b) + 1}.$$

(Otherwise, b would have period $z_3(b) - z_1(b)$, with $z_2(b)$ being the only zero in the period interval $(z_1(b), z_3(b))$. This would imply

$$z_2(b) + 1 = z_2(b) - z_0(b) = z_3(b) - z_1(b),$$

hence $b_{z_2(b)} + 1 = b_0$, a contradiction to the assumption above.)

Choose $\epsilon \in (0,1)$ with $z_4(b)+1+\epsilon < z_5(b)$. For $z_2(b)+1-\epsilon \leq t \leq z_4(b)+1+\epsilon$,

$$b_t \in -K \quad \text{if and only if} \quad t = z_3(b) + 1.$$

It follows that for every t as above,

$$b_t \neq b_{z_1(b)} + 1 \quad (\in -K).$$

Note $\overline{K} = \{\phi \in C : 0 \leq \phi\}$, and $\overline{K} \cap (-K) = \emptyset$. We conclude that

$$d := \mathrm{dist}(pb_{z_1(b)} + 1, p(\overline{W \cap K}) \cup \{pb_t : z_2(b)+1-\epsilon \leq t \leq z_4(b)+1+\epsilon\})$$

is strictly positive.

2. Corollary 13.4 and Corollary 13.3 permit us to find $j \in \mathbb{N}$ with

$$\|px_{z_{2\iota(j)+1}} + 1 - pb_{z_1(b)} + 1\| < \frac{d}{2}$$

and

$$\mathrm{dist}(px_t, \{pb_s : z_2(b)+1-\epsilon \leq s \leq z_4(b)+1+\epsilon\}) < \frac{d}{2}$$

for $z_{2\iota(j)+2} + 1 \leq t \leq z_{2\iota(j)+4} + 1$.

It follows that for $0 \leq t \leq 1$, the points

$$p \circ \xi_{\iota(j)+2}(t) = pF(t \cdot \{z_{2[\iota(j)+2]} - z_{2[\iota(j)+2]-2}, x_{z_{2[\ldots]-2}} + 1)$$

$$= px_t \cdot (z_{2\iota(j)+4} - z_{2\iota(j)+2}) + z_{2\iota(j)+2} + 1$$

belong to the $\frac{d}{2}$–neighborhood of

$$\{pb_s : z_2(b)+1-\epsilon \leq s \leq z_4(b)+1+\epsilon\},$$

so

$$\frac{d}{2} < \mathrm{dist}(p \circ \xi_{\iota(j)+2}([0,1]), pb_{z_1(b)} + 1).$$

Since $\xi_{\iota(j)+2}([-1,0]) \subset W \cap K$,

$$p \circ \xi_{\iota(j)+2}([-1,0]) \subset p(\overline{W \cap K});$$

therefore,

$$\frac{d}{2} < \mathrm{dist}(p \circ \xi_{\iota(j)+2}([-1,0]), pb_{z_1(b)} + 1).$$

Combining these we have

(13.4) $$\frac{d}{2} < \mathrm{dist}(|p \circ \xi_{\iota(j)+2}|, pb_{z_1(b)} + 1)$$

3. Corollary 12.1.4 says that the point

$$px_{z_{2\iota(j)+1}} + 1 \in \left\{ \chi \in L : \|\chi - pb_{z_1(b)} + 1\| < \frac{d}{2} \right\}$$

on $|p \circ \xi_{\iota(j)+1}|$ belongs to $\mathrm{ext}(p \circ \xi_{\iota(j)+2})$. Using (13.4) we infer that

$$pb_{z_1(b)} + 1 \in \mathrm{ext}(p \circ \xi_{\iota(j)+2}),$$

as well, and this is a contradiction to part 1 of Proposition 13.1.   □

In case $b \neq 0$ we set

$$\tau_b := z_2(b) + 1$$

and define a simple closed smooth curve $\beta : [0, \tau_b] \longrightarrow C$ by

$$\beta(t) := b_t.$$

In this case, $b^- = b(\cdot + z_1(b) + 1)$ has the same period τ_b, and for the simple closed curve $\beta^- : [0, \tau_b] \ni t \longrightarrow b_t^- \in C$,

$$|\beta^-| = |\beta|.$$

At the end of this section we shall prove assertion 3 of Theorem 5.4, i.e.,

$$(\overline{W} \setminus W) \cap p^{-1}(\text{int}(p \circ \eta)) = |\beta| \quad \text{if} \quad b \neq 0,$$

$$(\overline{W} \setminus W) \cap p^{-1}(\text{int}(p \circ \eta)) = \{0\} \quad \text{if} \quad b = 0,$$

in case that (11.1) holds.

The first, simple step towards this is the following.

PROPOSITION 13.3. 1. *In case $b = 0$, $pW \subset L \setminus \{0\}$.*
2. *In case $b \neq 0$, $pW \subset ext(p \circ \beta)$.*

PROOF. 1. The first part follows from $pW \subset S$, $0 \notin pS$.
2. Recall $|p \circ \beta| \subset \text{int}(p \circ \eta)$ (part 1 of Proposition 13.1). Hence

$$|p \circ \eta| \subset \text{ext}(p \circ \beta).$$

For each $\psi \in W$, a phase curve through ψ connects ψ in $W = W \setminus |\beta|$ to a point in the open neighborhood $p^{-1}(\text{ext}(p \circ \beta))$ of $|\eta|$. Apply p and deduce $p\psi \in \text{ext}(p \circ \beta)$.   □

PROPOSITION 13.4. 1. *In case $b \neq 0$,*

$$ext(p \circ \beta) \cap int(p \circ \eta) \subset \cup_{\mathbb{N}} ext(p \circ \xi_j).$$

2. *In case $b = 0$,*

$$int(p \circ \eta) \setminus \{0\} \subset \cup_{\mathbb{N}} ext(p \circ \xi_j).$$

PROOF. 1. The case $b \neq 0$. Let $\chi \in \text{ext}(p \circ \beta) \cap \text{int}(p \circ \eta)$ be given. Recall (4.2) which implies

$$L = (L \setminus pK) \cup (L \setminus p(-K)).$$

1.1. The case $\chi \in L \setminus pK$.
1.1.1. $L \setminus \{\chi\}$ is an open neighborhood of $|p \circ \beta|$. Corollary 13.3 implies that there exists $j_0 \in \mathbb{N}$ such that for all integers $j \geq j_0$,

$$p \circ \xi_{\iota(j)+1}([0, 1]) \subset L \setminus \{\chi\}.$$

We introduce homotopies

$$\text{hom}_j : [0, 1] \times [-1, 1] \longrightarrow L, \quad j \in \mathbb{N} \quad \text{and} \quad j \geq j_0$$

which deform $p \circ \xi_{\iota(j)+1}([-1,0])$ in the convex set pK into the line segment from $p\xi_{\iota(j)+1}(-1)$ to $p\xi_{\iota(j)+1}(0)$:

$$\text{hom}_j(s,t) := s \cdot (-t \cdot p\xi_{\iota(j)+1}(-1) + (1+t) \cdot p\xi_{\iota(j)+1}(0)) + (1-s) \cdot p\xi_{\iota(j)+1}(t)$$

$$\text{for} \quad t \le 0,$$

$$\text{hom}_j(s,t) := p\xi_{\iota(j)+1}(t) \quad \text{for} \quad 0 \le t.$$

Observe that hom_j is continuous, and that each $\text{hom}_j(s,\cdot)$ is a closed curve which is piecewise smooth. We have

$$\chi \notin \text{hom}_j([0,1] \times [-1,1])$$

since

$$\text{hom}_j([0,1] \times [-1,0]) \subset pK \subset L \setminus \{0\}, \quad \text{hom}_j([0,1] \times [0,1]) \subset L \setminus \{\chi\}.$$

It follows that the winding numbers of $p \circ \xi_{\iota(j)+1} = \text{hom}_j(0,\cdot)$ and $\text{hom}_j(1,\cdot)$ with respect to χ coincide.

1.1.2. The advantage of $\text{hom}_j(1,\cdot)$ over $p \circ \xi_{\iota(j)+1}$ is that the complex-valued integrands of the Riemann integrals over $[-1,1]$ which define

$$\text{wind}(\chi, \text{hom}_j(1,\cdot))$$

converge uniformly to the integrand of the integral which defines the winding number

$$\text{wind}(\chi, \overline{p \circ \beta})$$

where the curve

$$\overline{p \circ \beta} : [-1,1] \longrightarrow L$$

is given by

$$\overline{p \circ \beta}(t) = pb_0 \quad \text{for} \quad t \le 0, \quad \overline{p \circ \beta}(t) = p\beta(t \cdot \tau_b) \quad \text{for} \quad 0 \le t.$$

This uniform convergence is a consequence of the following simple facts. First,

$$\xi_{\iota(j)+1}(0) = x_{z_{2\iota(j)}+1} \quad \text{and} \quad \xi_{\iota(j)+1}(-1) = \xi_{\iota(j)+1}(1) = x_{z_{2\iota(j)+2}+1}$$

both converge to $b_0 = b_{z_2(b)+1}$ as $j \longrightarrow \infty$. This implies that uniformly for $t \in [-1,0]$,

$$\text{hom}_j(1,t) = -t \cdot p\xi_{\iota(j)+1}(-1) + (1+t) \cdot p\xi_{\iota(j)+1}(0) \longrightarrow pb_0 = \overline{p \circ \beta}(t)$$

as $j \longrightarrow \infty$, and

$$D_2\text{hom}_j(1,t) = p\xi_{\iota(j)+1}(0) - p\xi_{\iota(j)+1}(-1) \longrightarrow 0 = D\overline{p \circ \beta}(t)1$$

as $j \longrightarrow \infty$.

Secondly,

$$z_{2\iota(j)+2} - z_{2\iota(j)} \longrightarrow z_2(b) + 1 \quad \text{and} \quad x_{z_{2\iota(j)}+1} \longrightarrow b_0$$

and uniform continuity of $p \circ F$ on the compact set

$$[0, z_2(b) + 2] \times (\{b_0\} \cup \{x_{z_{2\iota(j)} + 1} : j \in \mathbb{N}\})$$

imply that uniformly for $t \in [0, 1]$,

$$\hom_j(1, t) = p\xi_{\iota(j)+1}(t) = pF(t \cdot (z_{2\iota(j)+2} - z_{2\iota(j)}), x_{z_{2\iota(j)} + 1})$$

$$\longrightarrow pF(t \cdot (z_2(b) + 1), b_0) = \overline{p \circ \beta}(t) \quad \text{as} \quad j \longrightarrow \infty.$$

Finally, observe that for $0 < t < 1$,

$$D_2 \hom_j(1, t) = pD\xi_{\iota(j)+1}(t)1$$

$$= p((z_{2\iota(j)+2} - z_{2\iota(j)}) \cdot (x')_{z_{2\iota(j)} + 1} + t \cdot (z_{2\iota(j)+2} - z_{2\iota(j)}))$$

and

$$D_2 \overline{p \circ \beta}(t)1 = pD\beta(t)1 = p((z_2(b) + 1) \cdot (b')_{t \cdot (z_2(b) + 1)});$$

eq. (1.1) yields

$$(x')_{...} = \mu x_{...} + f \circ x_{...-1}$$

$$= \mu \cdot F(t \cdot (z_{2\iota(j)+2} - z_{2\iota(j)}), x_{z_{2\iota(j)} + 1}) + f \circ F(t \cdot (z_{2\iota(j)+2} - z_{2\iota(j)}), x_{z_{2\iota(j)}})$$

and analogously

$$(b')_{t \cdot (z_2(b) + 1)} = \mu \cdot F(t \cdot (z_2(b) + 1), b_0) + f \circ F(t \cdot (z_2(b) + 1), b_{-1}).$$

Using uniform continuity as above, and in addition

$$x_{z_{2\iota(j)}} \longrightarrow b_{-1} \quad \text{as} \quad j \longrightarrow \infty,$$

we conclude that uniformly for $t \in (0, 1)$,

$$D_2 \hom_j(1, t)1 \longrightarrow D\overline{p \circ \beta}(t)1 \quad \text{as} \quad j \longrightarrow \infty.$$

1.1.3. The convergence of integrands obtained in part 1.1.2, and the fact that winding numbers are integers, yield finally that for j sufficiently large

$$0 = \text{wind}(\chi, p \circ \beta) \qquad \text{(by hypothesis)}$$

$$= \text{wind}(\chi, \overline{p \circ \beta}) = \text{wind}(\chi, \hom_j(1, \cdot)) = \text{wind}(\chi, p \circ \xi_{\iota(j)+1}),$$

i.e.,

$$\chi \in \text{ext}(p \circ \xi_{\iota(j)+1}).$$

1.2. The case $\chi \in L \setminus p(-K)$. We have

$$\chi \in \text{ext}(p \circ \beta) = \text{ext}(p \circ \beta^-)$$

since $|\beta| = |\beta^-|$. We work with the curves $\xi_{\iota(j)+1}^-$, instead of $\xi_{\iota(j)+1}$, argue as in case 1.1 and find, for some $j \in \mathbb{N}$,

$$\chi \in \text{ext}(p \circ \xi_{\iota(j)+1}^-).$$

Corollary 13.1 yields

$$\chi \in \text{ext}(p \circ \xi_k) \quad \text{for integers} \quad k > \iota(j) + 2.$$

2. The case $b = 0$. Let $\chi \in \text{int}(p \circ \eta) \setminus \{0\}$ be given. Choose a convex neighborhood N_L of 0 in L such that $\chi \notin N_L$.

2.1. The case $\chi \in L \setminus pK$. Corollary 13.2 implies that for $j \in \mathbb{N}$ sufficiently large,

$$p\xi_{\iota(j)+1}([0,1]) \subset N_L.$$

Also,

$$p\xi_{\iota(j)+1}([-1,0]) \subset pK \subset L \setminus \{\chi\}.$$

We infer

$$\chi \notin |p \circ \xi_{\iota(j)+1}|, \quad |p \circ \xi_{\iota(j)+1}| \subset N_L \cup pK.$$

Now, pK is a convex cone with $0 \in \overline{pK}$. Observe that $\chi \in L \setminus (N_L \cup pK)$ can be radially connected in $L \setminus (N_L \cup pK)$, i.e. without crossing $|p \circ \xi_{\iota(j)+1}|$, to points with arbitrarily large modulus. This implies $\chi \in \text{ext}(p \circ \xi_{\iota(j)+1})$.

2.2. The case $\chi \in L \setminus p(-K)$. We consider the curves $\xi_{\iota(j)+1}^-$, argue as before and find

$$\chi \in \text{ext}(p \circ \xi_{\iota(j)+1}^-) \quad \text{for some} \quad j \in \mathbb{N}.$$

Corollary 13.1 yields

$$\chi \in \text{ext}(p \circ \xi_k) \quad \text{for integers} \quad k > \iota(j) + 2. \quad \square$$

PROOF OF ASSERTION 3 OF THEOREM 5.4 IN CASE THAT (11.1) HOLDS. 1. Using Proposition 13.4 and Proposition 12.5 we infer that in case $b \neq 0$,

$$\text{ext}(p \circ \beta) \cap \text{int}(p \circ \eta) \subset pW \cap \text{int}(p \circ \eta)$$

while for $b = 0$,

$$\text{int}(p \circ \eta) \setminus \{0\} \subset pW \cap \text{int}(p \circ \eta).$$

Proposition 13.3 now implies

$$pW \cap \text{int}(p \circ \eta) = \text{ext}(p \circ \beta) \cap \text{int}(p \circ \eta) \quad \text{if} \quad b \neq 0,$$

$$pW \cap \text{int}(p \circ \eta) = \text{int}(p \circ \eta) \setminus \{0\} \quad \text{if} \quad b = 0.$$

2. Part 1 of Proposition 13.1 gives

$$|\beta| \subset (\overline{W} \setminus W) \cap p^{-1}(\text{int}(p \circ \eta)) \quad \text{if} \quad b \neq 0,$$

$$0 \in (\overline{W} \setminus W) \cap p^{-1}(\text{int}(p \circ \eta)) \quad \text{if} \quad b = 0.$$

It remains to show

$$(\overline{W} \setminus W) \cap p^{-1}(\text{int}(p \circ \eta)) \subset |\beta| \quad \text{if} \quad b \neq 0,$$

$$(\overline{W} \setminus W) \cap p^{-1}(\text{int}(p \circ \eta)) \subset \{0\} \quad \text{if} \quad b = 0.$$

3. The case $b \neq 0$. By Proposition 13.3, $pW \subset \text{ext}(p \circ \beta)$. Hence

$$p\overline{W} \subset \overline{pW} \subset \overline{\text{ext}(p \circ \beta)} = |p \circ \beta| \cup \text{ext}(p \circ \beta),$$

and therefore

$$p(\overline{W} \setminus W) \cap \text{int}(p \circ \eta) = (p\overline{W} \setminus pW) \cap \text{int}(p \circ \eta)$$

$$\subset ((|p \circ \beta| \cup \text{ext}(p \circ \beta)) \setminus pW) \cap \text{int}(p \circ \eta)$$

$$\subset (|p \circ \beta| \cup [\text{ext}(p \circ \beta) \cap \text{int}(p \circ \eta)]) \setminus pW$$

$$\subset |p \circ \beta| \quad \text{(see part 1 above)}.$$

It follows that

$$(\overline{W} \setminus W) \cap p^{-1}(\text{int}(p \circ \eta)) \subset |\beta|.$$

4. The case $b = 0$. Then

$$p(\overline{W} \setminus W) \cap \text{int}(p \circ \eta) = (p\overline{W} \setminus pW) \cap \text{int}(p \circ \eta)$$

$$\subset \text{int}(p \circ \eta) \setminus pW \subset (\{0\} \cup pW) \setminus pW \quad \text{(see part 1)}$$

$$= \{0\};$$

it follows that

$$(\overline{W} \setminus W) \cap p^{-1}(\text{int}(p \circ \eta)) \subset \{0\}. \quad \square$$

14. The Case of Eventually Monotone Solutions

In this section we assume that (11.2) holds, i.e. there exists a solution $x :$ $\mathbb{R} \longrightarrow \mathbb{R}$ of eq. (1.1) with phase curve in W and

$$(14.1) \qquad\qquad\qquad J(x) < \infty.$$

We shall prove that in this case

$$(14.2) \qquad\qquad \{0\} = (\overline{W} \setminus W) \cap p^{-1}(\text{int}(p \circ \eta)).$$

Together with the result of the preceding section this will complete the proof of part 3 of Theorem 5.4.

The first observations are the following. The stationary point $0 \in C$ of the semiflow F is stable and attractive (Remark 4.1, parts 3 and 2). We have

$$(14.3) \qquad\qquad x_t \notin |\eta| \quad \text{for all} \quad t \in \mathbb{R}$$

since otherwise x would be periodic and we would have a contradiction to existence and boundedness of zeros.

A remark in Section 3 yields

$$x(t) \longrightarrow 0 \quad \text{as} \quad t \longrightarrow \infty.$$

Hence

$$\{0\} = \omega(x).$$

By Theorem 5.1, $0 \in \text{int}(p \circ \eta)$; the open neighborhood $\text{int}(p \circ \eta)$ of 0 contains points px_t. Using (14.3) we infer

$$px_t \in \text{int}(p \circ \eta) \quad \text{for all} \quad t \in \mathbb{R}.$$

The latter implies that for some integer $j \leq J(x)$,

$$x_{z_j + 1} \in h((-\underline{a}, 0)).$$

Relabeling the zeros $z_k = z_k(x)$ of x if necessary we achieve

$$2 \leq J(x), \quad z_0 = -1, \quad x_0 = h(a) \quad \text{for some} \quad a \in (-\underline{a}, 0).$$

Now we are in the situation of Corollary 12.1, and Corollary 12.2 yields

$$\omega(x) \subset \overline{W} \setminus W.$$

Recall from Section 12 the simple closed curve $\xi_0 = \xi_{0,a}$ associated with a, and the homotopy $M = M_{0,a}$ of closed curves which deforms a reparameterization $M(a, \cdot)$ of ξ_0 in W into the reparameterization $M(0, \cdot)$ of η:

$$M(a,t) = \xi_0(2t - 1) \quad \text{and} \quad M(0,t) = \eta(t \cdot \tau) \quad \text{for} \quad 0 \leq t \leq 1.$$

The curves

$$p \circ F(j\tau, \cdot) \circ \xi_0, \quad j \in \mathbb{N},$$

are simple, closed and piecewise smooth.

PROPOSITION 14.1. *For every $j \in \mathbb{N}$,*

$$p(\overline{W} \setminus W) \cap int(p \circ \eta) \subset int(p \circ F(j\tau, \cdot) \circ \xi_0).$$

PROOF. Let $\chi \in p(\overline{W} \setminus W) \cap \text{int}(p \circ \eta)$. In particular, $\chi \notin pW$, and therefore

$$\chi \notin p \circ F(j\tau, \cdot) \circ M([a, 0] \times [0, 1]).$$

It follows that

$$0 \neq \text{wind}(\chi, p \circ \eta) = \text{wind}(\chi, p \circ F(j\tau, \cdot) \circ \eta) \quad \text{(by periodicity of } \eta),$$

or

$$\chi \in \text{int}(p \circ F(j\tau, \cdot) \circ \eta) = \text{int}(p \circ F(j\tau, \cdot) \circ M(0, \cdot))$$

(with $|\eta| = |M(0, \cdot)|$). By homotopy invariance,

$$0 \neq \text{wind}(\chi, p \circ F(j\tau, \cdot) \circ M(0, \cdot)) = \text{wind}(p \circ F(j\tau, \cdot) \circ M(a, \cdot)),$$

hence

$$\chi \in \text{int}(p \circ F(j\tau, \cdot) \circ M(a, \cdot)) = \text{int}(p \circ F(j\tau, \cdot) \circ \xi_0)$$

(with $|\xi_0| = |M(a, \cdot)|$). $\quad\square$

PROPOSITION 14.2. *For every solution $x^* : \mathbb{R} \longrightarrow \mathbb{R}$ of eq. (1.1) with phase curve in $W \cap p^{-1}(int(p \circ \eta))$,*

$$x^*(t) \longrightarrow 0 \quad as \quad t \longrightarrow \infty.$$

PROOF. Suppose there exists a solution $x^* : \mathbb{R} \longrightarrow \mathbb{R}$ of eq. (1.1) with phase curve in $W \cap p^{-1}(\text{int}(p \circ \eta))$ so that

$$0 < \limsup_{t \longrightarrow \infty} |x^*(t)|.$$

Then $J(x^*) = \infty$ (see Section 3), and

$$x_t^* \notin |\eta| \quad \text{for all} \quad t \in \mathbb{R}.$$

Let $(z_j^*)_{-\infty}^{\infty}$ denote the increasing sequence of zeros of x^*. We may assume

$$z_0^* = -1 \quad \text{and} \quad x_0^* = h(a^*) \quad \text{where} \quad \underline{a} < a^* < 0.$$

Then

$$\omega(x^*) \subset \overline{W} \setminus W \qquad \text{(Corollary 12.2)},$$

and

$$x_{z_{2j+1}^*}^* + 1 \in -K \quad \text{for all integers} \quad j.$$

There is a strictly increasing map $\nu : \mathbb{N} \longrightarrow \mathbb{N}$ such that the subsequence of points

$$x_{z_{2\nu(j)+1}^*}^* + 1, \quad j \in \mathbb{N},$$

converges to some

$$\phi^* \in (\overline{-K}) \cap \overline{W} \cap \omega(x^*).$$

We have

$$\phi^* \neq 0$$

since otherwise, attractivity of 0 would imply $x_t^* \longrightarrow 0$ as $t \longrightarrow \infty$ (see Remark 4.1, part 2), a contradiction to our assumption. Using

$$x_t \longrightarrow 0 \quad \text{as} \quad t \longrightarrow \infty, \quad x_t \longrightarrow |\eta| \subset W \quad \text{as} \quad t \longrightarrow -\infty,$$

$$x_t \in W \quad \text{for all} \quad t \in \mathbb{R}, \quad \text{and} \quad \phi^* \in \omega(x^*) \subset \overline{W} \setminus W,$$

we infer

$$0 < \operatorname{dist}(p\phi^*, \{px_t : t \in \mathbb{R}\}) =: d_x.$$

Furthermore,

$$0 < \operatorname{dist}(p\phi^*, |p \circ \eta|) =: d_\eta$$

and

$$p\phi^* \in \operatorname{int}(p \circ \eta),$$

as follows from

$$\phi^* \in \overline{W} \setminus W, \quad |\eta| \subset W; \quad p\phi^* \in p(\overline{W} \setminus W) = p\overline{W} \setminus pW, \quad |p \circ \eta| \subset pW,$$

$$p\phi^* = \lim_{j \longrightarrow \infty} px_{z_{2\nu(j)+1}^*}^* + 1 \in \overline{\operatorname{int}(p \circ \eta)} = (\operatorname{int}(p \circ \eta)) \cup |p \circ \eta|.$$

Set

$$d := \frac{1}{2} \cdot \min\{d_x, d_\eta\}.$$

Choose an integer j so large that for

$$j_1 := 2\nu(j) + 1 \quad \text{and} \quad j_2 := 2\nu(j+1) + 1,$$

$j_2 > j_1$, both points

$$px_{z_{j_1}^*}^* + 1 \quad \text{and} \quad px_{z_{j_2}^*}^* + 1 \quad \text{in} \quad p(-K) = -pK$$

belong to the open ball $N_L \subset L$ with center $p\phi^*$ and radius d. We define a closed curve

$$\xi^* : [-1, z_{j_2}^* - z_{j_1}^*] \longrightarrow C$$

by

$$\xi^*(t) := (-t) \cdot x^*_{z^*_{j_2} + 1} + (1+t) \cdot x^*_{z^*_{j_1} + 1} \quad \text{for} \quad -1 \le t \le 0,$$

$$\xi^*(t) := x^*_{t + z^*_{j_1} + 1} \quad \text{for} \quad 0 < t \le z^*_{j_2} - z^*_{j_1}.$$

Now, ξ^* is piecewise smooth. Observe that

$$\xi^*([-1,0]) \subset -K$$

since K is convex.

Recall from the proof of Theorem 5.1 the construction of the homotopy of closed curves in S which connects a reparameterization of the curve $\eta : [0, z_{2n} + 1] \longrightarrow C$ given by the periodic solution y to the curve $n \cdot c_r$ in $L \setminus \{0\}$;

$$\text{wind}(0, n \cdot c_r) = n.$$

Using that for $-1 \le t \le z^*_{j_1} + 1 - 1 - z^*_{j_1}$,

$$\xi^*(t) \in \{\phi \in C : \phi < 0 \quad \text{in} \quad (-1,0)\}$$

(this set corresponds to $-K_{k0}$ in the proof of Theorem 5.1) we obtain, by an obvious modification of the construction in the proof of Theorem 5.1, a homotopy of closed curves in S from a reparameterization of ξ^* to the closed curve

$$n \cdot (-c_r) \quad \text{where} \quad n := \frac{1}{2} \cdot (j_2 - j_1) \ge 1.$$

Applying the projection p we arrive at a homotopy in $L \setminus \{0\}$ from a reparameterization of $p \circ \xi^*$ to the curve $n \cdot (-c_r)$ in L. It follows that

$$\text{wind}(0, p \circ \xi^*) = n \cdot \text{wind}(0, -c_r) = n \ne 0;$$

$$0 \in \text{int}(p \circ \xi^*);$$

$\text{int}(p \circ \xi^*)$ is an open neighborhood of 0 in L. Consequently, for $t > 0$ sufficiently large,

(14.4) $px_t \in \text{int}(p \circ \xi^*).$

We have

$$|p \circ \xi^*| \subset \text{int}(p \circ \eta)$$

since $px^*_t \in \text{int}(p \circ \eta)$ for all $t \in \mathbb{R}$, and $p \circ \xi^*([-1,0])$ is the line segment from

$$px^*_{z^*_{j_2} + 1} \quad \text{to} \quad px^*_{z^*_{j_1} + 1}$$

in the convex set N_L which is disjoint with $|p \circ \eta|$. This implies

$$|p \circ \eta| \subset \text{ext}(p \circ \xi^*).$$

As $x_{-t} \longrightarrow |\eta|$ as $t \longrightarrow \infty$,

(14.5) $px_{-t} \in \text{ext}(p \circ \xi^*)$

for $t > 0$ sufficiently large. From (14.4) and (14.5) we obtain that for some $t \in \mathbb{R}$

$$px_t \in |p \circ \xi^*|; \quad x_t \in |\xi^*|,$$

which leads to

$$0 \neq \phi^* \in \omega(x),$$

a contradiction. $\quad\square$

PROOF OF (14.2). 1. Theorem 5.1 and

$$\{0\} = \omega(x) \subset \overline{W} \setminus W$$

give

$$0 \in (\overline{W} \setminus W) \cap p^{-1}(\mathrm{int}(p \circ \eta)).$$

2. It remains to show that, given an open ball N_L in L with center 0, we have

$$p(\overline{W} \setminus W) \cap \mathrm{int}(p \circ \eta) \subset N_L.$$

The stationary point $0 \in C$ is attractive in the sense of Remark 4.1, part 2:newline There exist positive ϵ, δ, c; $\epsilon \leq c$, so that for all $\phi \in C$ with $\|\phi\| < \epsilon$ and for all $t \geq 0$,

$$\|F(t, \phi)\| \leq c \cdot e^{-\delta t}.$$

For every $\psi \in |\xi_0|$ there are $t = t(\psi) > 0$ and a neighborhood U_ψ of ψ in C with

$$\|F(t, \psi^*)\| < \epsilon \quad \text{for all} \quad \psi^* \in U_\psi.$$

It follows that for $\psi^* \in U_\psi$ and for all $t > t(\psi) + \frac{1}{\delta} \log\left(\frac{c}{\epsilon}\right)$,

$$\|F(t, \psi^*)\| < \epsilon.$$

The compact set $|\xi_0|$ is covered by a finite collection $U_{\psi_1}, \ldots, U_{\psi_n}$ of such neighborhoods. Consequently,

$$\text{for} \quad t_0 := \max_{1,\ldots,n} t(\psi_n) \quad + \frac{1}{\delta} \log\left(\frac{c}{\epsilon}\right) \quad + 1 \quad \text{and for each} \quad \psi \in |\xi_0|,$$

$$\|F(t_0, \psi)\| < \epsilon.$$

For $t \geq t_0$ and $\psi \in |\xi_0|$,

$$\|F(t, \psi)\| \leq c \cdot e^{-\delta(t-t_0)}.$$

We infer that for $n \in \mathbb{N}$ sufficiently large,

$$|F(n\tau, \cdot) \circ \xi_0| \subset p^{-1}(N_L);$$

$$|p \circ F(n\tau, \cdot) \circ \xi_0| \subset N_L.$$

Proposition 14.1 gives

$$p(\overline{W} \setminus W) \cap \mathrm{int}(p \circ \eta) \subset \mathrm{int}(p \circ F(n\tau, \cdot) \circ \xi_0) \subset N_L. \quad\square$$

15. Outside the Projected Periodic Orbit

The proof of part 2 of Theorem 5.4, i.e. that

$$E := (\overline{W} \setminus W) \cap p^{-1}(\text{ext}(p \circ \eta))$$

is the orbit of a slowly oscillating periodic solution of eq. (1.1), is analogous but easier than the proof of part 3, because of the following simple observations.

PROPOSITION 15.1. $0 \notin E$.

PROOF. Otherwise, $0 = p0 \in \text{ext}(p \circ \eta)$, a contradiction to Theorem 5.1. $\square$

PROPOSITION 15.2. *For every solution* $x : \mathbb{R} \longrightarrow \mathbb{R}$ *of eq.* (1.1) *with phase curve in* $W \cap p^{-1}(\text{ext}(p \circ \eta))$,

$$J(x) = \infty.$$

PROOF. Suppose $J(x) < \infty$ for a solution $x : \mathbb{R} \longrightarrow \mathbb{R}$ of eq. (1.1) with phase curve in $W \cap p^{-1}(\text{ext}(p \circ \eta))$. Then $x(t) \longrightarrow 0$ as $t \longrightarrow \infty$ (Section 3), hence

$$\text{ext}(p \circ \eta) \ni px_t \longrightarrow p0 = 0 \quad \text{as} \quad t \longrightarrow \infty;$$

$$0 \in \overline{\text{ext}(p \circ \eta)} = |p \circ \eta| \cup \text{ext}(p \circ \eta),$$

a contradiction to Theorem 5.1. $\square$

Constructions like those in Sections 11–13 now yield the desired result. Proposition 15.2 guarantees that we do not have to deal with eventually monotone solutions (as in Section 14), and Proposition 15.1 shows that the analogue of the case $b = 0$ in Section 13 does not occur. We omit the details.

REFERENCES

1. R. Abraham and J. Robbin, *Transversal Mappings and Flows*, Benjamin, New York, 1967.
2. N. Angelstorf, *Spezielle periodische Lösungen einiger autonomer zeitverzögerter Differentialgleichungen mit Symmetrien*, Ph.D. Thesis, Universität Bremen.
3. Y. Cao, in preparation.
4. S. N. Chow, *Existence of periodic solutions of autonomous functional differential equations*, J. Differential Equations **15** (1974), 350–378.
5. S. N. Chow and H. O. Walther, *Characteristic multipliers and stability of periodic solutions of* $\dot{x}(t) = g(x(t-1))$, Trans. Amer. Math. Soc. **307** (1988), 127–142.
6. O. Diekmann, S. van Gils, S. Verduyn Lunel and H. O. Walther, *Delay equations: complex, functional and nonlinear analysis*, in preparation.
7. J. Dieudonné, *Foundations of Modern Analysis*, Academic Press, New York, 1960.
8. P. Dormayer, *The stability of special symmetric solutions of* $\dot{x}(t) = \alpha f(x(t-1))$ *with small amplitudes*, Nonlinear Anal. **14** (1990), 701–715.
9. P. Dormayer, *Smooth bifurcation of symmetric periodic solutions of functional differential equations*, J. Differential Equations **82** (1989), 109–155.
10. C. C. Fenske, Personal communication.
11. R. B. Grafton, *A periodicity theorem for autonomous functional differential equations*, J. Differential Equations **6** (1969), 87–109.
12. K. P. Hadeler, F. Tomiuk, *Periodic solutions of difference-differential equations*, Arch. Rat. Mech. Anal. **65** (1977), 87–95.
13. J. K. Hale, *Theory of Functional Differential Equations*, Springer, New York, 1977.
14. J. K. Hale and X. B. Lin, *Symbolic dynamics and nonlinear semiflows*, Ann. Mat. Pura Appl. (4) **144** (1986), 229–259.

15. D. Henry, *Geometric Theory of Semilinear Parabolic Equations*, Lect. Notes Math. Vol. 840, Springer, New York, 1981.

16. M. W. Hirsch and S. Smale, *Differential Equations, Dynamical Systems and Linear Algebra*, Academic Press, New York, 1974.

17. A. F. Ivanov, B. Lani-Wayda and H. O. Walther, *Unstable hyperbolic periodic solutions of differential delay equations*, Preprint.

18. J. L. Kaplan and J. A. Yorke, *Ordinary differential equations which yield periodic solutions of differential delay equations*, J. Math. Anal. Appl. **48** (1974), 317–324.

19. J. L. Kaplan and J. A. Yorke, *On the stability of a periodic solution of a differential delay equation*, SIAM J. Math. Anal. **6** (1975), 268–282.

20. J. L. Kaplan and J. A. Yorke, *On the nonlinear differential delay equation $x'(t) = -f(x(t), x(t-1))$*, J. Differential Equations **23** (1977), 293–314.

21. J. Mallet-Paret, *Morse decompositions for differential delay equations*, J. Differential Equations **72** (1988), 270–315.

22. J. Mallet-Paret and R. D. Nussbaum, *Global continuation and asymptotic behaviour for periodic solutions of a differential delay equation*, Ann. Mat. Pura Appl. (4) **145** (1986), 33–128.

23. J. Mallet-Paret and G. Sell, in preparation.

24. J. Mallet-Paret and H. L. Smith, *The Poincaré-Bendixson theorem for monotone cyclic feedback systems*, J. Dynamics Diff. Equations **2** (1990), 367–421.

25. A. Neugebauer, *Invariante Mannigfaltigkeiten und Neigungslemmata für Abbildungen in Banachräumen*, Diploma Thesis, Universität München.

26. R. D. Nussbaum, *Periodic solutions of some nonlinear autonomous functional differential equations II*, J. Differential Equations **14** (1973), 360–394.

27. R. D. Nussbaum, *Periodic solutions for some nonlinear autonomous functional differential equations*, Ann. Mat. Pura Appl. (4) **101** (1974), 263–306.

28. R. D. Nussbaum, *A global bifurcation theorem with applications to functional differential equations*, J. Functional Anal. **19** (1975), 319–339.

29. R. D. Nussbaum, *Uniqueness and nonuniqueness for periodic solutions of $x'(t) = -g(x(t-1))$*, J. Differential Equations **34** (1979), 25–54.

30. Ya. B. Peşin, *On the behavior of a strongly nonlinear differential equation with retarded argument*, Differentsialnye Uravnenija **10** (1974), 1025–1036.

31. R. A. Smith, *Existence of periodic orbits of autonomous retarded functional differential equations*, Math. Proc. Camb. Phil. Soc. **88** (1980), 89–109.

32. H. O. Walther, *A theorem on the amplitudes of periodic solutions of delay equations, with an application to bifurcation*, J. Differential Equations **29** (1978), 396–404.

33. H. O. Walther, *On instability, ω-limit sets and periodic solutions of nonlinear autonomous differential delay equations*, in Functional Differential Equations and Approximation of Fixed Points, Bonn 1978 (H.O. Peitgen and H.O. Walther, eds.), Lect. Notes Math. Vol. 730, Springer, New York, 1979, pp. 489-503..

34. H. O. Walther, *Density of slowly oscillating solutions of $\dot{x}(t) = -f(x(t-1))$*, J. Math. Anal. Appl. **79** (1981), 127–140.

35. H. O. Walther, *Bifurcation from periodic solutions in functional differential equations*, Math. Z. **182** (1983), 269–289.

36. H. O. Walther, *An invariant manifold of slowly oscillating solutions for $\dot{x}(t) = -\mu x(t) + f(x(t-1))$*, J. Reine Angew. Math. **414** (1991), 67–112.

37. H. O. Walther, *On Floquet multipliers of periodic solutions of delay equations with monotone nonlinearities*, Functional Differential Equations, Kyoto 1990 (J. Kato and T. Yoshizawa, eds.), World Scientific, Singapore, 1991, pp. 349–356.

38. X. Xie,, *Uniqueness and stability of slowly oscillating periodic solutions of delay equations with bounded nonlinearity*, J. Dynamics Diff. Equations **3** (1991), 515–540.

MATHEMATISCHES INSTITUT UNIVERSITÄT MÜNCHEN D 8000 MÜNCHEN 2 GERMANY

E-mail: Hans-Otto.Walther@@MATHEMATIK.UNI-MUENCHEN.DBP.DE

Contemporary Mathematics
Volume **129**, 1992

The Fourier Method for Partial Differential Equations with Piecewise Continuous Delay

J. WIENER and L. DEBNATH

ABSTRACT. Separation of variables is justified for general classes of partial differential equations with piecewise continuous delay, which enables us to find the solutions of appropriate boundary value problems and to investigate their asymptotic and oscillatory properties.

1. MAIN RESULTS

Functional differential equations (FDE) with delay provide a mathematical model for a physical or biological system in which the rate of change of the system depends upon its past history. The theory of FDE with continuous argument is well developed and has numerous applications in natural and engineering sciences. This paper continues our earlier work [1-5] in an attempt to extend this theory to differential equations with discontinuous argument deviations. In these papers, ordinary differential equations with arguments having intervals of constancy have been studied. Such equations represent a hybrid

1991 Mathematics Subject Classification. Primary 35A05, 35B25, 35L10, 34K25.
Supported by U.S. Army Grant DAALO3-89-G-0107 and by the University of Central Florida.

of continuous and discrete dynamical systems and combine
properties of both differential and difference equations. They
include as particular cases loaded and impulse equations, hence
their importance in control theory and in certain biomedical
models. Continuity of a solution at a point joining any two
consecutive intervals implies recursion relations for the values of
the solution at such points. Therefore, differential equations with
piecewise continuous argument (EPCA) are intrinsically closer to
difference rather than differential equations. In [6] boundary
value problems for some linear EPCA in partial derivatives have
been considered and the properties of their solutions studied. The
influence of certain discontinuous delays on the behavior of
solutions to some typical equations of mathematical physics has
been explored. In [7] initial value problems for EPCA in partial
derivatives have been investigated, closed form solutions found in
some classical cases, and a class of loaded equations that arise in
solving certain inverse problems handled within the general
framework of differential equations with piecewise continuous
delays. The purpose of this paper is to analyze boundary value
problems for EPCA in partial derivatives.

We consider the boundary value problem (BVP) consisting of
the equation

$$(1.1) \qquad \frac{\partial u(x,t)}{\partial t} + P\left(\frac{\partial}{\partial x}\right)u(x,t) = Q\left(\frac{\partial}{\partial x}\right)u(x,[t/h]h) \,,$$

where P and Q are polynomials of the highest degree m with
coefficients that may depend only on x, the boundary conditions

$$(1.2) \qquad L_j u = \sum_{k=1}^{m}\left(M_{jk}u^{(k-1)}(0) + N_{jk}u^{(k-1)}(1)\right) = 0 \,,$$

$$(M_{jk} \text{ and } N_{jk} \text{ are constants, } j = 1, ..., m)$$

and the initial condition

(1.3) $$u(x, 0) = u_0(x) .$$

Here [·] designates the greatest integer function, $(x, t) \epsilon [0, 1] \times [0, \infty)$, and $h = $ constant > 0. Equations (1.2) will be written briefly as

(1.4) $$Lu = 0 .$$

DEFINITION 1.1. A function $u(x, t)$ is called a solution of the above BVP if it satisfies the conditions: (i) $u(x, t)$ is continuous in $G = [0, 1] \times [0, \infty)$; (ii) $\partial u / \partial t$ and $\partial^k u / \partial x^k$ $(k = 0, 1..., m)$ exist and are continuous in G, with the possible exception of the points (x, nh), where one-sided derivatives exist $(n = 0, 1, 2..)$; (iii) $u(x, t)$ satisfies Eq. (1.1) in G, with the possible exception of the points (x, nh), and conditions (1.2) - (1.3).

Let $u_n(x, t)$ be the solution of the given problem on the interval $nh \leq t < (n + 1)h$, then

(1.5) $$\partial u_n(x, t) / \partial t + P u_n(x, t) = Q u_n(x) ,$$

where

$$u_n(x) = u_n(x, nh) .$$

Write

$$u_n(x, t) = w_n(x, t) + v_n(x) ,$$

which gives the equation

$$\partial w_n / \partial t + P w_n + P v_n(x) = Q u_n(x) ,$$

and require that

(1.6) $$\partial w_n / \partial t + P w_n = 0 ,$$

(1.7) $$Pv_n(x) = Qu_n(x) .$$

Assuming both w_n and v_n satisfy (1.4) leads to an ordinary BVP (1.4)-(1.7), whose solution is denoted by

$$v_n(x) = P^{-1}Qu_n(x) ,$$

and to BVP (1.4), (1.6), whose solution is sought in the form

(1.8) $$w_n(x,t) = T_n(t)X(x) .$$

Separation of variables produces the ODE $T'_n + \lambda T_n = 0$ with a solution $T_n(t) = e^{-\lambda(t-nh)}$, and the BVP

(1.9) $$P(d/dx)X - \lambda X = 0, \qquad LX = 0$$

where L is defined in (1.2) and (1.4). If BVP (1.9) has an infinite countable set of eigenvalues λ_j and corresponding eigenfunctions $X_j(x) \epsilon C^m[0,1]$, then the series

$$w_n(x,t) = \sum_{j=1}^{\infty} C_{nj} e^{-\lambda_j(t-nh)} X_j(x), \qquad C_{nj} = \text{constant}$$

represents a formal solution of problem (1.4)-(1.6) and

(1.10) $$u_n(x,t) = \sum_{j=1}^{\infty} C_{nj} e^{-\lambda_j(t-nh)} X_j(x) + P^{-1}Qu_n(x)$$

is a formal solution of (1.1)-(1.2). At $t = nh$ we have

(1.11) $$u_n(x) = \sum_{j=1}^{\infty} C_{nj} X_j(x) + P^{-1}Qu_n(x) .$$

Therefore, assuming the sequence $\{X_j\}$ is complete and orthonormal in $C^m[0,1]$ yields for the coefficients C_{nj} the formula

(1.12) $$C_{nj} = \int_0^1 X_j(x) \, (I - P^{-1}Q) \, u_n(x) dx, \qquad (n = 0, 1, 2...) .$$

Substituting the initial function $u_0(x) \epsilon C^m[0,1]$ in (1.12) produces the coefficients C_{0j}, and putting them together with $u_0(x)$ in (1.10) as $n = 0$ gives the solution $u_0(x,t)$ of BVP (1.1)-(1.3) on the interval $0 \leq t \leq h$. Since $u_0(x,h) = u_1(x,h) = u_1(x)$, we can find from (1.12) the numbers C_{1j} and then substitute them along with $u_1(x)$ in (1.10), to obtain the solution $u_1(x,t)$ on $h \leq t \leq 2h$. This method of steps allows to extend the solution to any interval $nh \leq t \leq (n+1)h$. Furthermore, continuity of the solution $u(x,t)$ implies

$$u_n(x,(n+1)h) = u_{n+1}(x,(n+1)h) = u_{n+1}(x),$$

hence, at $t = (n+1)h$ we get from (1.10) the recursion relations

$$(1.13) \qquad u_{n+1}(x) = \sum_{j=1}^{\infty} C_{nj} e^{-\lambda_j h} X_j(x) + P^{-1} Q u_n(x) .$$

This concludes the derivation of the following result:

THEOREM 1.1. Formula (1.10), with coefficients C_{nj} and functions $u_n(x)$ defined by recursion relations (1.12) and (1.13), represents a formal solution of BVP (1.1)-(1.3) in $[0,1] \times [nh,(n+1)h]$, for $n = 0,1,...$, if BVP (1.9) has a countable number of eigenvalues λ_j and a complete orthonormal set of eigenfunctions $X_j(x) \epsilon C^m[0,1]$ and the initial function $u_0(x) \epsilon C^m[0,1]$ satisfies (1.2).

Let

$$Py = \sum_{j=0}^{m} p_j y^{(m-j)} ,$$

where p_j are real-valued functions of classes C^{m-j} on $0 \leq x \leq 1$ and $p_0(x) \neq 0$ on $[0,1]$. Assuming $C^m[0,1]$ is embedded in $L^2[0,1]$ with the inner product

$$(y,z) = \int_0^1 y(x) z(x) dx ,$$

BVP (1.9) is called self-adjoint if

$$(Py,z) = (y,Pz) ,$$

for all $y, z \in C^m[0,1]$ that satisfy the boundary conditions

$$Ly = Lz = 0 .$$

If BVP (1.9) is self-adjoint, then all its eigenvalues are real and form at most a countable set without finite limit points. The eigenfunctions corresponding to different eigenvalues are orthogonal.

THEOREM 1.2. The BVP (1.1)-(1.3) has a solution in $[0,1] \times [nh, (n+1)h]$, for each $n = 0, 1, ...,$ given by formula (1.10) if the following hypotheses hold true.

(i) BVP (1.9) is self-adjoint, all its eigenvalues λ_j are positive.

(ii) For each λ_j, the roots of the equation $P(z) - \lambda_j = 0$ have non-positive real parts.

(iii) The initial function $u_0(x) \in C^m[0,1]$ satisfies (1.2).

PROOF. According to (1.7), we find the solution $v_0(x) = P^{-1}Qu_0(x)$ of the equation $Pv_0(x) = Qu_0(x)$ satisfying the boundary conditions $Lv_0 = 0$. Then the difference $u_0(x) - P^{-1}Qu_0(x) \in C^m[0,1]$ satisfies (1.4), and therefore we conclude from (1.11) that the Fourier series $\sum C_{0j} X_j(x)$ converges to it absolutely and uniformly on $[0,1]$, where $\{X_j(x)\}$ is the set of the orthonormal eigenfunctions of (1.9). Since $\lambda_j > 0$, the series in (1.10) also converges absolutely and uniformly on $[0,1] \times [0,h]$. Furthermore, the same is true on $[0,1]$ for the series in (1.13) at $n = 0$, and $u_1(x)$ satisfies (1.2). Hence, $u_1(x)$ should be used now to find the solution $v_1(x) = P^{-1}Qu_1(x)$ of the equation $Pv_1(x) = Qu_1(x)$ satisfying $Lv_1 = 0$, then to calculate the coefficients C_{1j} by (1.12) and the solution $u_1(x,t)$ of the given BVP on $[0,1] \times [h, 2h]$, according to (1.10). This procedure can be continued successively to construct the solution $u_n(x,t)$ for any $n \geq 0$. From (1.11) we conclude that all $u_n(x)$

satisfy (1.4). Differentiating (1.10) term by term with respect to t produces a series which converges to $\partial u_n/\partial t$ uniformly on $[0,1] \times [nh + \delta, (n+1)h]$, for sufficiently small $\delta > 0$, since $\lambda_j > 0$. Furthermore, it follows from (1.12) that

$$C_{nj} = \lambda_j^{-1} \int_0^1 (PX_j)(I - P^{-1}Q)u_n(x)\,dx$$

and since $X_j(x)$, $u_n(x)$, and $P^{-1}Qu_n(x)$ satisfy (1.4), then

$$C_{nj} = \lambda_j^{-1} \int_0^1 X_j(x)(P - Q)u_n(x)\,dx .$$

Hence,

$$(1.14) \quad |C_{nj}| \le \lambda_j^{-1} \left[\int_0^1 X_j^2\,dx\right]^{1/2} \left[\int_0^1 (Pu_n - Qu_n)^2\,dx\right]^{1/2} \le c_n\lambda_j^{-1} .$$

Let $p_0(x) = 1$ and $\lambda = \rho^m$, then in any domain W of the complex ρ-plane the equation

$$P(d/dx)y - \lambda y = 0$$

has m linearly independent solutions $y_1, ... y_m$ which are regular with respect to $\rho \epsilon W$, for sufficiently large $|\rho|$, and satisfy the relations

$$y_k^{(r-1)}(x) = \rho^{r-1}\exp(\rho\omega_k x)\left[\omega_k^{r-1} + O(\rho^{-1})\right]$$

where $(k, r = 1, ..., m)$ and $\omega_1, ..., \omega_m$ are the different m-order roots of unity [8]. Therefore, by virtue of condition (ii) and estimates (1.14), differentiating series (1.10) term by term r times $(r = 1, ..., m)$ with respect to x produces series that converge uniformly on $[0,1] \times [nh + \delta, (n+1)h]$, for sufficiently small δ and large λ_j. Letting $t = (n+1)h$ in each of these series and taking into account (1.13) shows that $u_{n+1}(x) \epsilon C^m[0,1]$ if $u_n(x) \epsilon C^m[0,1]$. By virtue of (iii), the proof is complete.

REMARK. We assumed in this theorem that $p_0(x) = 1$, where $p_0(x)$ is the leading coefficient of the operator $P(d/dx)$. If $p_0 = $ constant $\neq 1$, then dividing the equation $Py - \lambda y = 0$ by p_0 produces an equation whose leading coefficient is 1. If $p_0(x) \neq$ constant on $[0,1]$ and retains it sign, then we may assume $p_0(x) > 0$ and use the substitution [8]

$$x_1 = \int_0^x p_0^{-1/m}(s)ds \, / \int_0^1 p_0^{-1/m}(s)ds \, ,$$

to reduce the above equation to a new one in the interval $0 \leq x_1 \leq 1$, with a constant leading coefficient.

EXAMPLE 1.1. The equation $u_t = a^2 u_{xx} - bu$ describes heat flow in a rod with both diffusion $a^2 u_{xx}$ along the rod and heat loss (or gain) across the lateral sides of the rod. Measuring the lateral heat change at discrete moments of time leads to the EPCA

$$(1.15) \qquad u_t(x,t) = a^2 u_{xx}(x,t) - bu(x,[t/h]h) \, .$$

The solution $u_n(x,t)$ of Eq. (1.15) in $[0,1] \times [nh, (n+1)h]$, with the boundary conditions $u_n(0,t) = u_n(1,t) = 0$ and initial condition $u_n(x,nh) = u_n(x)$ is sought in form (1.8). Separation of variables produces

$$X_j(x) = \sqrt{2}\,\sin(\pi j x), \quad T'_{nj}(t) + a^2 \pi^2 j^2 T_{nj}(t) = -bT_{nj}(nh) \, ,$$

whence

$$T_{nj}(t) = C_{nj} e^{-a^2 \pi^2 j^2 (t - nh)} - \frac{b}{a^2 \pi^2 j^2} T_{nj}(nh) \, .$$

We put $t = nh$ in this equation and get

$$C_{nj} = \left(1 + \frac{b}{a^2 \pi^2 j^2}\right) T_{nj}(nh) \, ,$$

that is, $T_{nj}(t) = E_j(t - nh)T_{nj}(nh)$, where

(1.16) $\qquad E_j(t) = e^{-a^2\pi^2 j^2 t} - \dfrac{b}{a^2\pi^2 j^2}\left(1 - e^{-a^2\pi^2 j^2 t}\right).$

At $t = (n+1)h$ we have $T_{nj}((n+1)h) = E_j(h)T_{nj}(nh)$ and since $T_{nj}((n+1)h) = T_{n+1,j}((n+1)h)$, then

$T_{n+1,j}((n+1)h) = E_j(h)T_{nj}(nh)$ $\qquad$ and $\qquad$ $T_{nj}(nh) = E_j^n(h)T_{0j}(0).$

Therefore, $T_{nj}(t) = E_j(t - nh)E_j^n(h)T_{0j}(0)$ and

(1.17) $\qquad u_n(x, t) = \displaystyle\sum_{j=1}^{\infty} \sqrt{2}\, E_j^n(h)T_{0j}(0)E_j(t - nh)\sin(\pi jx).$

Putting $t = 0$, $n = 0$ gives

$$u_0(x) = \sum_{j=1}^{\infty} T_{0j}(0)\sqrt{2}\,\sin(\pi jx)$$

and

$$T_{0j}(0) = \sqrt{2}\int_0^1 u_0(x)\sin(\pi jx)\,dx .$$

If $|E_j(h)| < 1$, then solution (1.17) decays exponentially as $t \to \infty$, uniformly with respect to x. From (1.16) it follows that this is true if

$$-a^2\pi^2 < b < a^2\pi^2 \left(e^{a^2\pi^2 h} + 1\right) \big/ \left(e^{a^2\pi^2 h} - 1\right).$$

Furthermore, from the equations

$$T_{nj}(nh) = E_j^n(h)T_{0j}(0), \; T_{nj}((n+1)h) = E_j^{n+1}(h)T_{0j}(0)$$

we see that $T_{nj}(nh)T_{nj}((n+1)h) < 0$ if $E_j(h) < 0$. The latter inequality holds true if

(1.18) $\qquad\qquad b > a^2\pi^2 \big/ \left(e^{a^2\pi^2 h} - 1\right).$

Hence, under condition (1.18), each function $T_{nj}(t)$ $(j = 1, 2, ..)$ has a zero in the interval $[nh, (n+1)h]$, in sharp contrast to the functions $T_j(t)$ in the Fourier expansion for the solution of the equation $u_t = a^2 u_{xx} - bu$ without time delay. Moreover, the inequality $E_j(h) < 0$ takes place for sufficiently large j and any $b > 0$. Therefore, for $b > 0$ and sufficiently large j, the functions $T_{nj}(t)$ are oscillatory.

EXAMPLE 1.2. Separation of variables for the equation

$$\partial u(x,t)/\partial t = a^2 \partial^2 u(x,t)/\partial x^2 - b\partial^2(x,[t])/\partial x^2$$

with the boundary conditions $u(0,t) = u(1,t) = 0$ produces the eigenfunctions $X_j(x) = \sqrt{2}\,\sin(\pi j x)$ and the equation

$$T'(t) + a^2\pi^2 j^2 T(t) = b\pi^2 j^2 T([t]) \ ,$$

which on the interval $n \le t < n+1$ becomes

$$T'_{nj}(t) + a^2\pi^2 j^2 T_{nj}(t) = b\pi^2 j^2 T_{nj}(n) \ .$$

From here, $T_{nj}(t) = F_j(t-n)T_{nj}(n)$, where

$$F_j(t) = e^{-a^2\pi^2 j^2 t} + \left(1 - e^{-a^2\pi^2 j^2 t}\right) b/a^2 \ .$$

At $t = n+1$ we have $T_{nj}(n+1) = F_j(1)T_{nj}(n)$ and since $T_{nj}(n+1) = T_{n+1,j}(n+1)$, then $T_{n+1,j}(n+1) = F_j(1)T_{nj}(n)$ and $T_{nj}(n) = F_j^n(1)T_{0j}(0)$. Hence,

$$u_n(x,t) = \sum_{j=1}^{\infty} \sqrt{2}\, F_j^n(1)T_{0j}(0)F_j(t-n)\sin(\pi j x) \ ,$$

where

$$T_{0j}(0) = \sqrt{2} \int_0^1 u_0(x)\sin(\pi j x)\,dx \ .$$

The inequalities

$$-a^2\left(e^{a^2\pi^2} + 1\right) \big/ \left(e^{a^2\pi^2} - 1\right) < b < a^2$$

are equivalent to $|F_j(1)| < 1$ and ensure the exponential decay of $u_n(x, t)$ as $t \to \infty$, uniformly with respect to x. For

$$b < -a^2 / (e^{a^2 \pi^2} - 1),$$

each function $T_{nj}(t)$ has a zero in the interval $[n, n+1]$, which is impossible for the equation $u_t = (a^2 - b) u_{xx}$.

The Fourier method can be also used to find weak solutions of BVP (1.1)-(1.3) and it is easily generalized to similar problems in a Hilbert space. First, we recall a few well known definitions. Let H be a Hilbert space and let P be a linear operator in H (additive and homogeneous but, possibly, unbounded) whose domain $D(P)$ is dense in H, that is, $\overline{D(P)} = H$. The operator P is called symmetric if $(Pu, v) = (u, Pv)$, for any u, $v \epsilon D(P)$. If P is symmetric, then (Pu, v) is a symmetric bilinear functional and (Pu, u) is a quadratic form. A symmetric operator P is called positive if $(Pu, u) \geq 0$ and $(Pu, u) = 0$ if and only if $u = 0$. A symmetric operator P is called positive definite if there exists a constant $\gamma^2 > 0$ such that $(Pu, u) \geq \gamma^2 \| u \|^2$. With every positive operator P a certain Hilbert space H_P can be associated, which is called the energy space of P. It is the completion of $D(P)$, with the inner product $(u, v)_P = (Pu, v)$; $u, v \epsilon D(P)$. This product induces a new norm $\| u \|_P = (Pu, u)^{1/2}$, $u \epsilon D(P)$, and if P is positive definite, then $\| u \| \leq \gamma^{-1} \| u \|_P$. Since $D(P)$ is dense in H, it follows by using the latter inequality that the energy space H_P of a positive definite operator P is dense in the original space H.

Assuming P is positive definite, we may consider the solution $u(x, t)$ of BVP (1.1)-(1.3) for a fixed t as an element of H_P. If $D(Q) \subset H$, then $Qu(x, [t/h]h)$ may be treated as an abstract function $Qu([t/h]h)$ with values in H. Therefore, the given

BVP is reduced to the abstract Cauchy problem

(1.19) $\dfrac{du}{dt} + Pu = Qu([t/h]h), \quad t > 0, \quad u\,|_{\,t\,=\,0} = u_0 \,\epsilon\, H$.

If (1.19) has a solution, we multiply each term by an arbitrary function $g(t)\epsilon\,H_P$ in the sense of inner product in H and get on $nh \le t < (n+1)h$ the equation

(1.20) $\left(\dfrac{du}{dt}, g\right) + (u,g)_P = (Qu_n, g)$,

where $u_n = u(nh)$. Conversely, if $u\,\epsilon\,C^1((nh,(n+1)h);D(P))$ for all integers $n \ge 0$ and satisfies (1.20), then it also satisfies Eq. (1.19). Indeed, if $u\,\epsilon\,D(P)$, then $(u,g)_P = (Pu,g)$, and (1.20) can be written as

$$\left(\dfrac{du}{dt} + Pu - Qu_n,\, g\right) = 0, \qquad nh \le t < (n+1)h$$.

Since H_P is dense in H, then $u(t)$ is a solution of equation (1.19).

DEFINITION 1.2.   An abstract function $u(t)\colon [0,\infty) \to H$ is called a weak solution of problem (1.19) if it satisfies the conditions: (i) $u(t)$ is continuous for $t \ge 0$ and strongly continuously differentiable for $t > 0$, with the possible exception of the points $t = nh$ where one-sided derivatives exist; (ii) $u(t)$ is continuous for $t > 0$ as an abstract function with the values in H_P and satisfies Eq. (1.20) on each interval $nh \le t < (n+1)h$, for any function $g(t)\colon [0,\infty) \to H_P$; (iii) $u(t)$ satisfies initial condition (1.19), that is,

$$\lim_{t\,\to\,0} \|\,u(t) - u_0\,\|_H = 0$$.

Clearly, a weak solution $u(t)$ is also an ordinary solution if $u(t)\epsilon D(P)$, for any $t > 0$, and $u(x,t) \to u_0(x)$ as $t \to 0$ not only in the norm of H but uniformly as well.   It is said [9] that a symmetric operator P has a discrete spectrum if it has an infinite sequence $\{\lambda_j\}$ of eigenvalues with a single limit point at infinity and a sequence $\{X_j\}$ of eigenfunctions which is complete in H. Suppose the operator P in (1.20) is positive definite and has a discrete spectrum and assume existence of a solution $u(t) = u(x, t)$ to Eq. (1.20) with the condition $u(0) = u_0$. On the interval $nh \leq t < (n+1)h$ this solution can be expanded into the series

$$(1.21) \qquad u_n(x, t) = \sum_{j=1}^{\infty} T_{nj}(t) X_j(x) ,$$

where $T_j(t) = (u(t), X_j)$. To find the coefficients $T_j(t)$, we put $g(t) = X_k$ in (1.20) and since X_k does not depend on t, then

$$\left(\frac{du(t)}{dt}, X_k \right) = \frac{d}{dt} (u(t), X_k) = T_k'(t) ,$$

$$(u, X_k)_P = (Pu, X_k) = (u, PX_k) = \lambda_k(u, X_k) = \lambda_k T_k(t) ,$$

which leads to the equation

$$T_{nj}'(t) + \lambda_j T_{nj}(t) = (Qu_n, X_j) .$$

By selecting a proper space H, a weak solution corresponding to conditions (1.2) can be constructed. The proof of the following theorem is omitted.

THEOREM 1.3. If P and Q are linear operators in a Hilbert space and P is positive definite with a discrete spectrum, then there exists a unique weak solution of problem (1.19).

For the BVP consisting of the equation

$$(1.22) \qquad \frac{\partial u(x, t)}{\partial t} + P\left(\frac{\partial}{\partial x} \right) u(x, t) = Q\left(\frac{\partial}{\partial x} \right) u(x, [\mu t/h]h) ,$$

with constants $h > 0, 0 < \mu < 1$ and conditions (1.2)-(1.3), the difference $t - [\mu t/h]h$ is unbounded as $t \to \infty$. We introduce the following:

DEFINITION 1.3. A function $u(x, t)$ is called a solution of the above BVP if it satisfies the conditions: (i) $u(x, t)$ is continuous   in $G = [0,1] \times [0, \infty)$;   (ii) $\partial u / \partial t$ and $\partial^k u / \partial x^k (k = 0, 1, ..., m)$ exist and are continuous in G, with the possible exception of the points $(x, nh/\mu)$, where one-sided derivatives exist $(n = 0, 1, 2, ...)$; (iii) $u(x, t)$ satisfies Eq. (1.22) in G, with the possible exception of the points $(x, nh/\mu)$, and conditions (1.2), (1.3).

Let $u_n(x, t)$ be the solution of the given problem on the interval $nh/\mu \leq t < (n+1)n/\mu$, then

$$(1.23) \qquad \partial u_n(x, t) / \partial t + P u_n(x, t) = Q c_n(x) ,$$

where $c_n(x) = u(x, nh)$. We next write $u_n(x, t) = w_n(x, t) + v_n(x)$, which gives the equation $\partial w_n / \partial t + P w_n + P v_n(x) = Q c_n(x)$, and require (1.6) and

$$(1.24) \qquad P v_n(x) = Q c_n(x) .$$

Assuming both w_n and v_n satisfy (1.2) leads to an ordinary BVP (1.2), (1.24), whose solution is denoted by $v_n(x) = P^{-1} Q c_n(x)$, and to BVP (1.2), (1.6), whose solution is sought in form (1.8). Separation of variables produces a formal solution

$$(1.25) \quad u_n(x, t) = \sum_{j=1}^{\infty} C_{nj} e^{-\lambda_j (t - nh/\mu)} X_j(x) + P^{-1} Q c_n(x)$$

of (1.2), (1.22), where λ_j and X_j are the eigenvalues and eigenfunctions of (1.9). At $t = nh/\mu$ we have

$$s_n(x) = \sum_{j=1}^{\infty} C_{nj} X_j(x) + P^{-1} Q c_n(x) ,$$

where $s_n(x) = u_n(x, nh/\mu)$. Therefore, the coefficients C_{nj} are given by the formula

$$(1.26) \quad C_{nj} = \int_0^1 (s_n(x) - P^{-1} Q c_n(x)) X_j(x) \, dx , \quad (n = 0, 1, 2, ...) .$$

Since $s_0(x) = c_0(x) = u_0(x)$, substituting the initial function $u_0(x) \in C^m[0,1]$ in (1.26) produces the coefficients C_{0j}, and putting them together with $u_0(x)$ in (1.25) as $n = 0$ gives the solution $u_0(x,t)$ of BVP (1.2), (1.3), (1.22) on the interval $0 \le t < h/\mu$. Since $u_0(x,h) = c_1(x)$ and $u_0(x,h/\mu) = s_1(x)$, we can find from (1.26) the numbers C_{1j} and then substitute them along with $c_1(x)$ in (1.25) as $n = 1$, to obtain the solution of $u_1(x,t)$ on $h/\mu \le t \le 2h/\mu$.   This method of steps allows to extend the solution to any interval $nh/\mu \le t \le (n+1)h/\mu$.

THEOREM 1.4.   Under the conditions of Theorem 1.2, BVP (1.2), (1.3), (1.22) has a solution given by formula (1.25) in $[0,1] \times [nh/\mu, (n+1)h/\mu]$, for each $n = 0, 1, \dots$.

EXAMPLE 1.3.   The solution $u_n(x,t)$ of the equation

$$(1.27) \qquad u_t(x,t) = a^2 u_{xx}(x,t) - bu(x, [\mu t/h]h)$$

in $[0,1] \times [nh/\mu, (n+1)h/\mu]$, with the boundary conditions $u_n(0,t) = u_n(1,t) = 0$ and initial condition $u_n(x, nh/\mu) = s_n(x)$, is sought in the form (1.8).   Separation of variables gives $X_j(x) = \sqrt{2}\, \sin(\pi jx)$ and

$$(1.28) \quad T_{nj}(t) = T_{nj}(nh/\mu)e^{-a^2\pi^2 j^2(t - nh/\mu)}$$

$$- ba^{-2}\pi^{-2}j^{-2}(1 - e^{-a^2\pi^2 j^2(t - nh/\mu)})T_j(nh) ,$$

$$(nh/\mu \le t \le (n+1)h/\mu) .$$

Since $u_n(x,t)$ is the restriction of the solution $u(x,t)$ to the interval $[nh/\mu, (n+1)h/\mu]$ and the point $t = nh$ does not belong to this interval, the subindex n is omitted from the term $T_j(nh)$ in (1.28).   Furthermore, the condition $0 < \mu < 1$ implies that the difference $t - nh/\mu$ in (1.28) becomes infinite as $t \to \infty$.   At

$t = (n+1)h/\mu$ we get from (1.28)

(1.29) $T_{nj}(h(n+1)/\mu) = A_j t_{nj} - B_j s_{nj}$,

where $t_{nj} = T_{nj}(nh/\mu)$, $s_{nj} = T_j(nh)$ and $A_j = e^{-a^2\pi^2 j^2 h/\mu}$,

$B_j = ba^{-2}\pi^{-2}j^{-2}(1 - e^{-a^2\pi^2 j^2 h/\mu})$.

Since continuity of the solution implies $T_{nj}(h(n+1)/\mu) = T_{n+1,j}(h(n+1)/\mu)$, Eq. (1.29) becomes

(1.30) $t_{n+1,j} = A_j t_{nj} - B_j s_{nj}$.

The difference equation (1.30) with respect to t_{nj} is of unbounded order because it contains $s_{nj} = T_j(nh)$, where $nh \in [Nh/\mu, (N+1)h/\mu]$, that is, N is the integral part of $n\mu$.

THEOREM 1.5. If $|b| < \pi^2 a^2$, then all functions $T_j(t)$ in the expansion (1.21) for the solution of Eq. (1.27) with homogeneous boundary conditions exponentially tend to zero as $t \to +\infty$.

PROOF. Denote

$$M_n^{(j)} = \max |T_j(t)|, \quad t \in [(n-1)h, \infty),$$

$$q = e^{-a^2\pi^2 h/\mu} + \pi^{-2}a^{-2}|b|(1 - e^{-a^2\pi^2 h/\mu}).$$

Then $|s_{n-1,j}| \le M_n^{(j)}$, $|t_{n-1,j}| \le M_n^{(j)}$, $A_j + |B_j| \le q$, and from (1.30) we get $|t_{nj}| \le qM_n^{(j)}$, while the condition $|b| < \pi^2 a^2$ implies $q < 1$. By induction, we conclude from (1.30) that $|t_{n+i,j}| \le qM_n^{(j)}$, $i \ge 1$. Furthermore, since on every interval $[nh/\mu, (n+1)h/\mu]$ the function $|T_{nj}(t)|$ attains its maximum at an endpoint of this interval, the inequality $|t_{[n/\mu],j}| \le qM_n^{(j)}$ leads to $M_{[n/\mu]}^{(j)} \le qM_n^{(j)}$. Therefore, $M_{[n/\mu]}^{(j)} \le q^2 M_{[n\mu]}^{(j)}$ and the proof is completed by lowering the subindex successively $[1/\mu]$ times. We also note that the functions $T_j(t)$ decay slower for Eq. (1.27) than for the equation without delay

(1.31) $$u_t(x,t) = a^2 u_{xx}(x,t) - bu(x,t) .$$

THEOREM 1.6.   If $-\pi^2 a^2 < b < 0$, then the functions $T_j(t)$ tend to zero monotonically as $t \to \infty$, and none of them has a zero in $(0, \infty)$.

THEOREM 1.7.   For $b > 0$, each function $T_j(t)$ is bounded and oscillatory, that is, it has infinitely large zeros.

PROOF.   Assume   that   a   certain   function $T_j(t)$ is nonoscillatory, say, positive for large t.   Then t_{nj} and s_{nj} are positive for large n, and therefore it follows from (1.30) that $t_{n+1,j} < A_j t_{nj}$, with $0 < A_j < 1$.   Hence, t_{nj} tends to zero faster than A_j^n as $n \to \infty$, whereas s_{nj} decays at a slower rate of $A_j^{n\mu}$ as $n \to \infty$.   This   contradicts   (1.30)   and   proves   that $T_j(t)$ is oscillatory.   To show that $T_j(t)$ is bounded, assume the opposite: let the sequence $\{t_{nj}\}$ contain an unbounded subsequence in which $t_{nj} > 0$ and $t_{n+1,j} < 0$.   Then $|t_{n+1,j} - A_j t_{nj}| > |t_{n+1,j}|$,   and   keeping   in   mind   that $\max |T_j(t)|$ on $nh/\mu \le t \le (n+1)h/\mu$ is   either $|t_{nj}|$ or $|t_{n+1,j}|$,   we   conclude   that   (1.30)   fails   for   large n since $|t_{n+1,j}|$ grows faster than $|B_j s_{nj}|$.   This theorem reveals a striking difference between the behavior of the functions $T_j(t)$ for equations (1.27) and (1.31) when $b > 0$: for Eq. (1.31) without delay, the $T_j(t)$ are always nonoscillatory.

THEOREM 1.8.   If $b < -\pi^2 a^2 m^2$, then the functions $T_1(t), ..., T_m(t)$ are unbounded.

2. SEPARATION OF VARIABLES IN SYSTEMS OF PDE

Consider the BVP consisting of the equation

(2.1) $$U_t(x,t) = AU_{xx}(x,t) + BU_{xx}(x,[t]) ,$$

with the boundary conditions

(2.2) $U(0,t) = U(1,t) = O$,

and the initial condition

(2.3) $U(x,0) = U_0(x)$.

Here $U(x,t)$ and $U_0(x)$ are real $m \times m$ matrices, A and B are real constant $m \times m$ matrices, and $[\cdot]$ designates the greatest integer function.   Looking for a solution of (2.1)-(2.3) in the form $U(x,t) = T(t)X(x)$ gives

$$T'(t)X(x) = AT(t)X''(x) + BT([t])X''(x) ,$$

whence

$$(AT(t) + BT([t]))^{-1}T'(t) = X''(x)X^{-1}(x) = -P^2 ,$$

which generates the BVP

(2.4) $X''(x) + P^2X(x) = O$,

$$X(0) = X(1) = O$$

and the equation with piecewise constant argument

(2.5) $T'(t) = -AT(t)P^2 - BT([t])P^2$.

The general solution of Eq. (2.4) is

$$X(x) = \cos(xP)C_1 + \sin(xP)C_2 ,$$

where

$$\cos(xP) = \sum_{n=0}^{\infty} \frac{(-1)^n x^{2n} P^{2n}}{(2n)!}, \quad \sin(xP) = \sum_{n=0}^{\infty} \frac{(-1)^n x^{2n+1} P^{2n+1}}{(2n+1)!}$$

and C_1, C_2 are arbitrary constant matrices .   From $X(0) = O$ we conclude that $C_1 = O$, and the condition $X(1) = O$ enables us to choose $\sin P = O$ (although this is not the necessary consequence of the equation $(\sin P)C_2 = O$).   This can be written as

$e^{iP} - e^{-iP} = O$ or $e^{2iP} = I$.   Assuming that all eigenvalues $p_1, p_2, ..., p_m$ of P are distinct and $S^{-1}PS = D = \text{diag}$ $(p_1, p_2, ..., p_m)$ we have $e^{2iSDS^{-1}} = I$, whence $Se^{2iD}S^{-1} = I$ and $e^{2iD} = I$. Therefore, $D = \text{diag}(\pi j_1, \pi j_2, ..., \pi j_m)$, where the j_k are integers, and

$$P = SDS^{-1}, P^2 = SD^2S^{-1} = S\text{diag}(\pi^2 j_1^2, \pi^2 j_2^2, ..., \pi^2 j_m^2)S^{-1},$$
$$\sin(xP) = S\sin(xD)S^{-1} = S\text{diag}(\sin\pi j_1 x, ..., \sin\pi j_m x)S^{-1}.$$

Furthermore, we can put

(2.6) $P_j = \text{diag}(\pi(m(j-1)+1), ..., \pi m j), \; (j = 1, 2, ...)$

in (2.4) and obtain the following result:

THEOREM 2.1. There exists an infinite sequence of matrix eigenfunctions for BVP (2.4)

(2.7) $X_j(x) = \sqrt{2} \, \text{diag}(\sin\pi(m(j-1)+1)x, ..., \sin\pi m j x), \; (j = 1, 2, ...)$

which is complete and orthonormal in the space $L^2[0,1]$ of the $m \times m$ matrices, that is,

$$\int_0^1 X_j(x)X_k(x) \, dx = \begin{cases} O, & j \neq k \\ I, & j = k \end{cases}$$

where I is the identity matrix.

REMARK. The matrices $SX_j(x)S^{-1}$ satisfy Theorem 2.1 for any nonsingular matrix S.

THEOREM 2.2. Let $E(t)$ be the solution of the problem

(2.8) $T'(t) = -AT(t)P^2, \; T(0) = I$

and let

(2.9) $M(t) = E(t) + (E(t) - I)A^{-1}B$.

If the matrix A is nonsingular, then Eq. (2.5) with the initial condition $T(0) = C_0$ has on $[0, \infty)$ a unique solution

$$(2.10) \qquad T(t) = M(t - [t])M^{[t]}(1)C_0 .$$

PROOF. On the interval $n \leq t < n+1$, where $n \geq 0$ is an integer, Eq. (2.5) turns into

$$T'(t) = -AT(t)P^2 - BC_nP^2, \quad C_n = T(n)$$

with the general solution

$$T(t) = E(t - n)C - A^{-1}BC_n .$$

At $t = n$, we have $C_n = C - A^{-1}BC_n$, whence $C = (I + A^{-1}B)C_n$ and

$$T(t) = (E(t-n) + (E(t-n) - I)A^{-1}B)C_n ,$$

that is,

$$(2.11) \qquad T(t) = M(t - n)C_n .$$

At $t = n+1$ we have $C_{n+1} = M(1)C_n$ and

$$(2.12) \qquad C_n = M^n(1)C_0 .$$

Hence, $T(t) = M(t - n)M^n(1)C_0$, which is equivalent to (2.10).

THEOREM 2.3. If $\| M(1) \| < 1$, then $\| T(t) \|$ exponentially tends to zero as $t \to +\infty$.

THEOREM 2.4. The solution $T = O$ of Eq. (2.5) is globally asymptotically stable as $t \to \infty$ if and only if the eigenvalues λ_r of the matrix $M(1)$ satisfy the inequalities

$$(2.13) \qquad | \lambda_r | < 1, \quad r = 1, ..., m .$$

PROOF. There exists a nonsingular matrix Q such that $M(1) = QJQ^{-1}$, where J is a diagonal or Jordan matrix with the diagonal elements λ_r. Hence, from (2.12) it follows that

$$C_n = (QJQ^{-1})^n C_0 = QJ^n Q^{-1} C_0$$

or

$$C_n = \sum_{r=1}^{k} Q_r(n)\lambda_r^n, \quad k \leq m$$

where the entries of the matrices $Q_r(n)$ are polynomials of degree not exceeding $k-1$. This implies $C_n \to O$ as $n \to \infty$ if and only if (2.13) holds, and the conclusion of the theorem follows from (2.11).

LEMMA. All entries of every solution of the equation

$$(2.14) \qquad\qquad T'(t) = A_1 T(t) A_2 ,$$

with constant $m \times m$ matrices A_1 and A_2, are linear combinations of terms $t^k \exp{(\alpha_i^{(1)} \alpha_j^{(2)} t)}$, where $\alpha_i^{(1)}$ and $\alpha_j^{(2)}$ are eigenvalues of A_1 and A_2 and k is a non-negative integer.

PROOF. Assume, for simplicity, that the eigenvalues of A_1 and A_2 are distinct, and let

$$S_1^{-1} A_1 S_1 = D_1 = \mathrm{diag}(\alpha_1^{(1)}, \alpha_2^{(1)}, ..., \alpha_m^{(1)}) ,$$
$$S_2^{-1} A_2 S_2 = D_2 = \mathrm{diag}\,(\alpha_1^{(2)}, \alpha_2^{(2)}, ..., \alpha_m^{(2)}) .$$

Then the substitution $T = S_1 V$ changes (2.14) to

$$(2.15) \qquad\qquad V'(t) = D_1 V(t) A_2 ,$$

and the substitution $V = W S_2^{-1}$ transforms (2.15) into

$$W'(t) = D_1 W(t) D_2 .$$

The entries of $W(t)$ are $c_{ij} \exp(\alpha_i^{(1)} \alpha_j^{(2)} t)$, with arbitrary constants c_{ij}, and the completion of the proof follows from the formula $T = S_1 W S_2^{-1}$. If some of the eigenvalues are multiple, polynomial factors will replace the constants c_{ij}.

THEOREM 2.5. If all eigenvalues of A have positive real parts, $U_0(x) \in C^3[0,1]$, and $\| A^{-1}B \| < 1$, then BVP (2.1)-(2-3) has a solution (1.21). This series and all its term-by-term derivatives converge uniformly.

PROOF. The functions $T_j(t)$ satisfy Eq. (2.5), with P_j given by (2.6). Therefore, setting in the above lemma $A_1 = -A$, $A_2 = P_j^2$ and noting that Re $\alpha_i^{(1)} < 0$, $\alpha_j^{(2)} > 0$, we conclude that $E_j(t) \to O$ as $j \to \infty$ and all $t > 0$, where $E_j(t)$ is the solution of (2.8). Hence, (2.9) implies $M_j(t) \to -A^{-1}B$ as $j \to \infty$ and $\| M_j(t) \| < 1$, for sufficiently large j, by virtue of the condition $\| A^{-1}B \| < 1$. From the formula

$$T_j(t) = M_j(t - [t])M_j^{[t]}(1)C_{0j}$$

we note that $\| T_j(t) \| \to 0$ exponentially as $t \to \infty$, for sufficiently large j. Therefore, for any $\varepsilon > 0$, there exist $t = t_0 > 0$ and $j = N$ such that $\| R_N(t) \| < \varepsilon$, for $t > t_0$, where $R_N(t)$ is the remainder of series (1.21) after the Nth term. For $0 < t \leq t_0$ and large j, we have $\| T_j(t) \| < \| C_{0j} \|$, and in this case the uniform convergence of series (1.21), together with its respective derivatives in t and x, follows from the smoothness of the initial function $U_0(x)$ and the formula

$$T_j(0) = C_{0j} = \int_0^1 U_0(x)X_j(x)\,dx \;,$$

where $X_j(x)$ are given by (2.7).

REFERENCES

1. WIENER, J., Differential equations with piecewise constant delays, in *Trends in the Theory and Practice of Nonlinear Differential Equations,* Lakshmikantham, V. (editor), Marcel Dekker, New York, 1983, 547-552.

2. COOKE, K. L. and WIENER, J., Retarded differential equations with piecewise constant delays, *J. Math. Anal. Appl. 99* (1), (1984), 265-297.

3. SHAH, S. M. and WIENER, J., Advanced differential equations with piecewise constant argument deviations, *Internat. J. Math. & Math. Sci.* 6(4), (1983), 671-703.

4. COOKE, K. L. and WIENER, J., Neutral differential equations with piecewise constant argument, *Bolletino Unione Matematica Italiana 7* (1987), 321-346.

5. COOKE, K. L. and WIENER, J., An equation alternately of retarded and advanced type, *Proc. Amer. Math. Soc. 99* (1987), 726-732.

6. WIENER, J., Boundary-value problems for partial differential equations with piecewise constant delay, *Internat. J. Math. & Math. Sci. 14* (1991), 301-321.

7. WIENER, J. and DEBNATH, L., Partial differential equations with piecewise constant delay, *Internat. J. Math. & Math. Sci. 14* (1991), 485-496.

8. NAIMARK, M.A., Linear Differential Operators (Russian), Moscow, 1954.

9. MIKHLIN, S.G., Linear Equations in Partial Derivatives (Russian), Moscow, 1977.

Department of Mathematics, University of Texas-Pan American, Edinburg, TX 76384, U.S.A.

Department of Mathematics, University of Central Florida, Orlando, FL 32816, U.S.A.

Recent Titles in This Series

(*Continued from the front of this publication*)

(See the AMS catalogue for earlier titles)